Dr. John Chung's

SAT MATH

Good Luck

Made in the USA

Dear Students:

Achieving a perfect score on any math exam is quite simple. Though this may sound clichéd, all it takes is practice. Practice by taking as many mock tests as you can, and take the time to go through and correct all of your incorrect answers. Keep your mistakes in mind as you take your next mock test.

Since 1992, I have personally helped more than 50 students each year achieve perfect scores on the SAT Math, SAT II Math I & II, and AP Calculus AB & BC exams. As you might imagine, during my many years of teaching, I have gone through almost every single SAT Math test preparation book out there. I have come to realize that every book is loaded down with explanations and not enough tests! What a waste of money!

Therefore, it is my honor to introduce to you my first test preparation book, **Dr. John Chung's SAT Math.** There are no tricks or fast-track methods in this book. I have put together 20 mock exams, complete with answers and explanations, to help you PRACTICE your math test taking skills. These are the mock exams that I have used in my private tutoring sessions with my own students, most of whom have gone on to achieve perfect scores on the SAT Math exam.

Special thanks to my latest star students, Angela Lao, Priya Vohra, Devi Mehrotra, Donna Cheung, Jennifer Wong, Amos Han, and Shalini Pammal, who provided invaluable feedback on the format of this book and assisted in the final proofreading session. They all achieved a perfect score on the math section of the PSAT, SAT Math , and SAT II Math I and II.

I hope this book helps you as much as it has helped my students.

Dr. John Chung
President, NYEA

CONTENTS

Dr. John Chung's SAT Math

50 TIPS

50 TIPS

Tip 1	Absolute Value	Tip 26	Coordinates of a Circle
Tip 2	Ratio of Similar Figures	Tip 27	Paths in Grid
Tip 3	Combined Range	Tip 28	Transformation
Tip 4	Classifying a Group	Tip 29	The Least & Greatest Number
Tip 5	Direct Variation	Tip 30	Maximum & Minimum
Tip 6	Inverse Variation	Tip 31	Percentage
Tip 7	Special Triangles	Tip 32	Proportion & Ratios
Tip 8	Exponents	Tip 33	Probability
Tip 9	Geometric Probability	Tip 34	Number of Guarantee
Tip 10	Domain & Range	Tip 35	Midpoint & Distance
Tip 11	Linear Function	Tip 36	Odd & Even Numbers
Tip 12	Triangle Inequality	Tip 37	Inequality
Tip 13	Permutation & Counting	Tip 38	Solids
Tip 14	Handshake	Tip 39	Sequences & Series
Tip 15	Percent of Solution	Tip 40	Defined Operation
Tip 16	Slope of a Line	Tip 41	Functions as Models
Tip 17	Number of Factors	Tip 42	Data Interpretation
Tip 18	Composition of Functions	Tip 43	Expected Value
Tip 19	Consecutive Numbers	Tip 44	Counting Digits
Tip 20	Must be or Could be True	Tip 45	Counting Multiples
Tip 21	Sum of Consecutive Numbers	Tip 46	Average Speed
Tip 22	No Solution	Tip 47	Factorings
Tip 23	Identical Equations	Tip 48	Divisibility & Prime
Tip 24	Pythagorean Theorem	Tip 49	Rate of Work
Tip 25	Similar in Right Triangle	Tip 50	Parallel Lines

TIP 1 **Absolute Value**

The absolute value of x, denoted "$|x|$" (and which is read as "the absolute value of x"), is regarded as the distance of x from zero.

1) If $|x| = a$ and $a > 0$, then $x = a$ or $-a$

2) If $|x| < a$ and $a > 0$, then $-a < x < a$

3) If $|x| > a$ and $a > 0$, then $x > a$ or $x < -a$

4) $|x| < 5 \Leftrightarrow x^2 < 25 \Leftrightarrow -5 < x < 5$

5) $|x| > 5 \Leftrightarrow x^2 > 25 \Leftrightarrow x < -5$ or $x > 5$

6) $|x - 10| = |10 - x|$

---

1. If $|x| = 5$, what is the value of x?

(Sol)

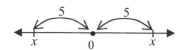

Ans: $x = 5$, $x = -5$

---

2. If $|x - 2| = 5$, what is the value of x?

(Sol)

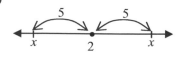

The distance of x from 2 is 5, therefore
$x = 2 + 5 = 7$ or $2 - 5 = -3$

Ans: $x = 7$ or $x = -3$

---

3. If $|x + 3| < 5$, what is the value of x?

(Sol)

$$|x + 3| = |x - (-3)| < 5$$

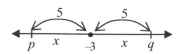

The distance of x from -3 is less than 5, therefore $p < x < q$.
$p = -3 - 5 = -8$ and $q = -3 + 5 = 2$

Ans: $-8 < x < 2$

Or

Algebraic Solution

$$-5 < x + 3 < 5$$
$$-3 \quad\quad -3\ -3$$
$$\overline{\quad\quad\quad\quad\quad\quad\quad}$$
$$-8 < \quad x \quad < 2$$

---

4. If $|x + 3| > 5$, what is the value of x?

(Sol)
$$|x - (-3)| > 5$$

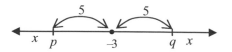

The distance of x from -3 is greater than 5, therefore $x < -8$ or $x > 2$
because $p = -8$ and $q = 2$.

Ans: $x < -8$ or $x > 2$

5. If $-8 < x < 2$, then express the interval using absolute value.

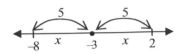

(Sol)
Step1) Find the midpoint between -8 and 2
$$\frac{-8+2}{2} = -3$$

Step2) Find the distance from the midpoint.
$$2 - (-3) = 5$$

From the figure above, the interval can be expressed with absolute value.

$$-8 < x < 2 \quad \longleftrightarrow \quad |x - (-3)| < 5$$
$$= |x + 3| < 5$$

Ans: $|x + 3| < 5$

SAT Practice

1. An art class of 20 students took a final exam and ten of the students scored between 78 and 86 in the exam. If s is defined as the scores of the ten students, which of the following describes all possible values of s ?

(A) $|s - 82| = 4$

(B) $|s + 82| = 4$

(C) $|s - 82| < 4$

(D) $|s + 82| < 4$

(E) $|s - 82| > 4$

$78 < s < 86$

$\frac{78 + 86}{2}$

82

$\cancel{78} - 82 =$

2. At a bottling company, a computerized machine accepts a bottle only if the number of fluid ounces is greater than or equal to $5\frac{3}{7}$, and less than or equal to $6\frac{4}{7}$. If the machine accepts a bottle containing f fluid ounces, which of the following describes all possible values of f ?

$\geq 5^{3}/_{7} \leq f \leq 6^{4}/_{7}$

(A) $|f - 6| < \frac{4}{7}$

(B) $|f - 6| \leq \frac{3}{7}$

(C) $|f + 6| > \frac{4}{7}$

(D) $|6 - f| \leq \frac{4}{7}$

(E) $|f + 6| \leq \frac{4}{7}$

3. At the O.K Daily Milk Company, machine X fills a box with milk, and machine Y eliminates milk-box if the weight is less than 450 grams, or greater than 500 grams. If the weight of the box that will be eliminated by machine Y is E, in grams, which of the following describes all possible values of E ?

$450 Y > 5.00$

(A) $|E - 475| < 25$

(B) $|E + 475| < 25$

(C) $|E - 500| > 450$

(D) $|475 - E| = 25$

(E) $|E - 475| > 25$

TIP 2   Ratio of Similar Figures

If the ratio of lengths is $a:b$,
the ratio of areas is $a^2:b^2$
the ratio of volumes is $a^3:b^3$

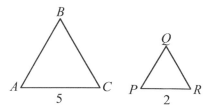

3. The figure above shows two similar triangles with a side 5 and a side 2 respectively. If the area of $\triangle ABC$ is 30, what is the area of $\triangle PQR$?

1. In the figures above, if the ratio of the diameter of circle O to the diameter of circle P is 5:3, what is the ratio of the area of circle O to the area of circle P ?

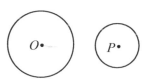

(Sol)

$$5^2:3^2 = 25:9$$

Ans: 25: 9

(Sol)
The ratio of area= $5^2:2^2 = 25:4$
The area of $\triangle ABC = 25k$ and the area of $\triangle PQR = 4k$.
$25k = 30 \rightarrow k = 1.2$. Therefore,
$4k = 4(1.2) = 4.8$

Ans: $\dfrac{24}{5}$ or 4.8

2. In the figure above, if the ratio of the circumference of circle O to the circumference of circle P is $4:3$, what is the ratio of the area of circle O to the area of circle P ?

4. In the figure above, the radius of a larger circle is $\dfrac{5}{2}$ times the radius of a smaller circle. What fraction of the larger is the shaded region?

(Sol)

$$4^2:3^2 = 16:9$$
Ans: 16:9

(Sol)
The ratio of the length $= 5:2$
The ratio of the area $= 5^2:2^2 = 25k:4k$
The area of e shaded region $= 25k - 4k = 21k$
The fraction will be $\dfrac{21k}{25k} = \dfrac{21}{25}$.

Ans: $\dfrac{21}{25}$ or .84

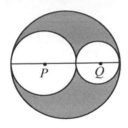

1. In the figure above, circle P and Q are inscribed in a circle. If the radius of circle P is 4, and the radius of circle Q is 2, what is the ratio of the shaded region to the area of the largest circle?

(A) $\dfrac{2}{9}$

(B) $\dfrac{4}{9}$

(C) $\dfrac{5}{9}$

(D) $\dfrac{5}{12}$

(E) $\dfrac{7}{12}$

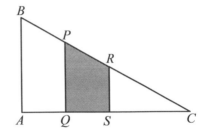

Note: Figure not drawn to scale.

2. In $\triangle ABC$ above, $AB \parallel PQ \parallel RS$ and the ratio of the lengths, $AQ:QS:SC = 2:2:3$. If the area of quadrilateral $PRSQ$ is 48, what is the area of $\triangle ABC$?

(A) 84
(B) 92
(C) 105
(D) 144
(E) 147

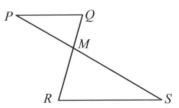

Note: Figure not drawn to scale.

3. In the figure above, $\overline{PQ}$ is parallel to $\overline{RS}$. If the ratio of the area of $\triangle PQM$ to the area of $\triangle SRM$ is 4:9 and the perimeter of $\triangle PQM$ is 15, what is the perimeter of $\triangle SRM$?

(A) $22\dfrac{1}{2}$

(B) $33\dfrac{3}{4}$

(C) $35\dfrac{1}{2}$

(D) $37\dfrac{1}{2}$

(E) $39\dfrac{1}{3}$

If $5 \le A \le 10$ and $2 \le B \le 5$, then

    (1) $7 \le A + B \le 15$

    (2) $10 \le A \times B \le 50$

    (3) $0 \le A - B \le 8$

    (4) $1 \le \dfrac{A}{B} \le 5$

Smallest Value $\le$ Combined Range $\le$ Largest Value

1. Given $2 \le P \le 8$ and $1 \le Q \le 4$. By how much is the maximum value of $\dfrac{P}{Q}$ greater than the minimum value of $\dfrac{P}{Q}$?

(Sol)

The maximum value is $\dfrac{8}{1} = 8$

The minimum value is $\dfrac{2}{4} = \dfrac{1}{2} = 0.5$

$8 - 0.5 = 7.5$

Ans: 7.5

2. If $-2 \le A \le 2$, and $-6 \le B \le -2$, and $C = (A - B)^2$, what is the smallest value of C ?

(Sol)
The minimum value is
$(-2) - (-2) = -2 + 2 = 0$
Ans: 0

3. If $1 \le P \le 6$, and $3 \le Q \le 10$, what is the smallest value of $P \times Q$?

(Sol)
    The smallest value is $1 \times 3 = 3$
    Ans: 3

SAT Practice

1. If $-2 < x < 4$ and $-3 < y < 2$, what are all possible values of $x - y$?

    (A) $-4 < x - y < 2$
    (B) $1 < x - y < 7$
    (C) $1 < x - y < 4$
    (D) $-4 < x - y < 7$
    (E) $-5 < x - y < 7$

2. The value of p is between 1 and 4, and the value of q is between 2 and 6. Which of the following is a possible value of $\dfrac{q}{p}$?

    (A) Between $\dfrac{1}{2}$ and $\dfrac{2}{3}$

    (B) Between $\dfrac{2}{3}$ and 2

    (C) Between $\dfrac{1}{2}$ and 6

    (D) Between 2 and 6

    (E) Between $\dfrac{1}{2}$ and $1\dfrac{1}{2}$

Classifying a Group into Two Different Ways.

In a certain reading group organized of only senior and junior students, $\frac{3}{5}$ of the students are boys, and the ratio of seniors to juniors is 4:5. If $\frac{2}{3}$ of girls are seniors, what fraction of the boys are juniors?

	Boys	Girls	
Seniors	A	B	$\frac{4}{9}$
Juniors	C	D	$\frac{5}{9}$
	$\frac{3}{5}$	$\frac{2}{5}$	

[Sol]

$B = \frac{2}{3}$ of girls $= \frac{2}{3} \times \frac{2}{5} = \frac{4}{15}$

$A + B = \frac{4}{9}$, $A = \frac{4}{9} - B = \frac{4}{9} - \frac{4}{15} = \frac{20-12}{45} = \frac{8}{45}$

$A + C = \frac{3}{5}$, $C = \frac{3}{5} - A = \frac{3}{5} - \frac{8}{45} = \frac{27-8}{45} = \frac{19}{45}$

$\frac{C}{3/5} = \frac{19/45}{3/5} = \frac{19}{27}$

Ans: $\frac{19}{27}$

1. On a certain college faculty, $\frac{4}{7}$ of the professors are male, and the ratio of the professors older than 50 years to the professors less than or equal to 50 years is 2:5. If $\frac{1}{5}$ of the male professors are older than 50 years, what fraction of the female professors are less than or equal to 50 years?

(A) $\frac{1}{7}$

(B) $\frac{1}{3}$

(C) $\frac{2}{5}$

(D) $\frac{3}{5}$

(E) $\frac{2}{3}$

2. Of the 24 company presidents attending a corporate meeting, $\frac{3}{4}$ of the presidents are male and $\frac{2}{3}$ of the presidents have children. If 2 female presidents do not have children, what is the number of the male presidents who have children?

(A) 6
(B) 8
(C) 10
(D) 12
(E) 14

3. In a certain group in Eton School, $\frac{3}{7}$ of the students are boys and the ratio of the students older than or equal to 15 years old to the students less than 15 years old is 3:8. If $\frac{3}{4}$ of girls are less than 15 years old, then what fraction of the boys are less than 15 years old?

(A) $\frac{1}{3}$

(B) $\frac{4}{11}$

(C) $\frac{5}{11}$

(D) $\frac{23}{33}$

(E) $\frac{40}{77}$

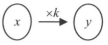

TIP 5 **Direct Variation**

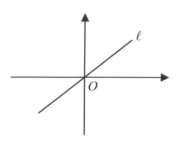

or

$$y = kx, \text{ where } k \text{ is a constant}$$

$$\frac{y}{x} = \frac{y_1}{x_1} = \frac{y_2}{x_2} = ... = k \quad \text{Proportion}$$

In xy-coordinate plane, $y = mx$, where m is slope. but y-intercept must be zero.

1. The value y changes directly proportional to the value of x. If $y = 15$ when $x = 5$, what is the value of y when $x = 12.5$?

(Sol)

$$\frac{y}{x} = k, \text{ Constant}$$

$$\frac{15}{5} = \frac{y}{12.5}, \quad 5y = 12.5 \times 15$$

$$y = \frac{12.5 \times 15}{5} = 37.5$$

Ans: 37.5

2. A group of workers can harvest all the grapes from 10 square meters of a vineyard in $\frac{1}{3}$ minutes. At this rate, how many minutes will the group need to harvest all the grapes from 300 square meters of this vineyard?

(Sol)
The area of the vineyard is directly proportional to time.

$$\frac{\text{area}}{\text{time}} = \frac{10}{\frac{1}{3}} = \frac{300}{x}, \qquad x = 10 \text{ minutes.}$$

Ans: 10 minutes

3. To make an orange dye, 5 parts of red dye are mixed with 3 parts of yellow dye. To make a green dye, 4 parts of blue dye are mixed with 2 parts of yellow dye. If equal amount of green and orange are mixed, what fraction of the new mixture is yellow dye?

(Sol)
Orange   5 red + 3 yellow = 8 parts
Green   4 blue + 2 yellow = 6 parts

To make equal amount, LCM= 24 parts

Orange   15 red + 9 yellow = 24 parts
Green   16 red + 8 yellow = 24 parts
            Total   48 parts

Therefore, there are 17 parts of yellow in the new mixture.

Ans: $\frac{17}{48}$

SAT Practice

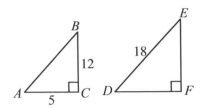

1. In the figure above, $\triangle ABC$ is similar to $\triangle DEF$. What is the length of $\overline{DF}$?

(A) 6

(B) $\frac{82}{13}$

(C) $\frac{90}{13}$

(D) 8

(E) $\frac{100}{13}$

x	y
1	a
a	$5a$

2. In the table above, y is directly proportional to x, where $a \neq 0$. Which of the following is the value of a ?

(A) 3
(B) 4
(C) 5
(D) 6
(E) 7

3. If y is directly proportional to x, which of the following could be the graph of $y = f(x)$?

(A)

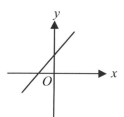

(B)

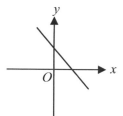

(C)

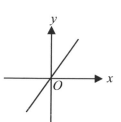

(D)

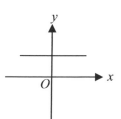

(E)

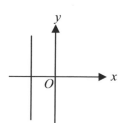

4. If y is directly proportional to x^2, which of the following could be the graph of $y = f(x)$?

(A)

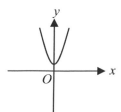

(B)

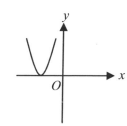

(C)

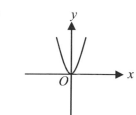

(D)

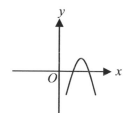

(E)

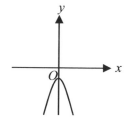

TIP 6	**Inverse Variation or Inverse Proportion**

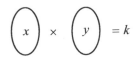

If y is inversely proportional to x, then

$$x \times y = k$$, where k is a nonzero constant.

or

$$x_1 \times y_1 = x_2 \times y_2 = \dots = k$$

In xy-coordinate plane, the graph of $y = f(x)$ is

1) $k > 0$

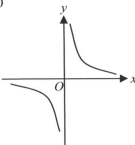

2) $k < 0$

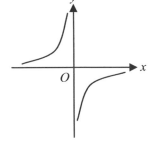

1. The cost of hiring a bus for a trip to Niagara Falls is $400. If 25 people go on the trip, what is the cost per person?

(Sol)
The cost per person varies inversely as the number of people who will go on the trip.

$$25 \times x = \$400, \quad x = \$16$$
Ans: $16

2. If 4 typists can complete the typing of a manuscript in 9 days, how long would it take 12 typists to complete the manuscript?

(Sol)
The number of days varies inversely as the number of typists.

$$4 \times 9 = 12 \times k$$
$$k = 3$$

Ans: 3 days

1. If a man can drive from his home to Albany in 5 hours at 45 miles per hour, how long would it take him if he drove at 50 miles per hour?

    (A) 4 hours
    (B) 4 hours 30 minutes
    (C) 5 hours
    (D) 5 hours 30 minutes
    (E) 6 hours

x	y
2	25
4	a
5	10
8	b

2. In the table above, y varies inversely as x. What is the value of $a+b$?

    (A) 16
    (B) 18
    (C) 18.75
    (D) 20.25
    (E) 23.75

3. If a job can be completed by 2 workers in 10 days, then what is the number of workers needed to complete the job in 5 days?

    (A) 1
    (B) 2
    (C) 3
    (D) 4
    (E) 5

4. The length of a rectangle varies inversely to the width. If the length is 10 when the width is 20, what is the length when the width is 40?

    (A) 2
    (B) 5
    (C) 10
    (D) 20
    (E) 40

5. If x and y are inversely proportional, which of the following graphs describes the relationship of x and y?

(A)
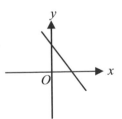

(B)
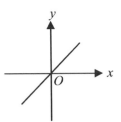

(C)
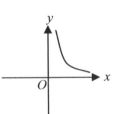

(D)
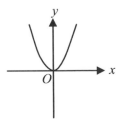

(E)
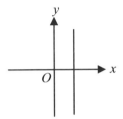

6. A certain job can be completed by p persons in h hours. How long would it take n persons, working at the same rate, to complete the same job?

(A) $\dfrac{hn}{p}$

(B) $\dfrac{n}{hp}$

(C) $\dfrac{hp}{n}$

(D) $\dfrac{np}{h}$

(E) $\dfrac{h}{np}$

7. If it takes 5 people d days to install the plumbing for a house, then how many days would it take two people to complete one-third of the same job?

(A) $d+1$

(B) $\dfrac{3d}{2}$

(C) $2d$

(D) $\dfrac{5d}{2}$

(E) $\dfrac{5d}{6}$

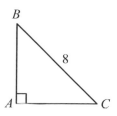

TIP 7 — Special Triangles

(A)

(B)

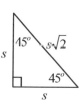

(C)

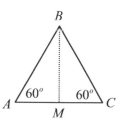

1. If $AC = 10$, what is the area of $\triangle ABC$?

(Sol)
$AM = 5$, $BM = 5\sqrt{3}$

The area of $\triangle ABC = \dfrac{1}{2} \times 10 \times 5\sqrt{3} = 25\sqrt{3}$

2. If $AB = AC$ and $\angle ABC = 45^{\circ}$, what is the area of $\triangle ABC$?

(Sol)
Special triangle in Tip 7(B)

$s\sqrt{2} = 8$, $s = \dfrac{8}{\sqrt{2}} = \dfrac{8\sqrt{2}}{\sqrt{2} \times \sqrt{2}} = 4\sqrt{2}$

The area of $\triangle ABC =$

$\dfrac{1}{2} \times 4\sqrt{2} \times 4\sqrt{2} = \dfrac{32}{2} = 16$

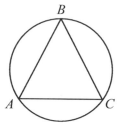

3. An equilateral triangle is inside a circle. If the radius of the circle is 10, what is the area of $\triangle ABC$?

(Sol)

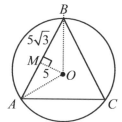

$BO = 10$, $BO = 2s$, $MB = s\sqrt{3}$,
$2s = 10$, $s = 5$ $\qquad MB = 5\sqrt{3}$
$AB = 10\sqrt{3}$, $MO = 5$

The area of

$\triangle ABC = 3 \times \left(\dfrac{1}{2} \times 10\sqrt{3} \times 5 \right) = 75\sqrt{3}$

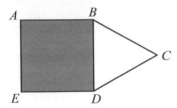

1. In the figure above, $ABDE$ is a square and $\triangle BCD$ is an equilateral triangle. If the area of $\triangle BCD$ is $16\sqrt{3}$, what is the area of the square?

(A) 32
(B) $32\sqrt{3}$
(C) 64
(D) $64\sqrt{2}$
(E) 72

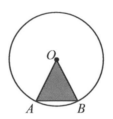

3. In the figure above, the radius of circle O is 8 and $\angle AOB = 60^o$. What is the area of $\triangle AOB$?

(A) $8\sqrt{3}$
(B) $10\sqrt{3}$
(C) $12\sqrt{2}$
(D) $15\sqrt{2}$
(E) $16\sqrt{3}$

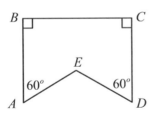

Note: Figure not drawn to scale.

2. In the figure above, $AB = BC = CD = 10$ and $AE = ED$. What is the perimeter of the figure?

(A) 40

(B) 50

(C) $30 + 10\sqrt{3}$

(D) $30 + 20\sqrt{3}$

(E) $30 + \dfrac{20\sqrt{3}}{3}$

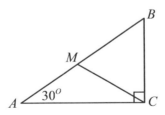

Note: Figure not drawn to scale.

4. In the figure above, M is the midpoint of $\overline{AB}$, $AM = MC$, and the length of $\overline{MC}$ is 20. What is the area of $\triangle ABC$?

(A) 200
(B) $100\sqrt{2}$
(C) $100\sqrt{3}$
(D) $200\sqrt{2}$
(E) $200\sqrt{3}$

TIP 8 — Exponents

$$5^2 = 25 \begin{cases} 5 = \text{base} \\ 2 = \text{exponent} \\ 25 = \text{power} \end{cases}$$

The Operations of exponents

(1) $a^m a^n = a^{m+n}$

(2) $\left(a^m\right)^n = a^{mn}$

(3) $\left(ab\right)^n = a^n b^n$

(4) $\left(\dfrac{a}{b}\right)^n = \dfrac{a^n}{b^n}$

(6) $\left(ab\right)^n = a^n b^n$

(7) $a^{-n} = \dfrac{1}{a^n}$

(8) $a^{\frac{m}{n}} = \sqrt[n]{a^m}$

SAT Practice

1. If $\left\{(-2)^3 \cdot 8^2\right\}^4 = \left(2^4\right)^n$, what is the positive value of n?

    (A) 6
    (B) 7
    (C) 8
    (D) 9
    (E) 10

2. If $4^3 + 4^3 + 4^3 + 4^3 = 2^n$, what is the value of n?

    (A) 2
    (B) 4
    (C) 6
    (D) 8
    (E) 10

3. If m and n are positive and $5m^5 n^{-3} = 20m^3 n$, what is the value of m in terms of n?

    (A) $\dfrac{1}{4n}$

    (B) $\dfrac{4}{n^2}$

    (C) $\dfrac{4}{n^3}$

    (D) $2n^2$

    (E) $4n^2$

4. If a and b are positive integers, $\left(a^{-4}b\right)^{-1} = 16$, and $b = a^2$, which of the following could be the value of a?

(A) 0
(B) 2
(C) 4
(D) 8
(E) 12

5. If $k^{-2} \times 2^3 = 2^7$, what is the value of k?

(A) 2

(B) 4

(C) 8

(D) $\dfrac{1}{4}$

(E) $\dfrac{1}{8}$

6. If p and q are positive integers, $p^{-3} = 2^{-6}$, and $q^{-2} = 4^2$, what is the value of pq?

(A) 1
(B) 2
(C) 3
(D) 4
(E) 5

7. If a and b are positive integers and $\left(a^6 b^4\right)^{\frac{1}{2}} = 675$, what is the value of $a+b$?

(A) 3
(B) 4
(C) 5
(D) 7
(E) 8

TIP 9 | Geometric Probability

"Geometric probability" is the probability dealing with the areas of regions instead of the "number" of outcomes. The equation becomes

$$\text{Probability} = \frac{\text{Favorable region}}{\text{Area of total region}}$$

A typical problem might be this: If you are throwing a dart at the rectangular target below and are equally likely to hit any point on the target, what is the probability that you will hit the small square?

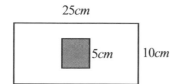

25cm

5cm   10cm

$$\text{Probablity} = \frac{\text{favorable}}{\text{total}} = \frac{25cm^2}{250cm^2} = \frac{1}{10}$$

This means that there is a 1 in 10 chance that a dart thrown at the rectangle will hit the small square.

SAT Practice

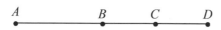

A   B   C   D

Note: Figure not drawn to scale.

1. In the figure above, B is the midpoint of $\overline{AD}$ and $\dfrac{BC}{CD} = \dfrac{2}{3}$. If a point will be chosen at random along the line segment, what is the probability that the point will be chosen from $\overline{BC}$?

(A) $\dfrac{1}{5}$

(B) $\dfrac{2}{5}$

(C) $\dfrac{1}{3}$

(D) $\dfrac{2}{3}$

(E) $\dfrac{3}{5}$

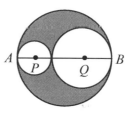

A — P — Q — B

2. In the figure above, the radius of the circle P is 2, the radius of the circle Q is 4, and AB is the diameter of the largest circle. If a dart is thrown at the circular target and is equally likely to hit any point on the target, what is the probability that the dart will hit the shaded region?

(A) $\dfrac{2}{9}$

(B) $\dfrac{1}{3}$

(C) $\dfrac{4}{9}$

(D) $\dfrac{5}{9}$

(E) $\dfrac{2}{3}$

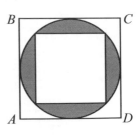

3. In the figure above, a circle is inside of and outside of a square. If a point is chosen at random from the square $ABCD$, what is the probability that the point is chosen from the shaded region?

(A) $\dfrac{1}{4}$

(B) $\dfrac{\pi - 50}{100}$

(C) $\dfrac{2\pi - 50}{100}$

(D) $\dfrac{\pi - 2}{8}$

(E) $\dfrac{\pi - 2}{4}$

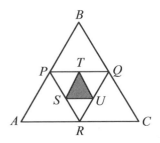

4. In the figure above, $\triangle ABC$, $\triangle PQR$, and $\triangle STU$ are equilateral triangles and $P, Q, R, S, T,$ and U are midpoints of $\overline{AB}, \overline{BC}, \overline{CA}, \overline{PR}, \overline{PQ},$ and $\overline{QR}$ respectively. If a point is chosen at random from $\triangle ABC$, what is the probability that the point is chosen from the shaded region?

(A) $\dfrac{1}{32}$

(B) $\dfrac{1}{24}$

(C) $\dfrac{1}{18}$

(D) $\dfrac{1}{16}$

(E) $\dfrac{1}{12}$

Domain and Range

The **domain** of a given function is the set of "input" values for which the function is defined. For instance, the domain of the square root would only be numbers greater than or equal to 0. In a representation of a function in a xy-coordinate system, the domain is represented <u>on the x axis</u> (or abscissa).

The **range** of a function is the set of all "output" values produced by that function. Sometimes it is called the image, or more precisely, the image of the domain of the function. Range is also occasionally used to indicate the difference between the largest and smallest numbers in a set of real-valued data. If f is a surjection then its range is equal to its codomain. In a representation of a function in a xy-coordinate system, the range is represented on the ordinate <u>(on the y axis)</u>.

(1)

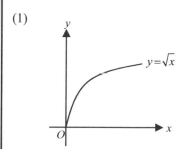

Domain: $x \geq 0$
Range : $y \geq 0$

(2)

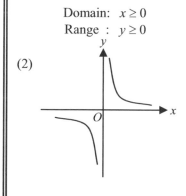

Domain: $x \neq 0$
Range : $y \neq 0$

1. If a function f is given by $f(x) = \dfrac{\sqrt{x}}{x-3}$, which of the following represents its domain?

    (A) $x \geq 0$
    (B) $x \neq 3$
    (C) $x \geq 3$
    (D) $x \geq 0$ and $x \neq 3$
    (E) all real x

2. If a function is given by $g(x) = \sqrt{x-2} - 5$, which of the following represents its range?

    (A) $y \geq 0$
    (B) $y \geq 2$
    (C) $y \geq 5$
    (D) $y \geq -5$
    (E) $y \leq -5$

SAT Practice

The functions are called "linear" because they are precisely the functions whose graph in the *xy*-coordinate plane is a straight line.
Such a function can be written as

$$f(x) = mx + b$$

m is the slope and *b* is the *y*-intercept.
The slope between any two points on the line is constant.

$$m = \frac{\Delta y}{\Delta x} = \frac{y_2 - y_1}{x_2 - x_1}$$

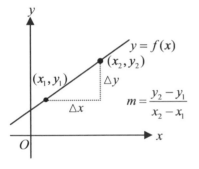

1. For a linear function f, $f(0) = 2$ and $f(3) = 5$. If $k = f(5)$, what is the value of k?

(A) 5
(B) 6
(C) 7
(D) 8
(E) 9

x	$f(x)$
0	a
1	12
2	b

2. The table above shows some values for the function f. If f is a linear function, what is the value of $a + b$?

(A) 24
(B) 36
(C) 48
(D) 60
(E) It cannot be determined from the information given.

3. A linear function is given by $ax + by + c = 0$ and $a > 0$, $b < 0$, and $c > 0$. Which of the following graphs best represents the graph of the function?

(A)

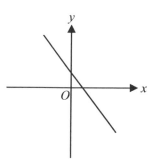

(B)

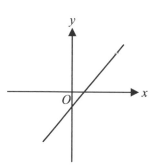

(C)

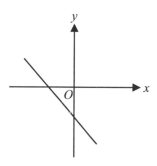

(D)

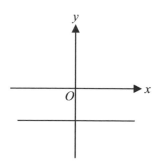

(E)

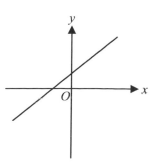

4. If f is a linear function and $f(3) = 2$ and $f(5) = 6$, what is the y-intercept of the graph of f?

(A) 4
(B) 2
(C) 0
(D) −2
(E) −4

5. If f is a linear function and $f(3) = -2$ and $f(4) = -4$, what is the x-intercept of the graph of f?

(A) 3
(B) 2.5
(C) 2
(D) 0
(E) -1

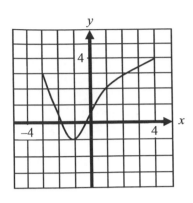

6. The figure above shows the graph of function f. Which of the following statements are true about the function?

    I. The domain of function f is $-3 \leq x \leq 4$.

    II. The range of function f is $-1 \leq f(x) \leq 4$.

    III. Function f has one zero.

(A) I only
(B) II only
(C) III only
(D) I and II only
(E) I, II, and III

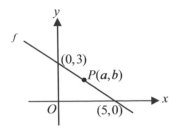

Note: Figure not drawn to scale.

7. The figure above shows the graph of function f. If $b = 2a$, what is the value of a?

(A) 2

(B) $\dfrac{5}{2}$

(C) $\dfrac{15}{13}$

(D) $\dfrac{5}{4}$

(E) $\dfrac{3}{2}$

t	$h(t)$
-1	6
0	4
1	2
2	0

8. The table above shows some values for the linear function h for selected values of t. Which of the following defines the function h?

(A) $h(t) = 4 - t$
(B) $h(t) = 4 - 2t$
(C) $h(t) = 4 + 2t$
(D) $h(t) = 4 + t$
(E) $h(t) = 2 - 0.5t$

9. Fahrenheit (F) and Celsius (C) are related by

$F = \dfrac{9}{5}C + 32$. If the Fahrenheit temperature

increased by 27 degrees, what is the degree increase in Celsius temperature?

(A) 15
(B) 20
(C) 32
(D) 59
(E) 81

10. In the formula $P = \dfrac{7}{12}Q + 60$, if P is

increased by 35, then what is the increase in Q ?

(A) 35
(B) 60
(C) 80
(D) 140
(E) 160

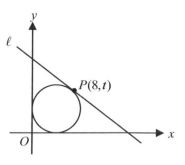

Note: Figure not drawn to scale.

11. In the figure above, a circle is tangent to line ℓ , x-axis, and y-axis. If the radius of the circle is 5, what is the value of t ?

(A) 7
(B) 8
(C) 9
(D) 10
(E) 11

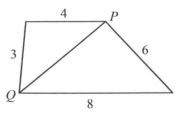

Note: Figure not drawn to scale.

TIP 12 — Triangle Inequality

Triangle Inequality Theorem

The sum of the measures of any two sides of any triangle is greater than the measure of the third side. That means that in a triangle, you can pick any two sides' measures, and when you add them together, the sum will be greater than the measure of the third side.

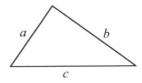

$$a+b>c,\ b+c>a,\ c+a>b$$
or
$$|a-b|<c<a+b$$

2. In the figure above, which of the following could be the length of $\overline{PQ}$?

    (A) 12
    (B) 10
    (C) 8
    (D) 7
    (E) 6

SAT Practice

3. Which of the following cannot be possible to construct a triangle with the given side lengths?

    (A) 6, 7, 11
    (B) 3, 6, 9
    (C) 28, 34, 39
    (D) 35, 120, 125
    (E) 40, 50, 60

1. If the lengths of the sides of $\triangle ABC$ is 3 , $x+3$, and 9, which of the following could be the value of x ?

    (A) 1
    (B) 2
    (C) 3
    (D) 4
    (E) 9

TIP 13 — Permutation, Counting

Permutation, also called an "**arrangement number**" or "**order**," is a rearrangement of the elements of an ordered list.

How many arrangements of the letters of the word CHEMISTRY are possible with R as the middle letter?

				R				

$8 \times 7 \times 6 \times 5 \times 1 \times 4 \times 3 \times 2 \times 1 = 8!$

└──▶ Number of possible choices

SAT Practice

1. If 11 square marbles, each red or white in color, are lined up side by side in a single row so that no two adjacent marbles are red, what is the minimum number of white marbles required?

(A) 3
(B) 4
(C) 5
(D) 6
(E) 7

2. In how many different ways can five students be arranged in a row?

(A) 60
(B) 80
(C) 120
(D) 160
(E) 240

3. How many distinct arrangements of the letters of the word LETTER are possible that begins and ends with a T?

(A) 6
(B) 12
(C) 24
(D) 120
(E) 720

4. A bag contains 8 white marbles, 8 blue marbles, 7 red marbles, and 6 yellow marbles. What is the least number of marbles that can be drawn from the bag, so that 3 of the same color marbles will be drawn?

(A) 6
(B) 9
(C) 12
(D) 13
(E) 15

5. How many arrangements of two letters and two numbers can be formed using the numbers and letters above, if each arrangement must start and end with a number, and no letter or number appears more than once in an arrangement?

(A) 360
(B) 120
(C) 36
(D) 18
(E) 12

6. If a fair die is thrown three times, what is the probability that a 5 comes up exactly two times?

(A) $\dfrac{5}{216}$

(B) $\dfrac{5}{72}$

(C) $\dfrac{1}{5}$

(D) $\dfrac{5}{24}$

(E) $\dfrac{1}{3}$

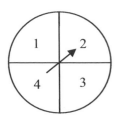

7. In the figure above, the arrow is spun twice on a wheel containing four equally likely regions, numbered 1 through 4. What is the probability that the first digit spun is larger than the second?

(A) $\dfrac{1}{8}$

(B) $\dfrac{1}{4}$

(C) $\dfrac{3}{8}$

(D) $\dfrac{1}{2}$

(E) $\dfrac{5}{8}$

8. A jar contains four white marbles and two blue marbles, all the same size. A marble is drawn at random and not replaced. A second marble is then drawn from the jar. What is the probability that one white and one blue marble are drawn?

(A) $\dfrac{8}{15}$

(B) $\dfrac{4}{15}$

(C) $\dfrac{1}{3}$

(D) $\dfrac{1}{2}$

(E) $\dfrac{2}{3}$

TIP 14 | Handshake

If there are five people in a room, and they shake each other's hands once and only once, how many handshakes are there altogether?

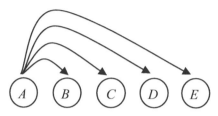

(A, B) (A, C) (A, D) (A, E)--------4 handshakes
(B, C) (B, D) (B, E) ---------------3 handshakes
(C, D) (C, E) ----------------------2 handshakes
(D, E) ------------------------------1 handshake

Therefore there will be
4+3+2+1= 10 handshakes

Or, $_5C_2 = \dfrac{5\times4}{2!} = 10$

SAT Practice

1. If you have 12 people in a group and each person shakes everyone else's hand only once, how many handshakes take place?

   (A) 132
   (B) 112
   (C) 88
   (D) 66
   (E) 36

2. At a party, everybody shakes hands with each other once. If there are 45 handshakes, how many people are there at the party?

   (A) 9
   (B) 10
   (C) 11
   (D) 12
   (E) 13

3. If there are five lines on a plane, what is the greatest number of possible intersection points?

   (A) 8
   (B) 9
   (C) 10
   (D) 11
   (E) 12

4. In the figure above, five points lie on the circle. If a line segment is formed between any two points, which of the following is the number of line segments?

   (A) 10
   (B) 9
   (C) 8
   (D) 7
   (E) 6

TIP 15 %of a solution

The percent of a solution is expressed as the percentage of solute over the total amount of solution.

$p\%$ of a solution is

$$\frac{\text{Solute}}{\text{Total amount of solution}} = \frac{p}{100}$$

or

$$\frac{\text{Solute}}{\text{Total amount of solution}} \times 100 = p\%$$

SAT Practice

1. How many gallons of water must be added to 40 gallons of 10% alcohol solution to produce a 8% alcohol solution?

   (A) 5
   (B) 8
   (C) 10
   (D) 12
   (E) 20

2. How many gallons of a 20% salt solution must be added to 10 gallons of a 50% salt solution to produce 30% salt solution?

   (A) 5 gallons
   (B) 10 gallons
   (C) 15 gallons
   (D) 20 gallons
   (E) 30 gallons

3. How many quarts of alcohol must be added to 10 quarts of a 25% alcohol solution to produce 40% alcohol solution?

   (A) 2.5 quarts
   (B) 8 quarts
   (C) 10 quarts
   (D) 15quarts
   (E) 20 quarts

4. How many gallons of acid must be added to G gallons of a $k\%$ acid solution to bring it up to an $m\%$ solution?

   (A) $\dfrac{G}{100-m}$

   (B) $\dfrac{Gm}{100-m}$

   (C) $\dfrac{G(m-k)}{100-m}$

   (D) $\dfrac{100-m}{G(m-k)}$

   (E) $\dfrac{G-m-k}{100-m}$

5. M gallons of a $p\%$ salt solution must be mixed up with G gallons of a $q\%$ salt solution to produce an $r\%$ solution. Which of the following best describes how to find the value of r?

   (A) $\dfrac{p+g}{M+G} = \dfrac{r}{100}$

   (B) $\dfrac{0.01p+0.01q}{M+G} = \dfrac{r}{100}$

   (C) $\dfrac{0.01p}{M} + \dfrac{0.01q}{G} = \dfrac{r}{100}$

   (D) $\dfrac{0.01M+0.01G}{M+G} = \dfrac{r}{100}$

   (E) $\dfrac{0.01pM+0.01qG}{M+G} = \dfrac{r}{100}$

TIP 16 — Slope of a Line

One of the most important properties of a straight line is its angle from the horizontal. This concept is called "slope". To find the slope, we need two points from the line.

From two points (x_1, y_1) and (x_2, y_2)

$$\text{Slope} \quad m = \frac{y_2 - y_1}{x_2 - x_1}$$

From slope-intercept form of a line

$$y = mx + b$$

$$m = \text{slope} \quad \text{and} \quad b = y-\text{intercept}$$

Ex) Find the slope of $2x + 3y = 4$.

$$3y = -2x + 4$$

$$y = \frac{-2}{3}x + \frac{4}{3}$$

$$\text{slope} = -\frac{2}{3} \quad \text{and} \quad y-\text{intercept} = \frac{4}{3}$$

The slope between any two points on the line is constant.

$$\frac{y_2 - y_1}{x_2 - x_1} = \frac{y_3 - y_2}{x_3 - x_2} = \frac{y_3 - y_1}{x_3 - x_1} = \dots = \text{constant}$$

SAT Practice

1. If f is a linear function and $f(3) = 6$ and $f(5) = 12$, what is the slope of the graph of f?

    (A) 2
    (B) 3
    (C) −2
    (D) −3
    (E) −4

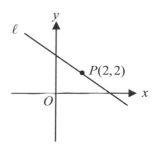

Note: Figure not drawn to scale.

2. In the figure above, line ℓ passes through point P and has a slope of $-\frac{1}{2}$. What is the $y-$intercept of line ℓ?

    (A) 6
    (B) 4
    (C) 3
    (D) 2
    (E) 1

x	$f(x)$
2	5
4	a
8	23

3. The table above gives values of the linear function f for selected values of x. What is the value of a?

    (A) 11
    (B) 14
    (C) 15
    (D) 16
    (E) 18

x	$f(x)$
2	a
5	6
8	b

4. The table above gives values of the linear function f for selected values of x. What is the value of $a+b$?

(A) 8
(B) 10
(C) 12
(D) 18
(E) 24

x	-1	1	0	2
$k(x)$	-0.5	-1.5	-1	-2

5. The table above gives values of the linear function k for selected values of x. Which of the following defines k?

(A) $k(x) = 2x - 1$

(B) $k(x) = -\dfrac{1}{2}x + 1$

(C) $k(x) = -\dfrac{1}{2}x - 1$

(D) $k(x) = -x - 1$

(E) $k(x) = -2x - 1$

6. Let the function F be defined by $F = \dfrac{9}{5}C + 32$. If C is increased by 20, how much of an increase in F will there be?

(A) 68
(B) 36
(C) 25
(D) 20
(E) 10

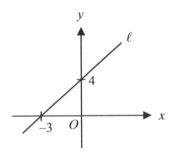

Note: Figure not drawn to scale.

7. In the figure above, a point $P(42, m)$ lies on line ℓ. What is the value of m?

(A) 39
(B) 42
(C) 45
(D) 52
(E) 60

TIP 17 Number of factors

The number of factors can be found by adding one to all exponents of prime factors and multiplying those results together.

$12 = 2^2 \times 3^1$
The number of factors $= (2+1)(1+1) = 6$

$$12 \rightarrow 1, 2, 3, 4, 6, 12$$

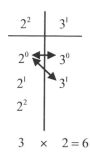

$$3 \quad \times \quad 2 = 6$$

These six products generate all the factors.

SAT Practice

1. Let a positive integer n be defined by $n = p^2 \times q^4$, where p and q are distinct prime numbers. How many factors does the number n have?

   (A) 6
   (B) 8
   (C) 12
   (D) 15
   (E) 20

2. Let a positive integer k be defined by $k = 24p^2$, where p is a prime number greater than 5. How many factors does the number k have?

   (A) 8
   (B) 16
   (C) 24
   (D) 32
   (E) It cannot be determined from the information given.

3. When two positive integers 12 and 24 are multiplied, how many factors does the resulting number have?

   (A) 18
   (B) 16
   (C) 12
   (D) 9
   (E) 6

Composition of Functions

Composition of functions is a process to combine two functions by adding, subtracting, multiplying, or dividing two given functions. There is another way which we substitute an entire function into another function.

Given $f(x) = 2x + 1$ and $g(x) = x^2$

$$f(g(x)) = 2g(x) + 1 = 2x^2 + 1$$
$$g(f(x)) = \left(f(x)\right)^2 = (2x+1)^2$$

Given $f(x) = 3x - 4$, let the function g be defined by $g(x) = 2f(x) + 3$. If $g(k) = 0$, what is the value of k?

(Sol)
$g(k) = 2f(k) + 3 = 0$,　substitute $f(k) = 3x - 4$
$2(3k - 4) = 0$
that is
$$k = \frac{4}{3}$$

SAT Practice

1. The function g is defined by $g(x) = 3f(x) - k$, and the function f is defined by $f(x) = 5x + 3$. If $g(2) = 25$, what is the value of k?

    (A) 18
    (B) 14
    (C) 12
    (D) 10
    (E) 8

2. The function g is defined by $g(x) = 2f(x) - 3$, and the function f is defined by $f(x) = ax + b$. If $g(1) = 3$ and $g(3) = 5$, what is the value of b?

    (A) $\dfrac{1}{2}$

    (B) $\dfrac{3}{2}$

    (C) $\dfrac{5}{2}$

    (D) $\dfrac{7}{2}$

    (E) 4

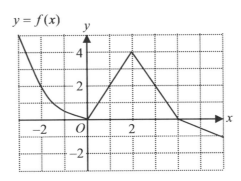

3. The figure above shows the graph of $y = f(x)$ for $-3 \le x \le 6$. If the function g is defined by $g(x) = 3f(x) - 1$, how many values of k are there in the function $g(k) = 8$?

    (A) None
    (B) One
    (C) Two
    (D) Three
    (E) Four

TIP 19 — Consecutive Numbers

Numbers which follow each other in order, without gaps, from smallest to largest.

12, 13, 14 and 15 are consecutive numbers.
14, 16, 18 are consecutive even numbers
13, 15, 17 are consecutive odd numbers.

For consecutive numbers (integers), if the first term is a_1 and the last term is a_n,

$$\text{Average (Arithmetic mean)} = \frac{a_1 + a_n}{2}$$

Median = Average

Ex) For the sequence of consecutive numbers

$$2, 3, 4, 5, 6, 7, 8, 9, 10$$

$$\text{Average} = \frac{2+3+4+5+6+7+8+9+10}{9} = 6$$

Or

$$\text{Average} = \frac{2+10}{2} = 6$$

That is

Median = Average = 6

SAT Practice

1. What is the sum of 11 consecutive integers if the middle one is 30?

    (A) 60
    (B) 120
    (C) 330
    (D) 660
    (E) 990

2. If the median of a list of 99 consecutive integers is 80, what is the greatest integer in the list?

    (A) 99
    (B) 128
    (C) 129
    (D) 157
    (E) 179

3. The median of a list of 10 consecutive even integers is 77. What is the sum of the integers?

    (A) 700
    (B) 770
    (C) 780
    (D) 800
    (E) 870

4. If the median of a list of 30 consecutive odd integers is 120, what is the greatest integer in the list?

    (A) 145
    (B) 147
    (C) 149
    (D) 151
    (E) 167

| TIP 20 | Must be true
Could be true. |
|---|---|

$$(a+b)^2 = (a-b)^2$$

The questions can be as follows
(1) If the statement above is true, which of the
    following must also be true (always true)?
                  or
(2) If the statement above is true, which of the
    following could be true (possibly true)?

(Sol)
$(a+b)^2 = a^2 + 2ab + b^2$
$(a-b)^2 = a^2 - 2ab + b^2$
Then
$\cancel{a}^2 + 2ab + \cancel{b}^2 = \cancel{a}^2 - 2ab + \cancel{b}^2$
$4ab = 0$, or $ab = 0$

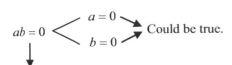

Must be true.

If the statement above is true, which of the following
must also be true?
        For this question, there is only one answer.
                  $ab = 0$
If the statement above is true, which of the following
could be true?
        For this question, there are two answers.
                  $a = 0$, or $b = 0$

Ex) If $a > b$ and $a(a-b) = 0$, which of the following
    must be true?

        I. $a = 0$
        II. $b < 0$
        III. $a + b < 0$

(Sol)
$a - b \neq 0$, therefore $a = 0$
From the given $a > b$, because $a = 0$, $b < 0$ "must be
true"
From III $a + b < 0$, $a = 0$, that is $b < 0$ "must be
true"

All of choices I, II, and III are "must be true".

1. If $k(a-b) = a - b$, which of the following
    could be true?

        I. $k = 1$
        II. $a = 2$ and $b = 2$
        III. $a = b$

    (A) I only
    (B) II only
    (C) III only
    (D) I and III only
    (E) I, II, and III

2. For real numbers a and b, $\sqrt{a-b} = \sqrt{a+b}$,
    where $a > b$. Which of the following must be
    true?

        I. $b = 0$
        II. $a > 0$
        III. $a = 0$

    (A) I only
    (B) II only
    (C) III only
    (D) I and II only
    (E) I and III only

3. If $a^2 + b^2 = 2ab$, which of the following must be true?

      I. $a = 1$
     II. $a = b$
    III. $a = 0$ and $b = 0$

(A) I only
(B) II only
(C) III only
(D) I and II only
(E) II and III only

4. If $\sqrt{a+b} = \sqrt{a} + \sqrt{b}$, where a and b are non-negative numbers, which of the following must be true?

      I. $a = 0$
     II. $b = 0$
    III. $ab = 0$

(A) I only
(B) II only
(C) III only
(D) I and II only
(E) I, II, and III

5. If $(k-2)(x-y) = (x-y)$, which of the following could be true?

      I. $k = 2$
     II. $x = y$
    III. $k - 3$

(A) I only
(B) II only
(C) III only
(D) I and II only
(E) II and III only

6. If $\sqrt{a^2 + b^2} = a - b$, where a is positive, which of the following must be true?

(A) $a = 0$
(B) $b = 0$
(C) $ab > 0$
(D) $a = 1$
(E) $a - b = 1$

Ex) The smallest integer of a set of consecutive integers is -10. If the sum of these integers is 23, how many integers are in this set?

(Sol)

$$-10, -9, -8 -7, ..., 0, ..., 7, 8, 9, 10, 11, 12$$

sum=0   sum=23

Therefore, the number of integers are

$$1, 2, 3, 4 ... 10 = 10 \text{ integers}$$
$$0 = 1 \text{ integer}$$
$$-1, -2, -3, ... -10 = 10 \text{ integers}$$

There are $10 + 1 + 10 + 2 = 23$ integers

SAT Practice

1. If the sum of the consecutive integers from -30 to x, inclusive, is 96, what is the value of x?

(A) 30
(B) 31
(C) 32
(D) 33
(E) 34

2. The smallest integer of a set of even consecutive integers is -20. If the sum of these integers is 72, how many integers are in the set?

(A) 24
(B) 25
(C) 43
(D) 44
(E) 45

3. The greatest integer of a set of consecutive integers is 61. If the sum of these integers is 61, how many integers are in this set?

(A) 2
(B) 61
(C) 121
(D) 122
(E) 125

No Solution

A system of linear equations means two or more linear equations. If two linear equations intersect, that point of intersection is called the solution to the system of linear equations.

▲ **The system has exactly one solution.**
When two lines have different slopes, the system has one and only one solution.

▲ **The system has no solution.**
When two lines are parallel and have different $y-$intercepts , the system has no solution.

▲ **The system has infinite solutions.**
When two lines are parallel and the lines have the same $y-$intercept.

Ex) For the system of equations

$$a_1 x + b_1 y = c_1$$
$$a_2 x + b_2 y = c_2$$

If $\dfrac{a_1}{a_2} \neq \dfrac{b_1}{b_2}$ One solution

If $\dfrac{a_1}{a_2} = \dfrac{b_1}{b_2} \neq \dfrac{c_1}{c_2}$ No solution

If $\dfrac{a_1}{a_2} = \dfrac{b_1}{b_2} = \dfrac{c_1}{c_2}$ Infinite solution

$$2x - 5y = 8$$
$$4x + ky = 17$$

1. For which of the following values of k will the system of equations above have <u>no</u> solution?

   (A) 10
   (B) 5
   (C) 0
   (D) −5
   (E) −10

$$5x - 2y = 3$$
$$ax + by = 6$$

2. For the system of equations above, the system has <u>infinite</u> solutions. What is the value of $a + b$?

   (A) 6
   (B) 4
   (C) 0
   (D) −4
   (E) −6

$$3x + by = 3$$
$$ax - 4y = 6$$

3. For which of the following values of $\{a, b\}$ will the system of equations above have <u>no</u> solution?

   (A) $\{-1, 2\}$
   (B) $\{1, 1\}$
   (C) $\{2, 1\}$
   (D) $\{3, -4\}$
   (E) $\{6, 2\}$

| TIP 23 | **Identical Equations** |

The two expressions (LHS and RHS) are always equal for <u>any value</u> we give to the variable. Equations that are true for any value of the variable are called **identical equations** or briefly an **identity.**

Ex)
If $2x+5 = ax+b$ for any value of x, what is the value of a and b?

(Sol)
For any value of x, the statement is true. The two expressions must be identical equations.
Therefore,
$$a = 2 \quad \text{and} \quad b = 5$$
LHS: left hand side   RHS: right hand side

SAT Practice

1. If $x(k-2) = 0$ for any value of x, what is the value of k?

    (A) 0
    (B) 2
    (C) 4
    (D) 6
    (E) 8

2. If $ax^2 + bx + c = 0$ for any value of x, what is the value of $a+b+c$?

    (A) 0
    (B) 1
    (C) 2
    (D) 3
    (E) It cannot be determined from the information given.

$$(k+1)x+5 = ax+k$$

3. If the two expressions above are true for any value of x, what is the value of a?

    (A) 6
    (B) 5
    (C) 2
    (D) 0
    (E) It cannot be determined from the information given.

$$a(x+1) + b(x-1) = 2x+4$$

4. If the two expressions above are true for any value of x, what is the value of a?

    (A) 0
    (B) 1
    (C) 2
    (D) 3
    (E) 4

TIP 24 — Pythagorean Theorem

In mathematics, the **Pythagorean Theorem** is a relation in Euclidean geometry among the three sides of a right triangle. The theorem is named after the Greek mathematician Pythagoras, who by tradition is credited with its discovery and proof, although knowledge of the theorem almost certainly predates him. The theorem is as follows:

In any right triangle, the area of the square whose side is the hypotenuse (the side opposite the right angle) is equal to the sum of the areas of the squares whose sides are the two legs (the two sides that meet at a right angle).

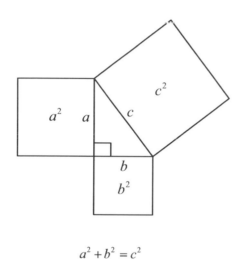

$$a^2 + b^2 = c^2$$

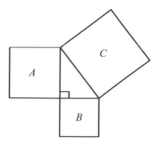

1. In the figure above, the area of the square A is 20 and the area of the square B is 16. What is the length of a side of the square C?

    (A) 4
    (B) 6
    (C) 8
    (D) 10
    (E) 36

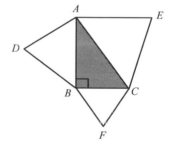

Note: Figure not drawn to scale.

2. In the figure above, $\triangle ABD$, $\triangle ACE$, and $\triangle BCF$ are equilateral triangles, and the ratio of BC to AB is $1:2$. If the area of $\triangle ACE$ is 20, what is the area of $\triangle ABD$?

    (A) 18
    (B) 16
    (C) 15
    (D) 14
    (E) 12

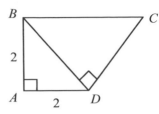

Note: Figure not drawn to scale.

3. In the figure above, $AB = AD = 2$, and $\triangle BCD$ is an isosceles triangle. What is the length of $\overline{BC}$?

(A) 3
(B) $3\sqrt{2}$
(C) 4
(D) $4\sqrt{2}$
(E) 6

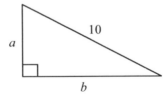

Note: Figure not drawn to scale.

4. Which of the following is true about the lengths a and b of the sides of the triangle above?

(A) $0 < (a+b)^2 \le 10$
(B) $10 < (a+b)^2 \le 40$
(C) $40 < (a+b)^2 \le 80$
(D) $80 < (a+b)^2 \le 100$
(E) $100 < (a+b)^2$

5. There are two joggers; one runs 8 miles north and then 5 miles east, and the other jogger runs 10 miles west and then 12 miles south. What is the shortest distance between these two joggers?

(A) 20
(B) 25
(C) 28
(D) 30
(E) 36

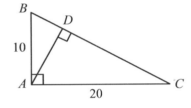

Note: Figure not drawn to scale.

6. In $\triangle ABC$ above, $AB = 10$ and $AC = 20$. What is the length of $\overline{AD}$?

(A) 5
(B) $5\sqrt{3}$
(C) $4\sqrt{5}$
(D) $6\sqrt{3}$
(E) $8\sqrt{5}$

TIP 25 — Similar in Right triangle

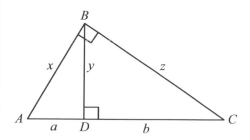

$\triangle ABD \sim \triangle BCD$ Similar triangles

$$\frac{BD}{CD} = \frac{AD}{BD} \quad \text{that is,} \quad BD^2 = AD \times CD$$

That results in $y^2 = ab$

From the same procedure,

$$x^2 = a(a+b)$$
$$y^2 = ab$$
$$z^2 = b(b+a)$$
$$xz = y(a+b)$$

SAT Practice

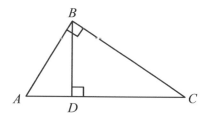

Note: Figure not drawn to scale.

1. In $\triangle ABC$ above, $AB = 6$ and $AD = 3$. What is the length of $\overline{CD}$?

Questions 2-5 refer to the following figure and information.

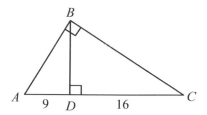

Note: Figure not drawn to scale.

In the figure above, $AD = 9$ and $CD = 16$.

2. What is the length of $\overline{AB}$?

(A) 12
(B) 13
(C) 14
(D) 15
(E) 18

3. What is the length of $\overline{BC}$?

(A) 18
(B) 20
(C) 25
(D) 28
(E) 30

4. What is the length of $\overline{BD}$?

(A) 10
(B) 11
(C) 12
(D) 13
(E) 14

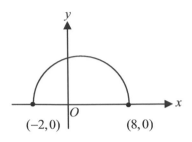

Note: Figure not drawn to scale.

2. In the semicircle above, which of the following are x-coordinates of two points on this semicircle whose y-coordinates are equal?

(A) −1 and 6
(B) 0 and 7
(C) 1 and 6
(D) 1 and 5
(E) 2 and 3

TIP 26 — Coordinates of a circle

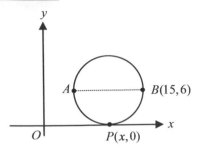

In the figure above, AB is a diameter of the circle and parallel to the x − axis. What is the value of x?

(Sol)
The radius of the circle = 6
Therefore $x = 15 - 6 = 9$

SAT Practice

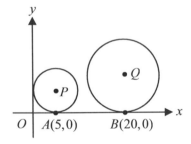

Note: Figure not drawn to scale.

1. In the xy-plane above, Points P and Q are the centers of the circles, which are tangent to the x-axis. If the radius of circle Q is twice the radius of circle P, what is the slope of line PQ (not shown)?

(A) $\dfrac{1}{4}$ (B) $\dfrac{1}{3}$ (C) $\dfrac{1}{2}$

(D) $\dfrac{2}{3}$ (E) $\dfrac{3}{4}$

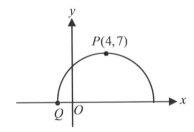

Note: Figure not drawn to scale.

3. In the xy-plane above, the semicircle has a maximum height at point P. What are the coordinates of point Q?

(A) $(-4, 0)$
(B) $(-3, 0)$
(C) $(-2, 0)$
(D) $(-0.5, 0)$
(E) $(-1, 0)$

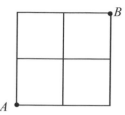

In the figure above, a path from point A to point B is determined by moving upward or to the right along the grid lines. How many different paths can be drawn from A to B?

(Sol)
At every intersection, add the number.

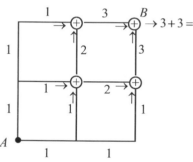

Therefore, there are six different paths.
For 3×3 grid lines,

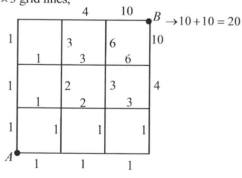

Therefore, there are 20 different paths from point A to B.

Or, in the figure above, it needs $3a$ and $3b$ to get B.

For example, *ababab*, *aaabbb*, *aababb*,...

Therefore, the number of paths is

$$\frac{6!}{3!3!} = 20 .$$

SAT Practice

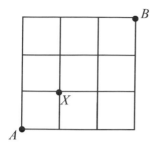

1. In the figure above, a path from point A to point B is determined by moving upward or to the right along the grid lines. How many different paths can be drawn from A to B that does <u>not</u> include point X?

    (A) 6
    (B) 8
    (C) 10
    (D) 16
    (E) 20

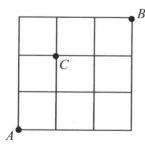

2. In the figure above, a path from point A to point B is determined by moving upward or to the right along the grid lines. How many different paths can be drawn from A to B that <u>must</u> include point C?

    (A) 4
    (B) 6
    (C) 9
    (D) 10
    (E) 12

SAT Practice

Types of transformations in math are

- Translation: involves "sliding" the object from one position to another.
- Reflection: involves "flipping" the object over a line called the line of reflection.
- Rotation: involves "turning" the object about a point called the center of rotation.
- Dilation: involves a resizing of the object. It could result in an increase in size (enlargement) or a decrease in size (reduction).

Translation of a graph

If the graph of
$$y = f(x)$$
Is translated a units horizontally and b units vertically, then the equation of the translated graph is

$$y - b = f(x - a)$$

Ex) The graph of $y = |x|$ is as follow.

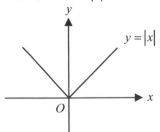

The graph of $y - 3 = |x + 5|$ or $y = |x + 5| + 3$ is

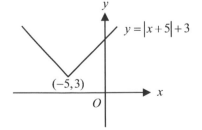

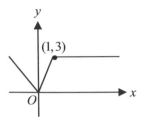

1. The graph of $y = f(x)$ is shown above. Which of the following could be the graph of $y = f(x + 2) - 2$?

(A)   (B)

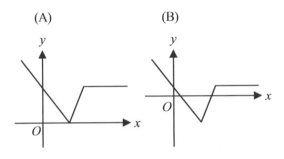

(C)   (D)

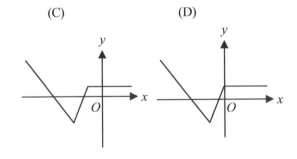

(E)

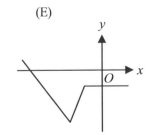

TIP 29 — The least, greatest number

Ex) If $0 \le x \le y$ and $(x+y)^2 - (x-y)^2 \ge 64$, what is the least possible value of y?

(Sol)

$$(x+y)^2 - (x-y)^2 = \cancel{x^2} + 2xy + \cancel{y^2} - \cancel{x^2} + 2xy - \cancel{y^2}$$

$$4xy \ge 64$$

To find the least value of y, the greatest value of x will be needed. That is $x = y$

$$4xy = 4y^2 \ge 64, \quad y^2 \ge \frac{64}{4} = 16, \quad y \ge 4$$

Therefore, the least value of y is 4.

Ex) If $x^2 - y^2 \ge 77$ and $x + y = 11$, what is the greatest possible value of y?

(Sol)

$$x^2 - y^2 = (x+y)(x-y) = 11(x-y) \ge 77$$
$$x - y \ge 7,$$
from the given $x + y = 11$, $x = 11 - y$
$$x - y = (11-y) - y = 11 - 2y \ge 7$$
$$2y \le 4, \quad y \le 2$$

Therefore, the greatest possible value of y is 2.

SAT Practice

1. If $a \ge 10$ and $a + 2b = 50$, what is the greatest possible value of b?

(A) 10
(B) 15
(C) 20
(D) 25
(E) 30

2. If $a + b = 30$ and $a > 12$, then which of the following must be true?

(A) $b > 18$
(B) $b < 18$
(C) $b = 18$
(D) $b > 0$
(E) $b < 30$

3. For positive integers x and y, $2x + y = 32$ and $x < 8$. What is the least possible value of y?

(A) 16
(B) 17
(C) 18
(D) 20
(E) 32

4. For positive integers a and b, $a + b < 1800$ and $\frac{a}{b} = 1.25$. What is the greatest possible value of b?

(A) 800
(B) 798
(C) 796
(D) 792
(E) 784

Maximum, Minimum

For the quadratic function, $f(x) = ax^2 + bx + c$

Axis of symmetry: $x = \dfrac{-b}{2a}$

If $a > 0$, minimum of $f = f\left(\dfrac{-b}{2a}\right)$

If $a < 0$, maximum of $f = f\left(\dfrac{-b}{2a}\right)$

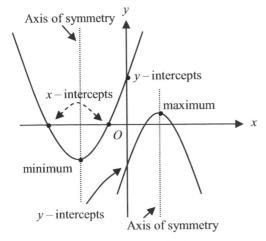

Ex) From the equation $f(x) = x^2 - 2x - 3$

$a = 1$, $b = -2$

Axis of symmetry: $x = \dfrac{-b}{2a} = \dfrac{-(-2)}{2(1)} = 1$

Because $a > 0$
Minimum of
$f = f(1) = (1)^2 - 2(1) - 3 = 1 - 2 - 3 = -4$

or
it can be factored to find the axis of symmetry.

$(x - 3)(x + 1) = 0$

The two x-intercepts are 3 and -1.

Axis of symmetry is $\dfrac{3 + (-1)}{2} = 1$

Ex) From the equation $f(x) = -x^2 + 2x + 3$

Axis of symmetry: $x = \dfrac{-b}{2a} = \dfrac{-2}{2(-1)} = 1$

Maximum of
$f = f(1) = (1)^2 - 2(1) - 3 = 1 - 2 - 3 = -4$

Questions 1-2 refer to the following information.

A ball is thrown straight up from the ground with an initial velocity of 256 feet per second. The equation $h = 256t - 16t^2$ describes the height the ball can reach in t seconds.

1. If the ball reaches its maximum height in k seconds, what is the value of k?

    (A) 4
    (B) 8
    (C) 12
    (D) 16
    (E) 24

2. What is the maximum height, in feet, that the ball will reach?

    (A) 360
    (B) 370
    (C) 384
    (D) 1024
    (E) 1200

Percentage

In mathematics, a **percentage** is a way of expressing a number as a fraction of 100 (*per cent* meaning "per hundred"). It is often denoted using the percent sign, "%". For example, 45% (read as "forty-five percent") is equal to 45 / 100, or 0.45.

When you learned how to **translate** simple English statements into mathematical expressions, you learned that "of" can indicate "times". This frequently comes up when using percentages.

- What percent of 20 is 30?

$$\frac{x}{100} \times 20 = 30, \ x = 150\%$$

- $\dfrac{\text{Amount of increase}}{\text{Original amount}} \times 100 = \%$ of increase

- $\dfrac{\text{Amount of decrease}}{\text{Original amount}} \times 100 = \%$ of decrease

Ex) The enrollment at a university increased from 15,000 students to 12,000 students over a period of 5 years. What is the <u>percent</u> increase in enrollment?

$$\frac{15,000 - 12,000}{12,000} \times 100 = 25\% \text{ increase}$$

Ex) What is the <u>percent</u> decrease on a DVD recorder that is marked down from $400 to $350?

$$\frac{350 - 400}{400} \times 100 = -12.5\%, \ 12.5\% \text{ decrease}$$

1. If 20 percent of 30 percent of a positive number is equal to 10 percent of k percent of the same number, what is the value of k?

(A) 80
(B) 60
(C) 40
(D) 15
(E) 10

2. The price of a music CD was first increased by 15 percent and then the new price was decreased by 30 percent. The final price was what percent of the initial price?

(A) 75%
(B) 78%
(C) 80.5%
(D) 82%
(E) 84.5%

3. If $2a + 3b$ is equal to 250 percent of $6b$, what is the value of $\dfrac{a}{b}$?

(A) $\dfrac{1}{6}$

(B) $\dfrac{1}{3}$

(C) 3

(D) 6

(E) 9

4. If 25 percent of m is 50, what is 15 percent of $2m$?

(A) 80
(B) 60
(C) 50
(D) 40
(E) 30

5. The cost of an automobile increases each year by 2.5 percent, and the cost this year is $20,000. If the cost c of the automobile is given by $c(n) = 20,000x^n$, what is the value of x?

(A) 2.5
(B) 1.25
(C) 1.025
(D) 0.25
(E) 0.025

6. Tom's salary was increased from $500 to $1000 this week. By what percent was his salary increased?

(A) 10%
(B) 50%
(C) 100%
(D) 200%
(E) 250%

7. If the price of a stock rises by 6 percent one day and falls by 5 percent next day, what was the change in the price of the stock after these two days?

(A) The price rose by 10%
(B) The price rose by 7%
(C) The price rose by 5%
(D) The price rose by 1%
(E) The price rose by 0.7%

8. If A is 25 percent of $2B$, then B is what percent of A?

(A) 50%
(B) 75%
(C) 100%
(D) 200%
(E) 250%

Proportion, Ratios

Ratio

A ratio is a comparison of two numbers. We can write this as 8:12 or as a fraction 8/12, and we say the ratio is *eight to twelve*.

Proportion

A proportion is an equation with a ratio on each side. It is a statement that two ratios are equal.
3/4 = 6/8 is an example of a proportion.

Rate

A rate is a ratio that expresses how long it takes to do something, such as traveling a certain distance. To walk 3 kilometers in one hour is to walk at the rate of 3 km/h. The fraction expressing a rate has units of distance in the numerator and units of time in the denominator.

Example:

Juan runs 4 km in 30 minutes. At that rate, how far could he run in 45 minutes?

Give the unknown quantity the name n. In this case, n is the number of km Juan could run in 45 minutes at the given rate. We know that running 4 km in 30 minutes is the same as running n km in 45 minutes; that is, the rates are the same. So we have the proportion
4km/30min = n km/45min, or 4/30 = n/45.
Finding the cross products and setting them equal, we get $30 \times n = 4 \times 45$, or $30 \times n = 180$. Dividing both sides by 30, we find that $n = 180 \div 30 = 6$ and the answer is 6 km.

1. If the cost of a 6-minute telephone call is $1.20, then at this rate, what is the cost of a 15-minute call?

   (A) $2.00
   (B) $3.00
   (C) $3.25
   (D) $3.75
   (E) $3.98

2. In 5 years the ratio of Julie's age to Song's age will be 3:5. In 10 years the ratio of Julie's age to Song's age will be 2:3. What is the sum of their current ages?

   (A) 15
   (B) 20
   (C) 30
   (D) 35
   (E) 40

3. If $\dfrac{x}{y} = \dfrac{2}{3}$ and $2x + 5y = 76$, what is the value of x?

   (A) 2
   (B) 4
   (C) 8
   (D) 16
   (E) 20

4. If a is divisible by 2, b is divisible by 5, and $\dfrac{a}{b} = \dfrac{7}{9}$, where a and b are positive numbers, and $a + b < 400$, what is one possible value of $a + b$?

Consider a *random sampling process* in which all the outcomes solely depend on the *chance*, i.e., each outcome is equally likely to happen. If the collection of all possible outcomes is U and the collection of desired outcomes is A, the **probability** of the desired outcomes is:

$$P(A) = \frac{\text{Number of } A}{\text{Number of } U} = \frac{n(A)}{n(U)}, \quad 0 \le P(A) \le 1$$

Accordingly, the probability of an unwanted outcome $\overline{A}$ is: $P(\overline{A}) = 1 - P(A)$

Law of addition

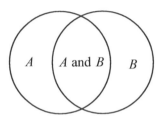

The probability of Event A or Event B (or both) occurring is given by

$$P(A \text{ or } B) = P(A) + P(B) - P(A \text{ and } B)$$

Example:
A bag contains 3 red marbles and 4 blue marbles. What is the probability that you select one red marble and one blue, at random, from the bag?

(Sol) There are two different selections.
RB or BR

$$P(\text{R and B}) = \frac{3}{7} \times \frac{4}{6} = \frac{12}{42},$$

$$P(\text{B and R}) = \frac{4}{7} \times \frac{3}{6} = \frac{12}{42}$$

The answer is $\dfrac{12}{42} + \dfrac{12}{42} = \dfrac{24}{42} = \dfrac{4}{7}$

1. A bag contains 3 red marbles and 3 blue marbles. What is the probability that you draw two red marbles without replacement?

(A) $\dfrac{1}{9}$

(B) $\dfrac{1}{6}$

(C) $\dfrac{1}{5}$

(D) $\dfrac{1}{3}$

(E) $\dfrac{2}{5}$

2. The three cards shown above were taken from a box of ten cards, each with a different integer on it from 1 to 10. What is the probability that the next two cards selected from the box will have both even integer on it?

(A) $\dfrac{10}{21}$

(B) $\dfrac{12}{23}$

(C) $\dfrac{4}{7}$

(D) $\dfrac{5}{7}$

(E) $\dfrac{6}{7}$

3. In a box, there are b blue marbles and g green marbles. If a person selects two marbles, what is the probability that both marbles are blue?

(A) $\dfrac{b}{b+g}$

(B) $\dfrac{b}{b+g+1}$

(C) $\dfrac{b \times b}{b+g}$

(D) $\dfrac{b \times b}{(b+g)(b+g-1)}$

(E) $\dfrac{b(b-1)}{(b+g)(b+g-1)}$

4. What is the probability that three quarters tossed in the air will land with only one head facing up?

(A) $\dfrac{1}{8}$

(B) $\dfrac{1}{4}$

(C) $\dfrac{3}{8}$

(D) $\dfrac{1}{2}$

(E) $\dfrac{2}{3}$

5. A traffic signal is green for 40 seconds, red for 35 seconds, and amber for 10 seconds. If this cycle continues, what is the probability that a driver will encounter a green signal?

(A) $\dfrac{6}{17}$

(B) $\dfrac{8}{17}$

(C) $\dfrac{10}{17}$

(D) $\dfrac{2}{5}$

(E) $\dfrac{3}{7}$

6. If a number is chosen at random from the set $\{-15,-10,-5,\ 0,\ 5,\ 10,\ 15,20\}$, what is the probability that it is a member of the solution set of $|x+2|<8$?

(A) 0

(B) $\dfrac{1}{4}$

(C) $\dfrac{2}{5}$

(D) $\dfrac{3}{8}$

(E) $\dfrac{2}{3}$

Socks of 4 different colors are in a drawer. A person wants to select at least three socks of the same color. What is the least number of socks he must select?

(A) 7
(B) 8
(C) 9
(D) 10
(E) 12

TIP 34 — Number of guarantee

Example:

There are 24 red marbles, 24 blue marbles, 24 green marbles, 24 white marbles, 24 black marbles, and 24 gray marbles in a box.

Question 1)
Tom wants to select at least <u>two marbles of the same color</u>. What is the least number of marbles he must select?

(Sol) Think about the worst case.

      R B G W B G
      ?

The next one will be one of these colors. Now two marbles are of same color. The answer is 7

Question 2)
Tom wants to select at least <u>four marbles of the same color</u>. What is the least number of marbles he must select?

      R B G W B G
      R B G W B G
      R B G W B G
      ?
The answer is 19.

SAT Practice

1. Marbles are randomly selected from a container. If there are 5 different colors and a person wants to make sure that at least 3 of the same color are selected, what is the minimum number of marbles he must select?

(A) 5
(B) 7
(C) 8
(D) 11
(E) 15

3. A box contains marbles of four different colors. What is the least number of marbles you would need to take from the box to make sure you have three of the same color?

(A) 9
(B) 10
(C) 11
(D) 12
(E) 15

The mid point formula:

Given the two end points (x_1, y_1) and (x_2, y_2)

The mid point of a segment $= \left(\dfrac{x_1 + x_2}{2}, \dfrac{y_1 + y_2}{2} \right)$

The distance formula:

The distance between these two points

$= \sqrt{(x_2 - x_1)^2 + (y_2 - y_1)^2}$

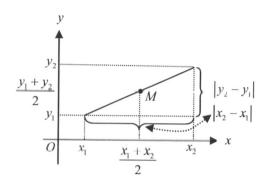

SAT Practice

1. In the *xy*-coordinate plane, the midpoint of $\overline{AB}$ is $(10,4)$. If the coordinates of point A are $(5,1)$, then what are the coordinates of point B?

    (A) $(5,3)$
    (B) $(6,4)$
    (C) $(15,5)$
    (D) $(7.5, 2.5)$
    (E) $(15,7)$

2. If point $M(a,b)$ is the midpoint of the line segment connecting point $A(2a,b)$ and point $B(x,y)$, what is the value of y?

    (A) 0
    (B) 1
    (C) a
    (D) b
    (E) $a+b$

3. In $\triangle ABC$ on a coordinate plane, the coordinates of A are $(-4, 4)$ and the coordinates of B are $(4,4)$. If the area of $\triangle ABC$ is 24, which of the following could be the coordinates of C?

    (A) $(3, 8)$
    (B) $(2, 10)$
    (C) $(0, -3)$
    (D) $(2, -5)$
    (E) $(-6, -4)$

4. If the distance between $(a,3)$ and $(b,8)$ is 13, then what is the value of $|a-b|$?

    (A) 3
    (B) 4
    (C) 8
    (D) 12
    (E) 16

Example:

If n is an odd number, which of the following must be even?

(A) $5n$ (B) n^2 (C) $2n - n$ (D) $n + 2$
(E) $(n+1)(n-2)$

(Sol) Algebraic solution by definitions above.
(A) $5n = odd \times odd = odd$
(B) $n^2 = n \times n = odd \times odd = odd$
(C) $2n - n = n = odd$
(D) $n + 2 = odd + even = odd$
(E) $(n+1)(n-2) = (odd - odd)(odd - even)$
$\qquad = (even)(odd) = even$

(Sol) Plug-in number, define $n = odd = 3$
(A) $5n = 5 \times 3 = odd$
(B) $n^2 = 3^2 = 9 = odd$
(C) $2n - n = 2(3) - 3 = 3 = odd$
(D) $n + 2 = 3 + 2 = 5 = odd$
(E) $(n+1)(n-2) = (3+1)(3-2) = 4 = even$

On this type of SAT question, using the '**plug-in number**' method is recommended.

Example:

If $a + 3$ is an odd integer, which of the following must be an even integer?

(A) $2a + 1$ (B) $4a$ (C) $\dfrac{a}{2}$ (D) $a - 1$ (E) $3a + 1$

(Sol)
$a + 3 = odd$, choose $a = 2$
(A) $2a + 1 = 2(2) + 1 = 5 = odd$
(B) $4a = 4(2) = 8 = even$

(C) $\dfrac{a}{2} = \dfrac{2}{2} = 1 = odd$ (D) $a - 1 = 2 - 1 = 1 = odd$

(E) $3a + 1 = 3(2) + 1 = 7 = odd$.
The answer is (B)

---

1. If exactly two of the three integers a, b, and c are even, which of the following must be odd?

   I. $a + b + c$
   II. $ab + c$
   III. $ab(2c)$

   (A) I only
   (B) II only
   (C) I and II only
   (D) I and III only
   (E) I, II, and III

---

2. If n is a positive integer such that n^2 is odd, then which of the following must be an odd integer?

   (A) $\dfrac{n}{2}$

   (B) $2n + n$

   (C) $2(n+1)$

   (D) $\dfrac{n+3}{3}$

   (E) $(n+1)(n-1)$

3. If a is an odd integer, which of the following is an even integer?

(A) $a-2$
(B) a^2
(C) a^2-2
(D) $(a-2)^2$
(E) a^2-a

5. If $p+q$ is an even integer, which of the following must be even?

(A) pq

(B) $2p+q$

(C) $(p+1)(q+1)$

(D) $\dfrac{p}{q}$

(E) p^2-q^2

4. If a and b are even integers, which of the following must be even?

    I. ab
    II. $a+b$
    III. $a(a^2-1)$

(A) I only
(B) II only
(C) III only
(D) I and II only
(E) I, II, and III

6. If k is a positive even integer, then $k(k+1)(k+2)$ could equal which of the following?

(A) 48
(B) 50
(C) 60
(D) 120
(E) 210

TIP 37   Inequalities

An inequality says that two values are not equal.
$a \neq b$ says that a is not equal to b.
There are other special symbols that show in what way things are not equal.

$a < b$ says that a is less than b.
$a > b$ says that a is greater than b.
(those two are known as strict inequality)

$a \leq b$ means that a is less than or equal to b.
$a \geq b$ means that a is greater than or equal to b.

• The properties of inequality

▲ If $a > b$ and $b > c$, then $a > c$

▲ If $a > b$, then $a + c > b + c$, $\ a - c > b - c$

▲ If $a > b$ and $c > 0$, then $ac > bc$, $\ \dfrac{a}{c} > \dfrac{b}{c}$

▲ If $a > b$ and $c < 0$, then $ac < bc$, $\ \dfrac{a}{c} < \dfrac{b}{c}$

▲ If $a > 0$ and $x^2 < a^2$, then $-a < x < a$

▲ If $a > 0$ and $x^2 > a^2$, then $x < -a$ or $x > a$

SAT Practice

1. If $a < b$, which of the following must be true?

   (A) $b < 0$
   (B) $a > 0$
   (C) $ab > 0$
   (D) $ab < 0$
   (E) $a - b < 3$

2. If $a > b > 0$, which of the following must be greater than $\dfrac{a}{b}$?

   (A) 1

   (B) $\dfrac{b}{a}$

   (C) $a - b$

   (D) $\dfrac{a}{2b}$

   (E) $\dfrac{2a}{b}$

3. If $s^3 t^4 u^3 w > 0$ and $w < 0$, which of the following must be true?

   (A) $s > 0$
   (B) $u < 0$
   (C) $su > 0$
   (D) $su < 0$
   (E) $t > 0$

4. If $b > a > 1$, which of the following must be true?

   (A) $\dfrac{1}{a} < \dfrac{1}{b}$

   (B) $a - b > 0$

   (C) $b^2 < ab$

   (D) $a + 2b > 3b$

   (E) $a^2 < b^2$

TIP 38   Solids

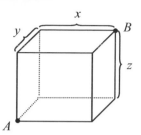

▲ Surface area $= 2(xy + yz + zx)$

▲ Volume $= xyz$

▲ Length of diagonal $= \sqrt{x^2 + y^2 + z^2}$

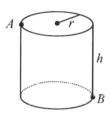

▲ Surface area $= 2\pi r^2 + 2\pi rh = 2\pi r(r + h)$

▲ Volume $= \pi r^2 h$

▲ Length of $\overline{AB} = \sqrt{(2r)^2 + h^2}$

SAT Practice

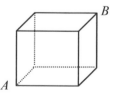

1.   In the figure above, if the volume of the cube is 64, what is the length of $\overline{AB}$ (not shown)?

(A) 4
(B) $4\sqrt{2}$
(C) $4\sqrt{3}$
(D) 8
(E) 12

2.   If the surface area of a cube is 96, what is the volume of the cube?

(A) 8
(B) 27
(C) 64
(D) 81
(E) 125

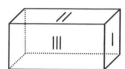

Note: figure not drawn to scale.

3.   In the rectangular solid above, the area of region I (side) is 8, the area of region II(top) is 10, and the area of region III (front) is 20. What is the volume of the solid?

(A) 40
(B) 60
(C) 80
(D) 100
(E) 200

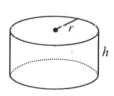

4.   The cylinder shown above has a radius of r and a height of h. If $r = h$, what is the surface area of the cylinder?

(A) $2\pi r^2$
(B) $2\pi r^3$
(C) $4\pi r^2$
(D) $4\pi r^3$
(E) $2\pi r^2 + \pi r$

Arithmetic Sequence
Arithmetic Progression
A sequence such as 1, 5, 9, 13, 17 or 12, 7, 2, –3, –8, –13, –18 which has a constant difference between terms. The first term is a_1, the common difference is d, and the number of terms is n.

nth term: $a_n = a_1 + (n-1)d$

Arithmetic Series
A series such as $3 + 7 + 11 + 15 + \cdots + 99$ or $10 + 20 + 30 + \cdots + 1000$ which has a constant difference between terms. The first term is a_1, the common difference is d, and the number of terms is n. The sum of an arithmetic series is found by multiplying the number of terms times the average of the first and last terms.

Sum: $S_n = n\left(\dfrac{a_1 + a_n}{2}\right)$ or $\dfrac{n\left[2a_1 + (n-1)d\right]}{2}$

Geometric Sequence
Geometric Progression
A sequence such as 2, 6, 18, 54, 162 or

$3, 1, \dfrac{1}{3}, \dfrac{1}{27}, \dfrac{1}{81}$ which has a constant ratio between

terms. The first term is a_1, the common ratio is r, and the number of terms is n.

nth term: $a_n = a_1 r^{n-1}$

Geometric Series
A series such as $2 + 6 + 18 + 54 + 162$ or

$3 + 1 + \dfrac{1}{3} + \dfrac{1}{27} + \dfrac{1}{81}$ which has a constant ratio

between terms. The first term is a_1, the common ratio is r, and the number of terms is n.

Sum: $S_n = \dfrac{a_1(r^n - 1)}{r - 1}$ or $\dfrac{a_1(1 - r^n)}{1 - r}$

Infinite Geometric Series
An infinite series that is geometric. An infinite geometric series converges if its common ratio r satisfies $-1 < r < 1$. Otherwise it diverges.

Sum: $S_\infty = \dfrac{a_1}{1 - r}$ as long as $-1 < r < 1$

$$-2, -1, \ 0, \ 1, \ 2$$

1. A sequence is formed by repeating the five numbers above in the same order indefinitely. What is the sum of the first 124 terms of the sequence?

    (A) –2
    (B) 0
    (C) 4
    (D) 124
    (E) 248

$$9, \ 27, \ 81, \ 243...$$

2. In the sequence above, the first term is 9 and each term after the first is 3 times the term before it. Which of the following is the expression for the 300th term of the sequence?

    (A) $3(299)$
    (B) $3(300)$
    (C) 3^{299}
    (D) 3^{300}
    (E) 3^{301}

$$5, -10, \ 20,...$$

3. In the geometric sequence above, what is the sum of the first eight terms of the sequence?

    (A) –425
    (B) –160
    (C) 120
    (D) 160
    (E) 425

4. If a certain salesman's salary increases 10% each year, approximately what is the percent increase in salary after 4 years?

(A) 40
(B) 44
(C) 46
(D) 50
(D) 146

5. Assume a ball bounces to a height of $\dfrac{3}{5}$ of the height from which it falls. If the ball is dropped from a height of 30 feet, after which bounce will the rebounded height be less than 4 feet?

(A) 3
(B) 4
(C) 5
(D) 6
(E) 7

$$a,\ 4a,\ 16a,\ 64a,\ldots$$

6. In the geometric sequence above, the first term is a, and the sum of the first 6 terms is 4095. What is the value of a?

(A) 3
(B) 4
(C) 5
(D) 6
(E) 9

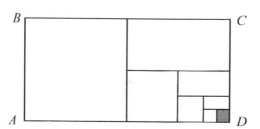

7. In the rectangle above, $BC = 2AB$, and it has been repeatedly divided in half resulting in the figure. What fraction of the area of rectangle $ABCD$ is the small shaded square?

(A) $\dfrac{1}{16}$

(B) $\dfrac{1}{32}$

(C) $\dfrac{1}{64}$

(D) $\dfrac{1}{128}$

(E) $\dfrac{1}{256}$

The **defined operations** are mathematical models (symbolic representations/notational systems/sign systems) of certain situations.

Example:

If the operation ▲ is defined by $▲a = a^a$, what is the value of $\dfrac{▲8}{▲4}$?

(Sol) $▲8 = 8^8 = \left(2^3\right)^8 = 2^{24}$

$▲4 = 4^4 = \left(2^2\right)^4 = 2^8$

The answer is $\dfrac{▲8}{▲4} = \dfrac{2^{24}}{2^8} = 2^{24-8} = 2^{16}$

SAT Practice

1. Let the operation $\odot$ be defined for all numbers by $a \odot b = \dfrac{a+b}{a-b}$. If $p \odot q = 3,$ what is the value of $\dfrac{p}{q}$?

(A) $\dfrac{1}{2}$

(B) 1

(C) $\dfrac{3}{2}$

(D) 2

(E) $\dfrac{5}{2}$

2. Let the operation $\triangle$ be defined by $a \triangle b = \dfrac{a}{b}$ for all positive numbers. If $4 \triangle (k \triangle 6) = 3$, what is the value of k?

(A) 4
(B) 8
(C) 12
(D) 20
(E) 36

3. Let the operation $n^{▲} = n(n-1)(n-2)(n-3).....(2)(1)$, where n is a positive integer. Which of the following is equivalent to $(n+1)^{▲}$?

(A) $n(n^{▲})$
(B) $(n+1)(n+1)^{▲}$
(C) $n(n-1)^{▲}$
(D) $(n+1)n^{▲}$
(E) $(n+1)(n-1)^{▲}$

Functions as Models
A function can serve as a simple kind of mathematical model, or a simple piece of a larger model. Remember that a function is just a rule. We can think of the rule (given in our model as a graph, a formula, or a table of values) as a representation of some natural cause and effect relationship.

SAT Practice

1. The total cost c, in dollars, of repairing shoes is given by the function $c(x) = \dfrac{200x\ \ 400}{x} + k$, where x is the number of repairing shoes and k is a constant. If 50 shoes were repaired at a cost of $300, what is the value of k?

(A) 100
(B) 108
(C) 126
(D) 150
(E) 300

2. The value of a computer decreases each year by 1.2 percent. This year the price of the computer was $1,200. If the price p of the computer n years from now is given by the function $p(n) = 1,200c^n$, what is the value of c?

(A) 0.012
(B) 0.88
(C) 0.988
(D) 1.012
(E) 1.12

3. Let the function m, average rate of change between a and b in the domain of the function, be defined by $m(x) = \dfrac{f(b) - f(a)}{b - a}$. If $f(x) = x^2$, what is the value of m between -2 and 3?

(A) -2
(B) -1
(C) 0
(D) 1
(E) 2

4. The present value p of a certain car that depreciates for a number of years is defined by $p(t) = k\left(1 - \dfrac{r}{100}\right)^t$, where k is the initial value of the car, r is the percent of depreciation per year, and t is the number of years. If a person purchases the car for $20,000 and the value of the car depreciates by 10% per year, how much will the value of the car be after three years from the date of purchase?

(A) $18,000
(B) $16,200
(C) $14,580
(D) $14,000
(E) $12,250

Data Interpretation

SAT Practice

Data interpretation problems usually require two basic steps. First, you have to read a chart or graph in order to obtain certain information. Then, second, you have to apply or manipulate the information in order to obtain an answer. Be sure to read all notes related to the data.

COST VS. WEIGHT
FOR 10 MEATS

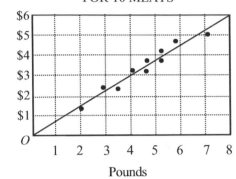

GEOMETRY TEST RESULTS
FOR 10 STUDENTS

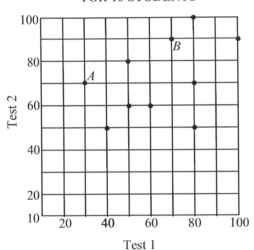

Test 1

In the scatter plot above, student *A* got 30 on test 1 and 70 on test 2. Student *B* got 70 on test 1 and 90 on test 2.

1) What is the median score on test 1 and 2?

(Sol) On test 1
30, 40, 50, 50, 60, 70, 80, 80, 80, 100

The median is $\dfrac{60+70}{2} = 65$

(Sol) On test 2
50, 50, 60, 60, 70, 70, 80, 90, 90, 100
The median is 70

2) What is the average (arithmetic mean) on test 1?
(Sol)

$$\frac{30+40+50+50+60+70+80+80+80+100}{10} = 64$$

The average is 64.

1. For 10 meats of different weights, the cost and weight of each are displayed in the scatter plot above, and the line of best fit for the data is shown. Which of the following is closest to the average (arithmetic mean) cost per pound for the 10 meats?

(A) $0.06
(B) $0.18
(C) $0.56
(D) $0.62
(E) $0.73

ITEMS PURCHASED
BY CUSTOMERS

Numbers of Customers	Number of Items
10	10
25	8
45	5
50	Fewer than 5

PEOPLE AT J.C HIGH SCHOOL

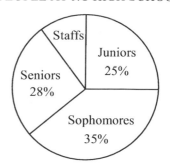

2. The table above shows the number of items 130 customers purchased from a stationery store during on Sunday. Which of the following can be obtained from the information in the table?

    I. The average (arithmetic mean) number of items
    II. The median number of items.
    III. The mode of the number of items.

(A) I only
(B) II only
(C) III only
(D) I and II only
(E) II and III only

3. In the circle graph above, there are 125 juniors in the school. How many people are staffs?

(A) 100
(B) 85
(C) 60
(D) 45
(E) 30

TIP 43   Expected value

The probability of an event is the ratio of the number of ways that an event can occur to the number of possible outcomes when each outcomes is equally likely to occur.
If

$n(E) =$ the number of ways that event E can occur.

$n(S) =$ the number of possible outcomes in the sample space S.

$P(E) =$ the probability of an event E.

then

$$P(E) = \frac{n(E)}{n(S)}$$

Expected value is merely an average. If we were to flip a coin 20 times, the frequency weighted average of Head will be 10 times. Because the probability of getting a head is $\frac{1}{2}$. Therefore

$$n(head) = p \times n = \frac{1}{2} \times 20 = 10$$

1. If the probability of a boy's being born is $\frac{1}{3}$, and if a family plans to have 6 children, what is the expected number of boys?

   (A) 1
   (B) 2
   (C) 3
   (D) 4
   (E) 6

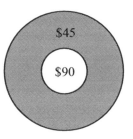

2. A carnival game consists of tossing a dart, which lands at a random spot within the larger circle. The shaded region loses $45 and the unshaded region wins $90. If the ratio of the radius of the smaller circle to the radius of the larger circle is 1:3, which of the following can be expected in this game?

   (A) lose $30
   (B) lose $10
   (C) make $10
   (D) make $20
   (E) make $30

Example:
How many 3-digit numbers have digits as odd integers?

$$\square\ \square\ \square$$

1	1	1
3	3	3
5	5	5
7	7	7
9	9	9

Hundreds place, tens place, and ones place have 5 possible odd integers each. Therefore,
$5 \times 5 \times 5 = 125$ numbers have digits as odd integers.

SAT Practice

1. How many 3-digit positive integers are odd and do not contain the digit 5?

(A) 64
(B) 288
(C) 360
(D) 400
(E) 420

2. For 4-digit numbers, the first digit is 8 and the third digit is 7. If the 4-digit numbers must have at least a 6 as a digit, and the numbers are even, how many numbers satisfy this condition?

(A) 10
(B) 12
(C) 14
(D) 16
(E) 18

3. How many 4 digit numbers between 5,000 and 10,000 are odd numbers?

(A) 200
(B) 400
(C) 1000
(D) 2500
(E) 3000

TIP 45 — Counting Multiples

Example:

In the first 1000 positive integers, how many numbers are multiples of 7?

Solution

$\left\lfloor \dfrac{1000}{7} \right\rfloor = 142$, $\lfloor x \rfloor$ = Greatest integer function.

142 is the greatest integer when you divide 1000 by 7.

Therefore, there are 142 numbers that are multiples of 7.

(*Answer*) 142

SAT Practice

2. For the first 1000 positive integers, how many integers are multiples of 3 or 4?

   (A) 470
   (B) 480
   (C) 500
   (D) 520
   (E) 550

1. Between 500 and 1000, how many integers are multiples of 5?

   (A) 99
   (B) 100
   (C) 150
   (D) 200
   (E) 300

3. Between 300 and 800, how many integers are multiples of 5 and 8?

   (A) 10
   (B) 12
   (C) 300
   (D) 799
   (E) 800

TIP 46   Average speed

Average speed is the total distance divided by the total time taken.

$$\text{Average speed} = \frac{\text{Total distance travelled}}{\text{Total time taken}}$$

Example:
If you travel from city A to city B at x miles per hour, and then you travel back at y miles per hour, what is the average speed for the whole trip?

(Sol)
Let the distance between city A and B be D.

Total distance $= 2D$, Total time $= \dfrac{D}{x} + \dfrac{D}{y}$

Average speed $= \dfrac{2D}{\dfrac{D}{x} + \dfrac{D}{y}} = \dfrac{2xy}{x + y}$

You can use any convenient number for D.

SAT Practice

1. A fellow travels from city A to city B in 3 hours. For the first hour, he drove at the constant speed of 40 miles per hour. Then he instantaneously increased his speed and, for the next 2 hours, kept it at 50 miles per hour. Find the average speed (mph) of the motion.

   (A) 45
   (B) $46\dfrac{2}{3}$
   (C) $46\dfrac{3}{4}$
   (D) 47
   (E) $48\dfrac{1}{3}$

2. A fellow travels from city A to city B. The first half of the way, he drove at the constant speed of 20 miles per hour. Then he increased his speed and traveled the remaining distance at 30 miles per hour. Find the average speed of the motion.

   (A) 24
   (B) 25
   (C) 26
   (D) 28
   (E) 30

3. Suppose we're driving a car from Amherst to Boston at a constant speed of 60 miles per hour. On the way back from Boston to Amherst, we drive a constant speed of 30 miles per hour. What is the average speed (mph) for the round trip?

   (A) 40
   (B) 43
   (C) 44
   (D) 45
   (E) 50

TIP 47 — Factoring

Factoring is to write an expression as a product of factors.
For SAT questions, the following factorings are needed.

(1) $a^2 + 2ab + b^2 = (a+b)^2$

(2) $a^2 - 2ab + b^2 = (a-b)^2$

(3) $a^2 - b^2 = (a+b)(a-b)$

(4) $a^2 - 2a - 3 = (a-3)(a+1)$

SAT Practice

1. If $(x-3)(x+3) = a$, then $(2x-6)(x+3) =$

(A) $2a$
(B) $3a$
(C) $4a$
(D) $5a$
(E) $6a$

2. Which of the following is equivalent to $\left(n - \dfrac{1}{n}\right)^2 + 4$?

(A) 4

(B) 8

(C) $\left(n + \dfrac{1}{n}\right)^2$

(D) $n^2 + \dfrac{1}{n^2}$

(E) $n^2 - \dfrac{1}{n^2}$

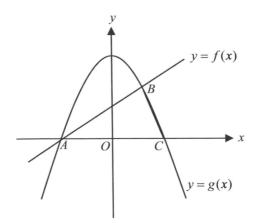

Note: Figure not drawn to scale.

3. In the figure above, the graph of a function f is defined by $f(x) = x + 2$ and the graph of a quadratic function g is defined by $g(x) = -x^2 + 4$. What is the area of $\triangle ABC$?

(A) 18
(B) 12
(C) 9
(D) 6
(E) 5

4. If $x^2 - y^2 = 24$, where x and y are positive integers $(x > y)$, what is one possible value of x?

TIP 48   **Divisiblity and Prime**

To determine if a number is prime or composite,

1. Find all of the factors of the number.
2. If the number has only two factors, 1 and itself, then it is prime.
3. If the number has more than two factors, then it is composite.

SAT Practice

1. If k is divisible by 2, 3, and 15, which of the following is also divisible by these numbers?

(A) $k+10$
(B) $k+15$
(C) $k+20$
(D) $k+30$
(E) $k+40$

2. If k is a positive integer divisible by 7, then which of the following must also divisible by 7?

(A) $\dfrac{k}{7}$

(B) $k+\dfrac{k}{7}$

(C) $2k+14$

(D) $\dfrac{77}{k}$

(E) $3k+10$

3. If n is divisible by 2, 3 and 5, which of the following is also divisible by these numbers?

(A) $n+12$
(B) $2n+15$
(C) $3n+20$
(D) $4n+45$
(E) $5n+60$

4. Which of the following numbers is divisible by the largest prime factor?

(A) 250
(B) 260
(C) 300
(D) 320
(E) 400

5. If a, b, and c are consecutive integers, which of the following must be true?

I. $a+b+c$ is divisible by 2
II. $a+b+c$ is divisible by 3
III. $a+b+c$ is divisible by 6

(A) I only
(B) II only
(C) III only
(D) I and II only
(E) II and III only

TIP 49 — Rate of Work

Let's assume we have two workers : A and B.
1) Worker A can finish a job in a hours when walking alone.

2) Worker B can finish a job in b hours when working alone.

If two workers are working together, the number of hours they need to complete the job is given by

Worker	Rate	Combined	Time
A	$1/a$	$\dfrac{1}{a}+\dfrac{1}{b}$	$\dfrac{1}{1/a + 1/b}$
B	$1/b$		

$$Rate = \frac{1}{Time} \quad \& \quad Time = \frac{1}{Rate}$$

$$Time = \frac{1}{\dfrac{1}{a}+\dfrac{1}{b}} = \frac{a+b}{ab}$$

SAT Practice

1. Worker A can finish a job in 5 hours. When worker A works together with worker B, they can finish the job in 4 hours. How long does it take for worker B to finish the job if he works alone?

(A) 3 hours
(B) 8 hours
(C) 12 hours
(D) 16 hours
(E) 20 hours

2. Raymond and Peter can paint a house in 20 hours when working together at the same time. If Raymond works twice as fast as Peter, how long would it take Peter to paint the house if he works alone?

(A) 10 hours
(B) 20 hours
(C) 30 hours
(D) 40 hours
(E) 60 hours

3. The swimming pool can be filled by pipe A in 5 hours and by pipe B in 8 hours. How long would it take to fill the pool if both pipes were used?

(A) $3\dfrac{1}{13}$ hours

(B) $5\dfrac{2}{3}$ hours

(C) 7 hours

(D) $8\dfrac{1}{3}$ hours

(E) 9 hours

4. If it takes 5 people 12 hours to paint 3 identical houses, then how many hours will it take 4 people working at the same rate to paint 5 identical houses?

(A) 15 hours
(B) 18 hours
(C) 19 hours
(D) 20 hours
(E) 25 hours

TIP 50 | Parallel lines

If a set of parallel lines are cut by a transversal, each of the parallel lines has 4 angles surrounding the intersection.

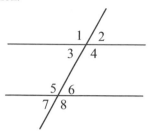

$\angle 1 \cong \angle 4$ and $\angle 2 \cong \angle 3$: Vertical angles

$\angle 2 \cong \angle 6$ and $\angle 4 \cong \angle 8$: Corresponding angles

$\angle 3 \cong \angle 6$ and $\angle 4 \cong \angle 5$: Alternate angles

$\angle 3 + \angle 5 = 180^o$ and $\angle 4 + \angle 6 = 180^o$:
Sum of the interior angles in the same side is 180°.

SAT Practice

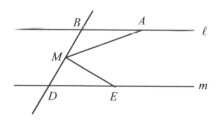

Note: Figure not drawn to scale.

1. In the figure above, If $AB = BM = DM = DE$ and $\ell \| m$, what is the measure of $\angle AME$?

    (A) 45°
    (B) 50°
    (C) 60°
    (D) 75°
    (E) 90°

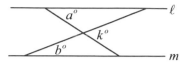

Note: Figure not drawn to scale.

2. In the figure above, $\ell \| m$, $a = 65$, and $b = 45$. What is the value of k ?

    (A) 70
    (B) 80
    (C) 90
    (D) 100
    (E) 110

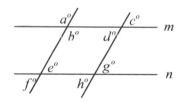

Note: Figure not drawn to scale.

3. In the figure above, line m is parallel to line n. Which of the following must be true?

    I. $a = c$
    II. $d = g$
    III. $b + e = 180$

    (A) I only
    (B) II only
    (C) III only
    (D) II and III only
    (E) I, II, and III

END.

NO MATERIAL ON THIS PAGE

50 TIPS ANSWER

TIP 1

1. (C)

Since $78 < s < 86$, the middle number is $\dfrac{78+86}{2} = 82$ and the distance is $86-82 = 4$. Therefore, $|s-82| < 4$.

2. (D)

The middle number is $\left(5\dfrac{3}{7}+6\dfrac{4}{7}\right) \div 2 = 6$, and the distance between $6\dfrac{4}{7}$ and 6 is $\dfrac{4}{7}$. Therefore,

$$|f-6| \le \dfrac{4}{7} \ \rightarrow \ |f-6| = |6-f| \le \dfrac{4}{7}.$$

3. (E)

Since $E < 450$ or $E > 500$, the middle number is 475 and the distance is 25. Therefore, $|E-475| > 25$.

TIP 2

1. (B)

The ratio of the lengths of the circles is 2:4:6=1:2:3. The ratio of the areas of the circles is $1:4:9 = k : 4k : 9k$. Thus the area of the shaded region is $9k-(k+4k) = 4k$. Therefore,

$$\frac{\text{area of shaded region}}{\text{area of the largest circle}} = \frac{4k}{9k} = \frac{4}{9}.$$

2. (E)

Those three triangles are similar.
The ratio of the lengths= 3:5:7.
The ratio of the areas=9:25:49=$9k : 25k : 49k$.

Since the area of $PRSQ$ is 48,
$(25k-9k) = 16k = 48$, or $k = 3$. Therefore, the area of $\triangle ABC$ is $49k = 49 \times 3 = 147$.

3. (A)

Since $\triangle PQM \sim \triangle SRM$, the ratio of the lengths is $\sqrt{4} : \sqrt{9} = 2 : 3 = 2k : 3k$

Thus $2k = 15$, then $k = 2.5$. Therefore, the perimeter of $\triangle SRM$ is $3k = 3(7.5) = 22.5$.

TIP 3

1. (D)

$$-2 < x < 4$$
$$-3 < y < 2$$

Largest of $(x-y)$ is $4-(-3) = 7$ and smallest of $(x-y)$ is $-2-2 = -4$. Therefore, $-4 < x-y < 7$.

2. (C)

$$2 < q < 6$$
$$1 < p < 4$$

$$\frac{2}{4} < \frac{q}{p} < \frac{6}{1}$$

TIP 4

1. (D)

	Male	Female	
50 ↑	$\frac{4}{35}$	$\frac{6}{35}$	$\frac{2}{7}$
50 ↓		$\frac{9}{35}$	$\frac{5}{7}$
	$\frac{4}{7}$	$\frac{3}{7}$	

Male & 50 $\uparrow = \frac{1}{5} \times \frac{4}{7} = \frac{4}{35}$. When you fill out

the blanks, female & 50 $\downarrow$ is $\frac{9/35}{3/7} = \frac{3}{5}$.

2. (D)

	M	F	
C	12	4	16
NC		2	8
	18	6	

From the table above, the answer is 12.

3. (D)

	Boys	Girls	
15 $\uparrow$	$10/77$	$1/7$	$3/11$
15 $\downarrow$	$23/77$	$3/7$	$8/11$
	$3/7$	$4/7$	

$(G)(15 \downarrow) = \frac{3}{4} \times \frac{4}{7} = \frac{3}{7}$. When you fill out the

blanks, $\frac{(B)(15 \downarrow)}{(B)} = \frac{23}{33}$.

1. (C)

Since $AB = 13$, $\frac{13}{18} = \frac{5}{DF}$. Therefore,

$DF = \frac{90}{13}$.

2. (C)

Direct proportion. $\frac{1}{a} = \frac{a}{5a} \rightarrow a = 5$.

3. (C)

Direct proportion: $y = kx$. y – intercept must
be zero. Therefore, the graph passes through the
origin.

4. (C)

Direction proportion: $y = k(x^2)$. The graph
passes through the origin.

1. (B)

Distance=Speed $\times$ Time. $45 \times 5 = 50 \times x$

$\rightarrow x = 4.5 \, \text{hr} = 4$ hour 30 minutes.

2. (C)

Inverse variation: $xy = k$ (constant). Since
$2 \times 25 = 4 \times a = 8 \times b$, $a = 12.5$ and $b = 6.25$.
Therefore, $a + b = 18.75$.

3. (D)

Job problem: days $\times$ workers $= k$ (constant) .
Thus, $10 \times 2 = 5 \times x$. Therefore, $x = 4$.

4. (B)

$Length \times Width = Area$ (Constant).
Therefore, $10 \times 20 = x \times 40,$ or $x = 5$.

5. (C)

Inverse variation: $xy = k$ (constant) $\rightarrow y = \frac{k}{x}$.

Therefore, the graph (C) is correct.

6. (C)

$p \times h = n \times x \rightarrow x = \frac{ph}{n}$.

7. (E)

To complete the same job, $5 \times d = 2 \times x$ or

$x = \frac{5d}{2}$ days. To complete $\frac{1}{3}$ of the job, it will

take $\frac{1}{3} \times \frac{5d}{2} = \frac{5d}{6}$ days.

1. (C)

The area of $\triangle BCD$ is

$\dfrac{2s \times s\sqrt{3}}{2} = 16\sqrt{3} \rightarrow s = 4$. $AB = 8$. Therefore, the area of the square is $8 \times 8 = 64$.

2. (E)

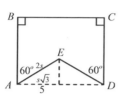

In the figure above, $s\sqrt{3} = 5$ or

$s = \dfrac{5}{\sqrt{3}} = \dfrac{5\sqrt{3}}{3}$. Therefore, $2s = \dfrac{10\sqrt{3}}{3}$, and

the perimeter is

$3 \times 10 + 2 \times \dfrac{10\sqrt{3}}{3} = 30 + \dfrac{20\sqrt{3}}{3}$.

3. (E)

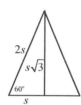

In the figure above, $2s = 8$ or $s = 4$. Therefore,

the area of AOB is $\dfrac{8 \times 4\sqrt{3}}{2} = 16\sqrt{3}$.

4. (E)

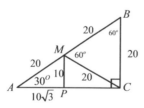

In the figure above, $\triangle AMC$ is an isosceles and $\triangle MBC$ is an equilateral triangle. Therefore, the

area of $\triangle ABC$ is $\dfrac{20\sqrt{3} \times 20}{2} = 2003$.

TIP 8

1. (D)

$\left[(-2)^3 (2^3)^2 \right]^4 = (-2)^{12} (2)^{24} = (2)^{12} (2)^{24} = 2^{36}$

Or, $2^{36} = 2^{4n}$. Therefore, $n = 9$.

2. (D)

$4^3 + 4^3 + 4^3 + 4^3 = 4 \times 4^3 = 4^4 = 2^8$. Therefore, $2^8 = 2^n$ or $n = 8$.

3. (D)

$5m^5 n^{-3} = 20m^3 n \rightarrow \dfrac{m^5}{m^3} = \dfrac{20n}{5n^{-3}} \rightarrow m^2 = 4n^4$.

Therefore, $m = 2n^2$.

4. (C)

$(a^{-4}b)^{-1} = a^4 b^{-1} = a^4 (a^7)^{-1} = a^4 a^{-2} - a^2$.

Since $a^2 = 16$, $a = 4$.

5. (D)

$k^{-2} = \dfrac{2^7}{2^3} = 2^4$. $\left(k^{-2} \right)^{-\frac{1}{2}} = \left(2^4 \right)^{-\frac{1}{2}}$. Therefore,

$k = 2^{-2}$ or $k = \dfrac{1}{4}$.

6. (A)

$\left(p^{-3} \right)^{-\frac{1}{3}} = \left(2^{-6} \right)^{-\frac{1}{3}} \rightarrow p = 2^2 = 4$.

And $\left(q^{-2} \right)^{-\frac{1}{2}} = \left(4^2 \right)^{-\frac{1}{2}} \rightarrow q = 4^{-1} = \dfrac{1}{4}$.

Therefore, $pq = 4 \times \dfrac{1}{4} = 1$.

7. (E)

$\left(a^6 b^4 \right)^{\frac{1}{2}} = a^3 b^2 = 3^3 5^2$. Therefore, $a = 3$ and $b = 5$.

$a + b = 8$.

TIP 9

1. (A)

Since $BC = 2k$, $AD = 10k$. Therefore,

$P = \dfrac{BC}{AD} = \dfrac{2k}{10k} = \dfrac{1}{5}$.

2. (C)

The ratio of the lengths of the three circles = 1:2:3,

and the ratio of the areas = $1:4:9 = k:4k:9k$.

The area of the shaded region =

$9k - (k + 4k) = 4k$.

Therefore, $P = \dfrac{4k}{9k} = \dfrac{4}{9}$.

3. (E)

If the length of a side of the smaller square is a,

The area of the smaller circle = a^2, the area of

the circle = $\pi\left(\dfrac{a\sqrt{2}}{2}\right)^2 = \dfrac{\pi a^2}{2}$, and the area of

the larger square = $\left(a\sqrt{2}\right)^2 = 2a^2$. Therefore,

$P = \dfrac{\text{the area of the shaded region}}{\text{the area of the larger square}} = \dfrac{\dfrac{\pi a^2}{2} - a^2}{2a^2}$

$= \dfrac{\pi - 2}{4}$.

4. (D)

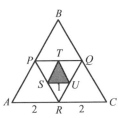

If $SU = 1$, then $Ar = 4$. The ratio of

$\dfrac{\text{the length of } SU}{\text{the length of } AC} = \dfrac{1}{4}$. Therefore, the ratio of

$\dfrac{\text{area of } \triangle STU}{\text{area of } \triangle ABC} = \dfrac{1^2}{4^2} = \dfrac{1}{16}$.

TIP 10

1. (D)

$\sqrt{x} \rightarrow x \geq 0$ and $x \neq 3$.

2. (D)

The graph will be as follows.

In the graph above, Range: $y \geq -5$.

TIP 11

1. (C)

$f(0) = 2 \rightarrow (0, 2)$, $f(3) = 5 \rightarrow (3, 5)$, and

$f(5) = k \rightarrow (5, k)$. If the function is linear, the

slope between any two points is constant.

Therefore, $\dfrac{5 - 2}{3 - 0} = \dfrac{k - 2}{5 - 0}$ or $k = 7$.

2. (A)

For three points of $(0, a)$ $(1, 12)$ $(2, b)$, the

slope is $\dfrac{12 - a}{1 - 0} = \dfrac{b - 12}{2 - 1}$ or $12 - a = b - 12$.

Therefore, $a + b = 24$.

3. (E)

Since $a > 0$, $b < 0$, $c > 0$, and $y = -\dfrac{a}{b}x - \dfrac{c}{b}$,

slope $= -\dfrac{a}{b} \rightarrow -\dfrac{(+)}{(-)} > 0$ and y-intercept

$= -\dfrac{b}{c} \rightarrow -\dfrac{(-)}{(+)} > 0$. Therefore, the graph has a

positive slope and a positive y-intercept.

4. (E)

For the given points $(3, 2)$ and $(5, 6)$, slope is

$\dfrac{6 - 2}{5 - 3} = 2$. Thus, the equation of the line is

$y = 2x + b$. Since point $(3, 2)$ lies on the line,

$2 = 2(3) + b$. Therefore, $b = -4$.

5. (C)

For the given points $(3, -2)$, $(4, -4)$, the

slope $= \dfrac{-4 - (-2)}{4 - 3} = -2$ and the equation is

$y = -2x + b$. Substitute $(3, -2)$ into the

equation, that is $-2 = -2(3) + b$ or $b = 4$.

Then $y = -2x + 4$.

To find x-intercept, $0 = -2x + 4 \rightarrow x = 2$.

6. (D)

The function has two x-intercepts (two zeros).

7. (C)

Slope $= \dfrac{0-3}{5-0} = \dfrac{-3}{5}$. slope of the points

(a, b) and $(5, 0)$ is $\dfrac{b-0}{a-5} = \dfrac{2a}{a-5} = \dfrac{-3}{5}$.

Therefore, $a = \dfrac{15}{13}$.

8. (B)

Substitute the values of t into the function given.

9. (A)

Slope $= \dfrac{9}{5} = \dfrac{\Delta F}{\Delta C} = \dfrac{27}{\Delta C}$. $\Delta C = \dfrac{27 \times 5}{9} = 15$.

10. (B)

Slope $= \dfrac{7}{12} = \dfrac{35}{\Delta Q}$. $\Delta Q = \dfrac{35 \times 12}{7} = 60$.

11. (C)

In the figure above, $t = 5 + 4 = 9$.

1. (D)

Triangle inequality: $9 - 3 < x + 3 < 9 + 3$.

Therefore, $3 < x < 9$.

2. (E)

Since $4 - 3 < PQ < 4 + 3$ and

$8 - 6 < PQ < 8 + 6$,

Thus, $2 < PQ < 7$.

3. (B)

$9 - 6 < 3 < 9 + 6 \rightarrow 3 < 3 < 15$ (False).

1. (C)

There are two possible arrangements for minimum number of white marbles as follows.

RWRWRWRWRWR $\rightarrow$ Five white marbles

WRWRWRWRWRW $\rightarrow$ Six white marbles

Therefore, minimum number is 5.

2. (C)

$5 \times 4 \times 3 \times 2 \times 1 = 120$.

3. (B)

T □ □ □ □ T
  L E E R

The number of ways to arrange 4 letters is $4 \times 3 \times 2 \times 1 = 24$. Because two letters E, E are the same, break the order by dividing by $2!$.

Therefore, $\dfrac{24}{2!} = 12$.

4. (B)

W B R Y

W B R Y

(W)

The ninth marbles gives the 3 of the same color.

5. (C)

○ □ □ ○

The arrangements of numbers $= 3 \times 2 = 6$.

The arrangements of letters $= 3 \times 2 = 6$.

Therefore, $6 \times 6 = 36$.

6. (B)

5	5	
5		5
	5	5

There are three possible ways as above.

Therefore, $P = \left(\dfrac{1}{6} \times \dfrac{1}{6} \times \dfrac{5}{6} \right) \times 3 = \dfrac{5}{72}$.

7. (C)

When spun twice, there are 16 possible outcomes.

The outcome that the first digit is greater than the second are

$(2,1)$ $(3,1)$ $(3,2)$ $(4,1)$ $(4,2)$ $(4,3)$.

Therefore, $P = \dfrac{6}{16} = \dfrac{3}{8}$.

8. (A)

Two possible outcomes: WB or BW.

For WR, $P = \dfrac{4}{6} \times \dfrac{2}{5} = \dfrac{8}{30}$ and for RW,

$P = \dfrac{2}{6} \times \dfrac{4}{5} = \dfrac{8}{30}$. Therefore, $\dfrac{8}{30} + \dfrac{8}{30} = \dfrac{8}{15}$.

TIP 14

1. (D)

For 12 people,

$11 + 10 + 9 + \dots + 2 + 1 = \left(\dfrac{11+1}{2}\right) \times 11 = 66$.

2. (B)

Since

$1 + 2 + 3 + 4 + 5 + 6 + 7 + 8 + \boxed{9} = 45$, there are

$\boxed{9} + 1 = 10$ people.

3. (C)

For 5 lines, $1 + 2 + 3 + 4 = 10$ intersection points.

4. (A)

For 5 points, $1 + 2 + 3 + 4 = 10$ lines.

TIP 15

1. (C)

Since $\dfrac{4}{40+x} = \dfrac{8}{100}$, $x = 10$.

2. (D)

	Mixture	Salt
Solution A	x	$0.2x$
Solution B	10	5

Since $\dfrac{0.2x+5}{x+10} = \dfrac{30}{100}$, cross multiplication

$\rightarrow 20x + 500 = 30x + 300 \rightarrow x = 20$.

3. (A)

	Mixture	Alcohol
Solution A	x	x
Solution B	10	25

Since $\dfrac{2.5+x}{10+x} = \dfrac{40}{100}$, $x = 2.5$.

4. (C)

	Mixture	Acid
Solution A	x	x
Solution B	G	$\dfrac{kG}{100}$

Therefore, $\dfrac{x + \dfrac{kG}{100}}{x + g} = \dfrac{m}{100}$.

Cross-multiplication $\rightarrow$

$1000x + kG = mx + mG \rightarrow x(100 - m)$

$= (m - k)G$

$\rightarrow x = \dfrac{(m-k)G}{100-m}$.

5. (E)

	Mixture	Salt
Solution A	M	$\dfrac{pM}{100}$
Solution B	G	$\dfrac{qG}{100}$

Then,

$\dfrac{\dfrac{pM}{100} + \dfrac{qG}{100}}{M + G} = \dfrac{r}{100} \rightarrow$

$\dfrac{0.01pM + 0.01qG}{M + G} = \dfrac{r}{100}$

Cross-multiplication $\rightarrow pM + qG = r(M + G)$.

Therefore, $r = \dfrac{pM + qG}{M + G}$

TIP 16

1. (B)

Since the points are $(3,6)$ and $(5,12)$, the slope is $\dfrac{12-6}{5-3} = 3$.

2. (C)

$y = -\dfrac{1}{2}x + b$ and $(2,2)$ lies on the line.

Therefore, $2 = -\dfrac{1}{2}(2) + b \rightarrow b = 3$.

3. (A)

The slope is the constant between any two points. $\dfrac{a-5}{4-2} = \dfrac{23-5}{6-2} \rightarrow a = 11$.

4. (C)

$\dfrac{6-a}{5-2} = \dfrac{b-6}{8-5}$ or $6-a = b-6$. Therefore, $a + b = 12$.

5. (C)

Substitute the values of x into the choices.

6. (B)

Slope $= \dfrac{\triangle F}{\triangle C} = \dfrac{9}{5} \rightarrow \dfrac{\triangle F}{20} = \dfrac{9}{5} \rightarrow \triangle F = 36$.

7. (E)

$\dfrac{m-4}{42-0} = \dfrac{4-0}{0-(-3)} \rightarrow m = 60$.

1. (D)

Since p and q are prime, the number of factors is $(2+1)(4+1) = 15$.

2. (C)

$24p^2 = 2^3 \times 3 \times p^2$. Since they are distinct prime, the number of factors is $(3+1)(1+1)(2+1) = 24$.

3. (A)

Since $12 \times 24 = 2^5 \cdot 3^2$, the number of factors is $(5+1)(2+1) = 18$.

1. (B)

Since $g(2) = 3f(2) - k = 25$ and $f(2) = 13$, $3(13) - k = 25$. Therefore, $k = 14$.

2. (C)

$g(1) = 2f(1) - 3 = 3 \rightarrow f(1) = 3$
$\rightarrow$ point$(1,3)$.
$g(3) = 2f(3) - 3 = 5 \rightarrow f(3) = 4$
$\rightarrow$ point$(3,4)$.
For $f(x) = ax + b$,
$3 = a + b$ and $4 = 3a + b$. Therefore,
$a = \dfrac{1}{2}$ and $b = \dfrac{5}{2}$.

3. (D)

$g(k) = 3f(k) - 1 = 8$ or $f(k) = 3$. Therefore, there are three values of k which have y-coordinate of 3.

1. (C)

Middle number is equal to the average. Therefore, the sum is $30 \times 11 = 330$.

2. (C)

Since the 50th number is 80, the 99th number is $80 + (99 - 50) = 129$.

3. (B)

Since the average is 77, the sum is $77 \times 10 = 770$.

4. (C)

The middle term is $\dfrac{1+30}{2} = 15.5$ or 16th term is 121. Therefore, the 30th term is $121 + 2 \times (30 - 16) = 149$.

1. (E)

Since $(a-b)(k-1)=0$, $a=b$ or $k=1$ are the answers for "could be true." Also " $a=2$ and $b=2$ " belongs to " $a=b$".

2. (D)
Square both sides. $a-b=a+b \rightarrow b=0$.
And also, when $b=0$, $a>b \rightarrow a>0$.

3. (B)
$a^2+b^2-2ab=0 \rightarrow (a-b)^2=0$. Therefore, $a=b$. III is the answer for "could be true."

4. (C)
Square both sides. $\not{a}+\not{b}=\not{a}+\not{b}+2\sqrt{ab}$.
$ab=0$.

5. (E)
$(k-2)(x-y)=x-y \rightarrow (k-3)(x-y)=0$.
Therefore, $k=3$ or $x=y$.

6. (B)

$\left(\sqrt{a^2+b^2}\right)^2=(a-b)^2 \rightarrow a^2+b^2=a^2-2ab+b^2$

Therefore, $ab=0$. But $a>0$, then $b=0$.

TIP 21

1. (D)

$-30,-29,-28,....-1, 0, 1,........30, 31, 32, 33$

$\underbrace{\qquad\qquad\qquad\qquad}_{\text{Sum}=0} \underbrace{\qquad}_{\text{Sum}=96}$

Therefore, $x=33$.

2. (A)

$-20,-18,-16,...,-2,0,2,....20,22,24,26.$

$\underbrace{\qquad\qquad\qquad\qquad}_{\text{Sum}=0} \underbrace{\qquad}_{\text{Sum}=72}$

Therefore, there are 24 even integers.

3. (D)
$-60,-59,...-1,0,1,...60,61$

$\underbrace{\qquad\qquad\qquad}_{\text{Sum}=0}$

From the list above, there are 122 integers.

TIP 22

1. (E)
Since no solution, $\dfrac{2}{4}=\dfrac{-5}{k}\neq\dfrac{8}{17}$. $k=-10$.

2. (A)
Since infinite solution, $\dfrac{5}{a}=\dfrac{-2}{b}=\dfrac{3}{6}$.
$a=10$ and $b=-4$. Therefore,
$a+b=10-4=6$.

3. (D)
$\dfrac{3}{a}=\dfrac{b}{-4}\neq\dfrac{3}{6}$. Therefore, $ab=-12$. Therefore, the product of a and b should be -12.

TIP 23

1. (B)
Since $x\times 0=0$, k should be 0.

2. (A)
$ax^2+bx+c=0 \rightarrow$ In order to be 0 for any x, $a=b=c=0$. Therefore, $a+b+c=0$.

3. (A)
Since they are identical, $k+1=a$ and $k=5$.
Therefore, $a=6$.

4. (D)
$a(x+1)+b(x-1)=(a+b)x+(a-b)=2x+4$.
Therefore, $a+b=2$ and $a-b=4$. When solving the system of equations, $a=3$ and $b=-1$.

1. (B)

Pythagorean Theorem: Since
$a^2 + b^2 = c^2$, $c = \sqrt{20 + 16} = 6$.

2. (B)
$AC = \sqrt{2^2 + 1^2} = \sqrt{5}$.

	$\triangle BCF$	$\triangle ADB$	$\triangle ACE$
The ratio of lengths	1	2	$\sqrt{5}$
The ratio of areas	1	4	5

or the areas of the three triangles $= k : 4k : 5k$.
Since $5k = 20$, $k = 4$. Therefore, the area of
$\triangle ADB = 4k = 4(4) = 16$.

3. (C)
Since $AD = CD = 2\sqrt{2}$, $BC = 2\sqrt{2} \times \sqrt{2} = 4$.

4. (E)
Since $a^2 + b^2 = 100$,
$(a+b)^2 = a^2 + b^2 + 2ab > 100$.

5. (B)

In the figure above, $\overline{AJ1} = 8 + 12 = 15$ and
$\overline{AJ2} = 20$. Therefore,
$\overline{J1J2} = \sqrt{15^2 + 20^2} = 25$.

6. (C)

$BC = \sqrt{10^2 + 20^2} = 10\sqrt{5}$.

The area of $\triangle ABC$ is $\dfrac{10 \times 20}{2} = 100$

and another way , the area is

$\dfrac{10\sqrt{5} \times x}{2} = 100$. Therefore, $x = 4\sqrt{5}$.

1. 9

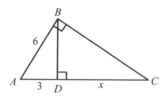

In the figure above,
$6^2 = 3(3 + x) \;\rightarrow\; 36 = 9 + 3x \;\rightarrow\; x = 9$.

2. (D)
$AB^2 = 9(9 + 16) = 225 \;\rightarrow\; AB = 15$.

3. (B)
$BC^2 = 16(16 + 9) = 400 \;\rightarrow\; BC = 20$.

4. (C)
$BD^2 = 9 \times 16 = 144 \;\rightarrow\; BD = 12$.

1. (B)

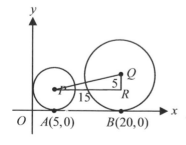

In the figure above, $PR = 20 - 5 = 15$ and $QR = 10 - 5 = 5$. Therefore, the slope is

$$\frac{5}{15} = \frac{1}{3}.$$

2.   (D)

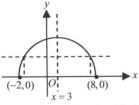

Since the midpoint of 8 and -2 is 3,

(D) : $\dfrac{1+5}{2} = 3$, is correct.

3.   (B)

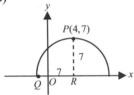

In the figure above, $PR = 7$ and $QR = 7$.
Therefore, x – coordinate of Q is $4 - 7 = -3$.

TIP 27

1.   (B)

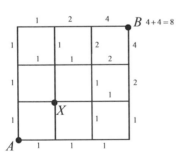

2.   (D)

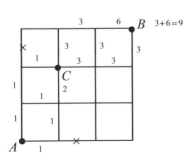

TIP 28

1.   (C)
Translate: Move to the left by 2 and move down by -2.

TIP 29

1.   (C)
Since $a = 50 - 2b \geq 10$, $2b \leq 40 \rightarrow b \leq 20$.
Therefore, the greatest one is 20.

2.   (B)
Since $a = 30 - b > 12$, $b < 18$.

3.   (C)
Since x is an integer, the maximum of x is 7.
When $x = 7$, $2(7) + y = 32$. Therefore the minimum of y is 18.

4.   (C)
To get the greatest value of b, a should be the smallest number. Since a and b are integers, $a = 1.25b$, but b must be a multiple of 4.

($\because a = \dfrac{5}{4}b$)

Since $1.25b + b = 2.25b < 1800$, $b < 800$.
Therefore, $b = 796$ and $a = 1.25b = 995$.
Both a and b are integers.

TIP 30

1.   (B)
For $y = ax^2 + bx + c$, axis of symmetry $= \dfrac{-b}{2a}$.

$$t = \frac{-b}{a} = \frac{-256}{-32} = 8$$

Or, $16t(16 - t) = 0$, Two zeros are $t = 0$, and $t = 16$.

Axis of symmetry $= k = \dfrac{0 + 16}{2} = 8$.

2.   (D)
When $t = 8$, h has the maximum height.
Therefore, $h(8) = 256 \times 8 - 16 \times 8^2 = 1024$.

1. (B)

Since $\dfrac{20}{100} \times \dfrac{30}{100} x = \dfrac{10}{100} \times \dfrac{k}{100} x$, $600 = 10k$.

Therefore, $k = 60$.

2. (C)

$(1 + 0.15) \times (1 - 0.3)x = 0.805x \rightarrow 80.5\%$ of original price.

3. (D)

Since $2a + 3b = 2.5(6b) = 15b$,

$2a = 12b \rightarrow a = 6b$.

Therefore, $\dfrac{a}{b} = \dfrac{6b}{b} = 6$.

4. (B)

Since

$\dfrac{25}{100} m = 50 \rightarrow m = 200$,

$\dfrac{15}{100} \times 2m = \dfrac{15}{100} \times 400 = 60$.

5. (C)

Since $c(n) = 20,000x^n = 20,000(1 + 0.025)^n$,

$x = 1.025$.

6. (C)

% of increase: $\dfrac{1000 - 500}{500} \times 100 = 100\%$.

7. (E)

$(1 + 0.06)(1 - 0.95)x = 1.007x = (1 + 0.007)x$.

Therefore, $0.007 \times 100 = 0.7\%$.

8. (D)

Since $A = 0.25 \times 2B = 0.5B$,

$\dfrac{B}{A} = \dfrac{B}{0.5B} = 2 \rightarrow 200\%$.

1. (B)

Proportion: $\dfrac{6}{1.2} = \dfrac{15}{x} \rightarrow x = \$3$.

2. (C)

Make a table.

	Now	$\rightarrow 5$	$\rightarrow 5$
Julie	10	3k	$3k + 5$
Song	20	5k	$5k + 5$

In 10 years, $\dfrac{3k + 5}{5k + 5} = \dfrac{2}{3}$. Thus $k = 5$.

Therefore, their current ages are,

Julie $= 15 - 5 = 10$ and Song $= 25 - 5 = 20$.

Sum $= 10 + 20 = 30$.

3. (C)

$3x = 2y \rightarrow y = \dfrac{3x}{2}$. Substitute into y,

$2x + 5\left(\dfrac{3x}{2}\right) = 76 \rightarrow x = 8$.

4. 160 or 320

Since $\dfrac{a}{b} = \dfrac{7k}{9k}$ and $b = 9k$ are divisible by 5,

k must be a multiple of 5. Thus

$b = 9k = 45, 90, 135, 180, 225, 270....$ .Then

$a = 7k = 35, 70, 105, 140, 175, 210.....$

Because a is even, and $a + b < 400$,

$(a, b) = (70, 90), (180, 140), (270, 210)....$

Therefore, $a + b = 160, 320$.

1. (C)

$$P = \begin{matrix} R \\ \dfrac{3}{6} \end{matrix} \times \begin{matrix} R \\ \dfrac{2}{5} \end{matrix} = \dfrac{1}{5}.$$

2. (A)

Since 3, 5, and 9 were taken already, seven numbers are left as follows.

1, 2, 4, 6, 7, 8, 10 (5 even out of 7)

$P(eve, even) = \dfrac{5}{7} \times \dfrac{4}{6} = \dfrac{10}{21}$

3. (E)

Since there are b blue marbles out of $(b+g)$ marbles. The probability of selling a blue one first is $\dfrac{b}{b+g}$. Now we have $(b-1)$ blue marbles out of $(b+g-1)$. Therefore,

$$P = \left(\frac{b}{b+g}\right)\left(\frac{b-1}{b+g-1}\right).$$

4. (C)

There are three different possible outcomes as follows.

$$H\ T\ T,\ T\ H\ T,\ \ T\ T\ H$$

Each probability is $\dfrac{1}{2} \times \dfrac{1}{2} \times \dfrac{1}{2} = \dfrac{1}{8}$.

Therefore, $P = 3 \times \dfrac{1}{8} = \dfrac{3}{8}$.

5. (B)

Since $G = 40$, $R = 35$, and $A = 10$, The probability of $Green = \dfrac{40}{85} = \dfrac{8}{17}$.

6. (D)

$|x+2| < 8 \ \rightarrow -8 < x+2 < 8 \ \rightarrow -10 < x < 6$.

Therefore, $(-5, 0, 5)$ belongs to this interval.

$P = \dfrac{3}{8}$.

TIP 34

1. (D)

$$A\ B\ C\ D\ E$$
$$A\ B\ C\ D\ E$$
$$(?)$$

In the figure above, the 11th marble will guarantee 3 of any same color.

2. (C)

$$A\ B\ C\ D$$
$$A\ B\ C\ D$$
$$A\ B\ C\ D$$
$$A\ B\ C\ D$$
$$(?)$$

The 21st number will guarantee 5 of the same color.

3. (A)

$$A\ B\ C\ D$$
$$A\ B\ C\ D$$
$$(?)$$

The 9th marble.

TIP 35

1. (E)

Let the coordinates of point B be (x, y).

Then, $\dfrac{5+x}{2} = 10$ and $\dfrac{1+y}{2} = 4$. Therefore, $x = 15$ and $y = 7$.

2. (D)

Midpoint: $\dfrac{x+2a}{2} = a \ \rightarrow x = 0$.

$\dfrac{b+y}{2} = b \ \rightarrow y = b$.

3. (B)

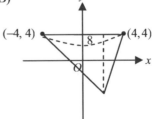

Since the area of $\triangle = \dfrac{1}{2} \times 8 \times h = 24$,

$h = 6$. Therefore, y-coordinate must be $4 - 6 = -2$ (downward) or $4 + 6 = 10$ (upward).

(B) $(2, 10)$ has y-coordinate of 10.

4. (D)

Distance $=\sqrt{(a-b)^2+(3-8)^2}=13$. Therefore,

$(a-b)^2=13^2-(3-8)^2=144 \rightarrow |a-b|=12$.

TIP 36

1. (A)

Use convenient numbers.

$(a,b,c) \rightarrow (2,2,1),(1,2,2),(2,1,2)$

   I. $a+b+c=5$ (always *odd*)

   II. $ab+c=odd, even, even$

   III. $ab(2c)=even$

2. (B)

Since $n^2=odd$, then $n=odd$. Use convenient numbers such as, $n=1,3,5..$

(A) $\dfrac{n}{2}=\dfrac{1}{2},\dfrac{3}{2}$.. not integers

(B) $2n+n=3n=odd$

(C) *even*

(D) $\dfrac{n+3}{3}=\dfrac{4}{3}$.. not integer

(E) $(n+1)(n-1)=even$

3. (E)

Let $a=1,3,5$ Only (E) $a^2-a=$ even for $a=1,3,5$.

4. (E)

Let $a=2$ and $b=4$. I, II, and III are all even.

5. (E)

Since $p+q=$ even, let $p=2$, $q=4$ or $p=1$ and $q=3$.

(A) $pq=even$ or *odd*.

(B) $2p+q=even$ or *odd*.

(C) $(p+1)(q+1)=odd$ or *even*.

(D) $\dfrac{p}{q}=$ not integer

(E) $p^2-q^2=even$ and *even*.

6. (D)

The product of three consecutive integers. If $k=2,4,6,8,10...$, then

$k(k+1)(k+2)=24,120,336...$

TIP 37

1. (E)

Since $a<b<b+3$, then $a<b+3$.

2. (E)

Since $a>b>0$, then a and b are both positive.

When divided by b, then $\dfrac{a}{b}>1$. Therefore,

$\dfrac{2a}{b}>2 \rightarrow \dfrac{2a}{b}$ is always greater than 1.

3. (D)

Since s^2,t^4, and u^2 are always positive,

$\dfrac{s^3t^4u^3w}{s^2t^4u^2}>0 \rightarrow suw>0$. Because $w<0$,

$su<0$.

4. (E)

Let $a=2$ and $b=3$.

(A) $\dfrac{1}{2}<\dfrac{1}{3}$ (false)

(B) $a-b=2-3>0$ (false)

(C) $b^2<ab \rightarrow 9<2\times3$ (false)

(D) $a+2b>3b \rightarrow 2+2(3)>3(3)$ (false)

(E) $2^2<b^2$ (true)

TIP 38

1. (C)

Let x be the length of an edge. Then

$x^3=64 \rightarrow x=4$. Therefore,

$AB=\sqrt{x^2+x^2+x^2}=\sqrt{48}=4\sqrt{3}$.

2. (C)

Let $x=$ the length of an edge. Since $6x^2=96$, then $x=4$. Therefore, volume$=4^3=64$.

3. (A)

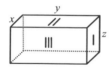

Volume $= xyz$.

The areas of the faces: $xz = 8$, $xy = 10$, and $yz = 20$. Thus, $(xy)(yz)(xz) = (10)(20)(8)$.

$(xyz)^2 = 1600 \rightarrow xyz = 40$.

4. (C)

The area of two circles $= 2\pi r^2$. The lateral surface area $= 2\pi r \times h = 2\pi r^2$. Therefore, $2\pi r^2 + 2\pi r^2 = 4\pi r^2$.

TIP 39

1. (A)

$$-2, -1, 0, 1, 2$$
$$-2, -1, 0, 1, 2$$
$$-2, -1, 0, 1, 2$$
$$.....................$$
$$.....................$$
$$-2, -1, 0, 1$$

Since sequence is repeating every 5 terms, the sum of the sequence up to 120 terms is 0. Therefore, the sum of next four terms is -2.

2. (E)

Since $a_1 = 9$ and common ratio is 3, then
$a_{300} = 9 \times 3^{300-1} = 3^2 \times 3^{299} = 3^{301}$.

3. (A)

Since $a_1 = 5$ and $r = -2$, the sum is

$$S_8 = \frac{5\left((-2)^8 - 1\right)}{(-2) - 1} = \frac{1275}{-3} = -425.$$

Or, $a_1 + a_2 + ... + a_8 = 5 + (-10) + .. + (-640)$
$\quad\quad = -42.5$

4. (C)

Let x be a salary.

$(1 + 0.1)^4 x = 1.4641x = (1 + 0.4641)x$.

Therefore, 46% increased.

5. (B)

After 1st bounce: $h = \frac{3}{5} \times 30 = 18$.

After 2nd bounce: $h = \frac{3}{5} \times 18 = 10.8$.

After 3rd bounce: $h = \frac{3}{5} \times 10.8 = 6.48$.

After the 4th bounce:

$h = \frac{3}{5} \times 6.48 = 3.888 < 4\,ft$

Therefore, the 4th bounce is correct.

6. (A)

$S = a(1 + 4 + 16 + 64 + 256 + 1024) = 4095$.
Therefore, $1365a = 4095 \rightarrow a = 3$.

7. (D)

The areas of the squares are changing with a constant ratio as follows.

$$1, \frac{1}{2}, \frac{1}{4}, \frac{1}{8}, \frac{1}{32}, \frac{1}{64}, \frac{1}{128}$$

TIP 40

1. (D)

Since $\frac{p+q}{p-q} = 3$, then $p = 2q$. Therefore,

$\frac{p}{q} = \frac{2q}{q} = 2$.

2. (B)

Since $4\Delta\left(\frac{k}{6}\right) = \frac{4}{k/6} = \frac{24}{k} = 3$, then $k = 8$.

3. (D)

$(n+1)^{\blacktriangle} = \underbrace{(n+1)n(n-1)(n-2)....1}_{n^{\blacktriangle}}$

Therefore, $(n+1)^{\blacktriangle} = (n+1)n^{\blacktriangle}$.

TIP 41

1. (B)

Since $300 = \frac{200 \times 50 - 400}{50} + k$, then $k = 108$.

2. (C)

Since $p(n) = 1200(1 - 0.012)^n$, c is equal to 0.988.

3. (D)

Since $f(-2) = 4$ and $f(3) = 9$,

$$m = \frac{9-4}{3-(-2)} = 1.$$

4. (C)

$$p(3) = 20000\left(1 - \frac{10}{100}\right)^3 = 20000 \times 0.9^3 = 14580.$$

TIP 42

1. (E)

The average cost per pound is equal to the slope of the line. Therefore, the slope of the line is

$\frac{6-0}{8-0} \cong 0.75$. The closest number is 0.73.

2. (B)

I. No exact information for 50 customers.
II. The median is 5 items.
III. Because not enough information for 50 customers. If all 50 customers bought 5 items, then the mode can be 5 items.

3. (C)

Let x = number of Staff.
Since the percent of Staff is 12%,

$$x = 125 \times \frac{12}{25} = 60.$$

TIP 43

1. (B)

The expected number $= 6 \times \frac{1}{3} = 2$ (boys)

2. (A)

The ratio of the lengths is 1:3
The ratio of the areas is 1:9. Therefore,

Probability (landing in shaded region) $= \frac{8}{9}$

Probability (landing unshaded region) $= \frac{1}{9}$

The expected amount $=$

$$\frac{8}{9} \times (-{}^\$45) + \frac{1}{9} \times ({}^\$90) = -{}^\$30$$

TIP 44

1. (B)
Multiply the possible choices.

1	0	1
2	1	3
3	2	~~5~~
4	3	7
~~5~~	4	9
6	~~5~~	
7	6	
8	7	
9	8	
	9	

Therefore, $8 \times 9 \times 4 = 288$, the answer is 288 numbers.

2. (C)

8		7	

(1) One 6
If you have a 6 in the second place, the last digits will be 0, 2, 4, 8--- 4 numbers
If you have a 6 in the last place, the second digits will be 0,1,2,3,4,5,7,8,9 --- 9 numbers
(2) Two 6's
Second and last places, --- 1 number
Total $= 4 + 9 + 1 = 14$

3. (D)

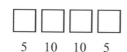

5	10	10	5

The first digits have 5 choices: 5, 6, 7, 8, 9
The second and third digits have 10 choices each.
The last digits must have 1, 3, 5, 7, 9: 5 choices.
Therefore, $5 \cdot 10 \cdot 10 \cdot 5 = 2500$

1. (A)

$\left\lfloor \dfrac{999}{5} \right\rfloor = 199$ and $\left\lfloor \dfrac{500}{5} \right\rfloor = 100$. Therefore, $199 - 100 = 99$.

2. (C)

$\left\lfloor \dfrac{1000}{3} \right\rfloor = 333$: There are 333 multiples of 3.

$\left\lfloor \dfrac{1000}{4} \right\rfloor = 250$: There are 250 multiples of 4.

$\left\lfloor \dfrac{1000}{12} \right\rfloor = 83$: There are 83 multiples of 12.

Therefore,

n(multiple 3 or 4)=n(multiple 3)+n(multiple 4)

　　　　　　-n(multiple 12)=333+250-83

　　　　　　=500 integers.

3. (B)

n(5 and 8) = n(40)

$n(40) = \left\lfloor \dfrac{799}{40} \right\rfloor - \left\lfloor \dfrac{300}{40} \right\rfloor = 19 - 7 = 12$

Therefore, 12 is the answer.

1. (B)

Distance traveled for the first hour = (40)(1) = 40miles

Distance traveled for the next two hours = (50)(2) = 100miles

Average speed

$= \dfrac{\text{Total distance}}{\text{Total hour}} = \dfrac{40+100}{1+2} = 46\dfrac{2}{3} mph$

2. (A)

Distance: 　1　　　　　1

Time 　$\dfrac{1}{20}$ 　+　 $\dfrac{1}{30}$

Average speed

$= \dfrac{2}{\dfrac{1}{20} + \dfrac{1}{30}} = 24mph$

3. (A)

Let distance from city A to B be 1.

Average speed $= \dfrac{2}{\dfrac{1}{60} + \dfrac{1}{30}} = 40mph$.

1. (A)

$(2x-3)(x+3) = 2(x-3)(x+3) = 2a$.

2. (C)

$\left(n - \dfrac{1}{n}\right)^2 + 4 = n^2 + 2 + \dfrac{1}{n^2}$

$= \left(n + \dfrac{1}{n}\right)^2$

3. (D)

At point B, $f(x) = g(x)$.

$x + 2 = -x^2 + 4 \rightarrow x^2 + x - 2$

$= (x+2)(x-1) = 0$

$f(1) = 1 + 2 = 3$.

Therefore, $A = (-2, 0)$ and $B = (1, 3)$.

At point C, $g(x) = 0$. $C = (2, 0)$. $AC = 4$ and height =3.

The area of $\triangle ABC = \dfrac{4 \times 3}{2} = 6$.

4. 5 or 7

$(x+y)(x-y) = 24$

$\begin{cases} x+y = 24 & 12 & 8 & 6 \\ x-y = 1 & 2 & 3 & 4 \end{cases}$

Therefore, $2x = 14, 10 \rightarrow x = 5$ or 7.

1. (D)

Least common multiple = 30. Therefore, the number must be a multiple of 30. $k + 30 = 60$.

2. (C)

Let $k = 7, 14, 21...$

(A) $\dfrac{k}{7} = 1, 2, 3$ are not multiples of 7.

(C) $2(k + 7) =$ multiple of 7.

3. (E)

LCM=30.

Only (E) is a multiple of 30.

4. (B)

(A) $250 = 2 \times 5^3$

(B) $260 = 2^2 \times 5 \times 13$

(C) $300 = 2^2 \times 3 \times 5^2$

(D) $320 = 2^6 \times 5$

(E) $400 = 2^2 \times 5^2$

Therefore, 13 is the largest prime number.

5. (B)

Let $\begin{cases} a = 1 \\ b = 2 \\ c = 3 \end{cases}$ or $\begin{cases} a = 2 \\ b = 3 \\ c = 4 \end{cases}$

Then, $a + b + c = 6$ or 9.

Only II is correct.

TIP 49

1. (E)

The working rate must be equal.

Let x be the hours for worker B.

$\dfrac{1}{4} = \dfrac{1}{5} + \dfrac{1}{x} \rightarrow \dfrac{1}{x} = \dfrac{1}{4} - \dfrac{1}{5} = \dfrac{1}{20}$

Therefore, $x = 20$ hours.

2. (E)

If Raymond takes x hours, then Peter will take $2x$ hours. Therefore, the rate of work is equal.

$\dfrac{1}{x} + \dfrac{1}{2x} = \dfrac{1}{20} \rightarrow x = 30$. $2x = 60$ hours.

3. (A)

The rate is equal. Let $x =$ hours when working together.

$\dfrac{1}{5} + \dfrac{1}{8} = \dfrac{1}{x} \rightarrow \dfrac{13}{40} = \dfrac{1}{x} \rightarrow x = 3\dfrac{1}{13}$ hours.

4. (E)

For one house, five people need 4 hours.

Four people need 5 hours to paint one house as follows.

$5 \times 4 = 4 \times x \rightarrow x = 5$ hours for one house.

To 5 identical houses, they need

$5 \times 5 = 25$ hours.

TIP 50

1.

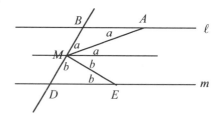

$\triangle ABM$ and $\triangle DEM$ are isosceles.

In the figure above, $2(a + b) = 180$.

Therefore, $a + b = 90$

2. (E)

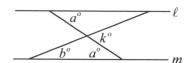

In the figure, the exterior angle

$k = a + b = 45 + 65 = 110$.

3. (D)

Because the other two lines may not be parallel, a is not always equal to b.

End

NO MATERIAL ON THIS PAGE

Dr. John Chung's SAT Math

TEST
1

ANSWER SHEET   TEST #:

SECTION 3

#		#		#		#	
1	Ⓐ Ⓑ Ⓒ Ⓓ Ⓔ	11	Ⓐ Ⓑ Ⓒ Ⓓ Ⓔ	21	Ⓐ Ⓑ Ⓒ Ⓓ Ⓔ	31	Ⓐ Ⓑ Ⓒ Ⓓ Ⓔ
2	Ⓐ Ⓑ Ⓒ Ⓓ Ⓔ	12	Ⓐ Ⓑ Ⓒ Ⓓ Ⓔ	22	Ⓐ Ⓑ Ⓒ Ⓓ Ⓔ	32	Ⓐ Ⓑ Ⓒ Ⓓ Ⓔ
3	Ⓐ Ⓑ Ⓒ Ⓓ Ⓔ	13	Ⓐ Ⓑ Ⓒ Ⓓ Ⓔ	23	Ⓐ Ⓑ Ⓒ Ⓓ Ⓔ	33	Ⓐ Ⓑ Ⓒ Ⓓ Ⓔ
4	Ⓐ Ⓑ Ⓒ Ⓓ Ⓔ	14	Ⓐ Ⓑ Ⓒ Ⓓ Ⓔ	24	Ⓐ Ⓑ Ⓒ Ⓓ Ⓔ	34	Ⓐ Ⓑ Ⓒ Ⓓ Ⓔ
5	Ⓐ Ⓑ Ⓒ Ⓓ Ⓔ	15	Ⓐ Ⓑ Ⓒ Ⓓ Ⓔ	25	Ⓐ Ⓑ Ⓒ Ⓓ Ⓔ	35	Ⓐ Ⓑ Ⓒ Ⓓ Ⓔ
6	Ⓐ Ⓑ Ⓒ Ⓓ Ⓔ	16	Ⓐ Ⓑ Ⓒ Ⓓ Ⓔ	26	Ⓐ Ⓑ Ⓒ Ⓓ Ⓔ	36	Ⓐ Ⓑ Ⓒ Ⓓ Ⓔ
7	Ⓐ Ⓑ Ⓒ Ⓓ Ⓔ	17	Ⓐ Ⓑ Ⓒ Ⓓ Ⓔ	27	Ⓐ Ⓑ Ⓒ Ⓓ Ⓔ	37	Ⓐ Ⓑ Ⓒ Ⓓ Ⓔ
8	Ⓐ Ⓑ Ⓒ Ⓓ Ⓔ	18	Ⓐ Ⓑ Ⓒ Ⓓ Ⓔ	28	Ⓐ Ⓑ Ⓒ Ⓓ Ⓔ	38	Ⓐ Ⓑ Ⓒ Ⓓ Ⓔ
9	Ⓐ Ⓑ Ⓒ Ⓓ Ⓔ	19	Ⓐ Ⓑ Ⓒ Ⓓ Ⓔ	29	Ⓐ Ⓑ Ⓒ Ⓓ Ⓔ	39	Ⓐ Ⓑ Ⓒ Ⓓ Ⓔ
10	Ⓐ Ⓑ Ⓒ Ⓓ Ⓔ	20	Ⓐ Ⓑ Ⓒ Ⓓ Ⓔ	30	Ⓐ Ⓑ Ⓒ Ⓓ Ⓔ	40	Ⓐ Ⓑ Ⓒ Ⓓ Ⓔ

SECTION 5

#		#		#		#	
1	Ⓐ Ⓑ Ⓒ Ⓓ Ⓔ	11	Ⓐ Ⓑ Ⓒ Ⓓ Ⓔ	21	Ⓐ Ⓑ Ⓒ Ⓓ Ⓔ	31	Ⓐ Ⓑ Ⓒ Ⓓ Ⓔ
2	Ⓐ Ⓑ Ⓒ Ⓓ Ⓔ	12	Ⓐ Ⓑ Ⓒ Ⓓ Ⓔ	22	Ⓐ Ⓑ Ⓒ Ⓓ Ⓔ	32	Ⓐ Ⓑ Ⓒ Ⓓ Ⓔ
3	Ⓐ Ⓑ Ⓒ Ⓓ Ⓔ	13	Ⓐ Ⓑ Ⓒ Ⓓ Ⓔ	23	Ⓐ Ⓑ Ⓒ Ⓓ Ⓔ	33	Ⓐ Ⓑ Ⓒ Ⓓ Ⓔ
4	Ⓐ Ⓑ Ⓒ Ⓓ Ⓔ	14	Ⓐ Ⓑ Ⓒ Ⓓ Ⓔ	24	Ⓐ Ⓑ Ⓒ Ⓓ Ⓔ	34	Ⓐ Ⓑ Ⓒ Ⓓ Ⓔ
5	Ⓐ Ⓑ Ⓒ Ⓓ Ⓔ	15	Ⓐ Ⓑ Ⓒ Ⓓ Ⓔ	25	Ⓐ Ⓑ Ⓒ Ⓓ Ⓔ	35	Ⓐ Ⓑ Ⓒ Ⓓ Ⓔ
6	Ⓐ Ⓑ Ⓒ Ⓓ Ⓔ	16	Ⓐ Ⓑ Ⓒ Ⓓ Ⓔ	26	Ⓐ Ⓑ Ⓒ Ⓓ Ⓔ	36	Ⓐ Ⓑ Ⓒ Ⓓ Ⓔ
7	Ⓐ Ⓑ Ⓒ Ⓓ Ⓔ	17	Ⓐ Ⓑ Ⓒ Ⓓ Ⓔ	27	Ⓐ Ⓑ Ⓒ Ⓓ Ⓔ	37	Ⓐ Ⓑ Ⓒ Ⓓ Ⓔ
8	Ⓐ Ⓑ Ⓒ Ⓓ Ⓔ	18	Ⓐ Ⓑ Ⓒ Ⓓ Ⓔ	28	Ⓐ Ⓑ Ⓒ Ⓓ Ⓔ	38	Ⓐ Ⓑ Ⓒ Ⓓ Ⓔ
9	Ⓐ Ⓑ Ⓒ Ⓓ Ⓔ	19	Ⓐ Ⓑ Ⓒ Ⓓ Ⓔ	29	Ⓐ Ⓑ Ⓒ Ⓓ Ⓔ	39	Ⓐ Ⓑ Ⓒ Ⓓ Ⓔ
10	Ⓐ Ⓑ Ⓒ Ⓓ Ⓔ	20	Ⓐ Ⓑ Ⓒ Ⓓ Ⓔ	30	Ⓐ Ⓑ Ⓒ Ⓓ Ⓔ	40	Ⓐ Ⓑ Ⓒ Ⓓ Ⓔ

Grid-in questions: 9, 10, 11, 12, 13, 14, 15, 16, 17, 18

SECTION 7	1 Ⓐ Ⓑ Ⓒ Ⓓ Ⓔ	11 Ⓐ Ⓑ Ⓒ Ⓓ Ⓔ	21 Ⓐ Ⓑ Ⓒ Ⓓ Ⓔ	31 Ⓐ Ⓑ Ⓒ Ⓓ Ⓔ
	2 Ⓐ Ⓑ Ⓒ Ⓓ Ⓔ	12 Ⓐ Ⓑ Ⓒ Ⓓ Ⓔ	22 Ⓐ Ⓑ Ⓒ Ⓓ Ⓔ	32 Ⓐ Ⓑ Ⓒ Ⓓ Ⓔ
	3 Ⓐ Ⓑ Ⓒ Ⓓ Ⓔ	13 Ⓐ Ⓑ Ⓒ Ⓓ Ⓔ	23 Ⓐ Ⓑ Ⓒ Ⓓ Ⓔ	33 Ⓐ Ⓑ Ⓒ Ⓓ Ⓔ
	4 Ⓐ Ⓑ Ⓒ Ⓓ Ⓔ	14 Ⓐ Ⓑ Ⓒ Ⓓ Ⓔ	24 Ⓐ Ⓑ Ⓒ Ⓓ Ⓔ	34 Ⓐ Ⓑ Ⓒ Ⓓ Ⓔ
	5 Ⓐ Ⓑ Ⓒ Ⓓ Ⓔ	15 Ⓐ Ⓑ Ⓒ Ⓓ Ⓔ	25 Ⓐ Ⓑ Ⓒ Ⓓ Ⓔ	35 Ⓐ Ⓑ Ⓒ Ⓓ Ⓔ
	6 Ⓐ Ⓑ Ⓒ Ⓓ Ⓔ	16 Ⓐ Ⓑ Ⓒ Ⓓ Ⓔ	26 Ⓐ Ⓑ Ⓒ Ⓓ Ⓔ	36 Ⓐ Ⓑ Ⓒ Ⓓ Ⓔ
	7 Ⓐ Ⓑ Ⓒ Ⓓ Ⓔ	17 Ⓐ Ⓑ Ⓒ Ⓓ Ⓔ	27 Ⓐ Ⓑ Ⓒ Ⓓ Ⓔ	37 Ⓐ Ⓑ Ⓒ Ⓓ Ⓔ
	8 Ⓐ Ⓑ Ⓒ Ⓓ Ⓔ	18 Ⓐ Ⓑ Ⓒ Ⓓ Ⓔ	28 Ⓐ Ⓑ Ⓒ Ⓓ Ⓔ	38 Ⓐ Ⓑ Ⓒ Ⓓ Ⓔ
	9 Ⓐ Ⓑ Ⓒ Ⓓ Ⓔ	19 Ⓐ Ⓑ Ⓒ Ⓓ Ⓔ	29 Ⓐ Ⓑ Ⓒ Ⓓ Ⓔ	39 Ⓐ Ⓑ Ⓒ Ⓓ Ⓔ
	10 Ⓐ Ⓑ Ⓒ Ⓓ Ⓔ	20 Ⓐ Ⓑ Ⓒ Ⓓ Ⓔ	30 Ⓐ Ⓑ Ⓒ Ⓓ Ⓔ	40 Ⓐ Ⓑ Ⓒ Ⓓ Ⓔ

Math Scoring Worksheet

A. Section 3   _____10_____   _____7_____
     numer of correct   *number of incorrect*

     +   +

B. Section 5 (1-8)   _____4_____   _____1_____
     numer of correct   *number of incorrect*

     +

C. Section 5 (9-18)   _____
     numer of correct

     +   +

D. Section 7   _____   _____
     numer of correct   *number of incorrect*

     =   =

E. Total Unrounded Raw Score   _____ − _____ ÷4 = _____
     numer of correct   *number of incorrect*

F. Total Rounded Raw Score   _____ (See table)

     Math Score Range = | —— |

Math Conversion Table

Raw Score	Scaled Score	Raw Score	Scaled Score
54	800	23	490-550
53	780-800	22	480-540
52	760-800	21	470-530
51	740-800	20	460-520
50	720-780	19	450-510
49	700-760	18	450-510
48	690-750	17	440-500
47	680-740	16	430-490
46	670-730	15	420-480
45	660-720	14	420-480
44	650-710	13	410-470
43	650-710	12	400-460
42	640-700	11	390-450
41	630-690	10	380-440
40	620-680	9	390-430
39	610-670	8	380-420
38	610-670	7	370-410
37	600-660	6	360-400
36	590-650	5	340-380
35	580-640	4	320-370
34	570-630	3	310-360
33	560-620	2	300-350
32	560-620	1	270-320
31	550-610	0	240-300
30	540-600	-1	200-290
29	530-590	-2	200-270
28	530-590	-3	200-260
27	520-580	-4	200-240
26	510-570	-5	200-220
25	500-560	-6 and below	200
24	500-560		

SECTION 3
Time- 25 minutes
20 Questions

Turn to Section 3 (Page 1) of your answer sheet to answer the questions in this section.

Directions: For this section, solve each problem and decide which is the best of the choices given. Fill in the corresponding circle on the answer sheet. You may use any available space for scratchwork.

Notes

1. The use of a calculator is permitted.
2. All numbers used are real numbers.
3. Figures that accompany problems in this test are intended to provide information useful in solving the problems. They are drawn as accurately as possible EXCEPT when it is stated in a specific problem that the figure is not drawn to scale. All figure lie in a plane unless other indicated.
4. Unless otherwise specified, the domain of any function f is assumed to be set of all real numbers x for which $f(x)$ is a real number.

Reference Information

$A = \pi r^2$
$C = 2\pi r$
$A = \ell w$
$A = \frac{1}{2}bh$
$V = \ell wh$
$V = \pi r^2 h$
$c^2 = a^2 + b^2$
Special Right Triangles

The numbers of degrees of arc in a circle is $360°$.

The sum of the measures in degrees of the angles is $180°$.

---

1. If $x - 2y = 10$, $y = z + 1$, and $z = 2$, what is the value of x?

    (A) 10
    (B) 12
    (C) 14
    (D) 16
    (E) 18

4, 8, 12, 16,

2. The first term in the sequence is 4, and each term after the first is determined by adding 4. What is the value of 20th term?

    (A) 20
    (B) 40
    (C) 60
    (D) 80
    (E) 100

GO ON TO THE NEXT PAGE

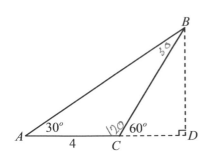

Note: Figure not drawn to scale.

3. In the figure above, $\angle BAC = 30^o$, $\angle BCD = 60^o$, and the length of $\overline{AC}$ is 4. What is the area of $\triangle ABC$?

(A) 4
(B) $4\sqrt{3}$
(C) 8
(D) $8\sqrt{3}$
(E) 16

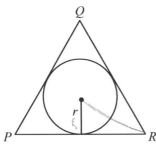

4. In the figure above, a circle is tangent to the side of equilateral triangle PQR and the radius r equals 5. What is the perimeter of $\triangle PQR$?

(A) 20
(B) 30
(C) $30\sqrt{3}$
(D) 35
(E) $40\sqrt{2}$

5. A bag contains 3 red marbles and 6 black marbles. What is the probability that you select two red marbles if you reach in the bag and randomly grab two marbles without replacement?

(A) $\dfrac{2}{9}$

(B) $\dfrac{2}{81}$

(C) $\dfrac{1}{36}$

(D) $\dfrac{1}{18}$

(E) $\dfrac{1}{12}$

6. Kimberly earns k dollars per week. At this rate how many weeks will it take her to earn p dollars?

(A) $10p$

(B) kp

(C) $\dfrac{k}{p}$

(D) $\dfrac{100k}{p}$

(E) $\dfrac{p}{k}$

GO ON TO THE NEXT PAGE

7. Which of the following graphs shows a relationship in which y is directly proportional to x?

(A)
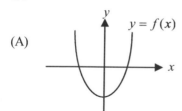

(B)
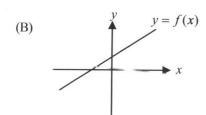

(C)
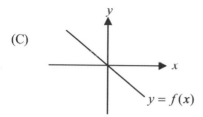

(D)
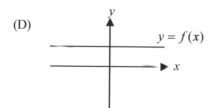

(E)
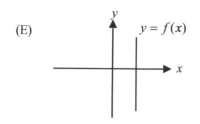

8. If $rs^2t^3u^3 > 0$ and $u < 0$, which of the following must be true?

    (A) $rt < 0$
    (B) $urt < 0$
    (C) $r > 0$
    (D) $t > 0$
    (E) $st > 0$

9. To arrive on time, a ship needs 5 hours to complete a voyage. If the ship must arrive in 4 hours, by what percent must the speed of the ship be increased?

    (A) 15%
    (B) 20%
    (C) 25%
    (D) 27%
    (E) 30%

GO ON TO THE NEXT PAGE

10. If $(p-3) \times \dfrac{p}{(p+5)(p-3)} = 0$, What is the value of p?

(A) 0
(B) 3
(C) 0 and 3
(D) 0 and -5
(E) No solution

$p = p + 5$

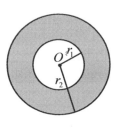

12. In the figure above, r_1 and r_2 is the radius of the circles which have the same center. If the area of the shaded region is 12 and the ratio of r_1 to r_2 is 1:2, what is the area of the smaller circle?

(A) 2
(B) 4
(C) 6
(D) 8
(E) 10

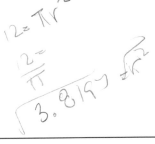

$12 = \pi r^2$

$\dfrac{12}{\pi} =$

$3.8197\ \pi r^2$

11. A certain job can be done in 20 hours by 4 people. How many people are needed to do the same job in 10 hours?

(A) 2
(B) 4
(C) 6
(D) 8
(E) 10

x	$f(x)$
a	b
2	7
3	9

13. The table above shows some values for the function f. If f is a linear function, what is the value of b in terms of a?

(A) $b = \dfrac{7}{2}a$

(B) $b = 3a$

(C) $b = a + 5$

(D) $b = a + 6$

(E) $b = 2a + 3$

$\dfrac{9-7}{3-2} = \dfrac{2}{1}$

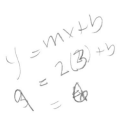

$y = mx + b$
$9 = 2(3) + b$
$9 = b$

GO ON TO THE NEXT PAGE

- 100 -

14. In the xy – plane, the equation of line ℓ is $x + 3y = 5$. If line m is perpendicular with line ℓ, what is a possible equation of line m?

(A) $y = -\dfrac{1}{3}x + 2$

(B) $y = \dfrac{1}{3}x - 1$

(C) $y = \dfrac{3}{2}x - 3$

(D) $y = -3x + 1$

(E) $y = 3x + \dfrac{5}{2}$

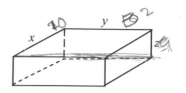

16. The figure above shows a rectangular solid with width x, length y, and height z. If $xy = 20$, $yz = 10$, and $zx = 18$, what is the volume of the rectangular solid?

(A) 60
(B) 70
(C) 80
(D) 90
(E) 100

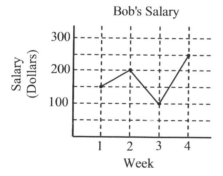

Bob's Salary

15. The graph above shows Bob's weekly salary in dollars for the first four weeks working at a public library. Bob's salary at week 4 is what percent greater than his salary at week 1?

(A) $33\dfrac{1}{3}\%$

(B) 40%

(C) 50%

(D) $66\dfrac{2}{3}\%$

(E) 100%

17. If exactly two of the three integers a, b, and c are odd, which of the following must be odd?

I. $a + bc$

II. $\dfrac{ab}{c}$

III. $a(b + c)$

(A) I only
(B) II only
(C) III only
(D) I and II only
(E) II and III only

GO ON TO THE NEXT PAGE

18. There are two different means, the arithmetic, $A = \dfrac{a+b}{2}$ and the harmonic, $H = \dfrac{2ab}{a+b}$. If the arithmetic mean is equal to the harmonic mean, which of the following must be true?

(A) $a = 0$ and $b = 0$

(B) $ab = 0$

(C) $a = 1$ and $b = 1$

(D) $a = b$

(E) $a + b = 0$

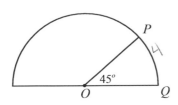

19. In the figure above, PQ is the arc of a circle with center O. If the length of arc PQ is 4, what is the area of sector OPQ?

(A) 54π

(B) $\dfrac{48}{\pi}$

(C) 32π

(D) $\dfrac{32}{\pi}$

(E) 16π

20. In a certain class, $\dfrac{4}{7}$ of the students are boys, and the ratio of the students older than or equal to 10 years to the students less than 10 years old is 2:3. If $\dfrac{2}{3}$ of the girls are less than 10 years old, what fraction of the boys are older than or equal to 10 years?

(A) $\dfrac{9}{20}$

(B) $\dfrac{11}{20}$

(C) $\dfrac{2}{3}$

(D) $\dfrac{14}{15}$

(E) $\dfrac{9}{35}$

STOP

If you finish before time is called, you may check your work on this section only.
Do not turn to any other section in the test.

SECTION 5
Time- 25 minutes
18 Questions

Turn to Section 5 (Page 1) of your answer sheet to answer the questions in this section.

Directions: For this section, solve each problem and decide which is the best of the choices given. Fill in the corresponding circle on the answer sheet. You may use any available space for scratchwork.

Notes

1. The use of a calculator is permitted.
2. All numbers used are real numbers.
3. Figures that accompany problems in this test are intended to provide information useful in solving the problems. They are drawn as accurately as possible EXCEPT when it is stated in a specific problem that the figure is not drawn to scale. All figure lie in a plane unless other indicated.
4. Unless otherwise specified, the domain of any function f is assumed to be set of all real numbers x for which $f(x)$ is a real number.

Reference Informatiom

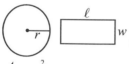   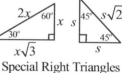

$A = \pi r^2$
$C = 2\pi r$
$A = \ell w$
$A = \frac{1}{2}bh$
$V = \ell w h$
$V = \pi r^2 h$
$c^2 = a^2 + b^2$
Special Right Triangles

The numbers of degrees of arc in a circle is $360°$.

The sum of the measures in degrees of the angles is $180°$.

1. If $|k - 5| \le 8$, which of the following CANNOT be the value of k?

   (A) 10
   (B) 8
   (C) 5
   (D) −3
   (E) −4

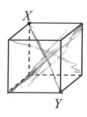

2. In the figure above, $\overline{XY}$ is the diagonal of the cube (not drawn in the figure). How many different diagonals of the same length as $\overline{XY}$ are there?

   (A) 1
   (B) 2
   (C) 3
   (D) 4
   (E) 5

GO ON TO THE NEXT PAGE

Votes For Favorite Professor

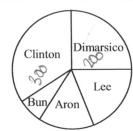

3. The circle graph above shows the results of a vote for favorite professors in JFK University. If 800 votes were cast, how many possible numbers of votes did Professor Lee receive?

(A) 150
(B) 200
(C) 250
(D) 400
(E) 500

4. In the fraction $\frac{a}{2b}$, a is 5 less than two times b.

If the fraction is equal to $\frac{1}{2}$, what is the denominator of this fraction?

(A) 2
(B) 4
(C) 6
(D) 8
(E) 10

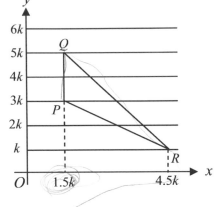

5. In the figure above, k is a positive number. If the area of $\triangle PQR$ is 12, what is the value of k?

(A) 1
(B) 2
(C) $2\sqrt{2}$
(D) 4
(E) $4\sqrt{2}$

6. If x, y, and z are positive and $\frac{x^2 y^{-3}}{z^2} = 9y$, which of the following is equal to x^{-1}?

(A) $\frac{1}{3y^2 z}$

(B) $\frac{3z}{2}$

(C) $\frac{1}{3z}$

(D) $\frac{yz}{3}$

(E) $\frac{9y}{z^2}$

GO ON TO THE NEXT PAGE

7. Allen and Barbra leave their house at the same time. Allen walks due east for 3 hours and due north for $3\frac{1}{2}$ hours at the average rate of 2 kilometers per hour, and Barbra walks due south for 3 hours and due west for 2 hours at the average of 3 kilometers per hour. At the end of their walk, what is the straight-line distance between them, in kilometers?

(A) 25
(B) 20
(C) $8\sqrt{2}$
(D) 8
(E) $6\sqrt{2}$

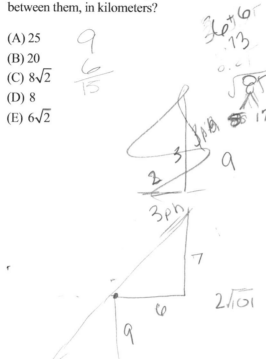

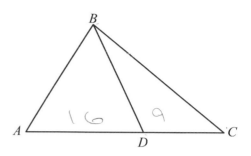

Note: Figure not drawn to scale.

8. In the figure above, the area of $\triangle ABD$ is 16 and the area of $\triangle BCD$ is 9. What is the ratio of the length of AD to the length of DC?

(A) 2:1
(B) 3:1
(C) 4:3
(D) 16:9
(E) 256:81

GO ON TO THE NEXT PAGE

Directions: For Students-Produced Response questions 9-18, use the grid at the bottom of the answer sheet page on which you have answered questions 1-8.

Each of the remaining 10 questions requires you to solve the problem and enter your answer by making the circles in the special grid, as shown in the examples below. You may use any available space for scratchwork.

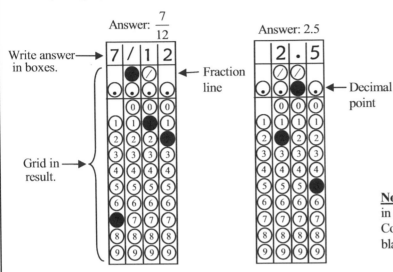

Answer: $\frac{7}{12}$

Write answer in boxes. → ← Fraction line

Grid in result. →

Answer: 2.5

← Decimal point

Answer: 201
Either position is correct.

Note: You may start your answers in any column, space permitting. Columns not needed should be left blank.

- Mark no more than one circle in any column.

- Because the answer sheet will be machine-scored, **you will receive credit only if the circles are filled in correctly.**

- Although not required, it is suggested that you write your answer in the boxes at the top of the columns to help you fill in the circles accurately.

- Some problems may have more than one correct answer. In such cases, grid only one answer.

- No question has a negative answer.

- **Mixed numbers** such as $3\frac{1}{2}$ must be gridded as 3.5 or 7/2. (If [3 1 / 2] is gridded, it will be interpreted as $\frac{31}{2}$, not $3\frac{1}{2}$.)

- **Decimal Answers:** If you obtain a decimal answer with more digits than the grid can accommodate, it may be either rounded or truncated, but it must fill the entire grid. For example, if you obtain an answer such as 0.6666…, you should record your result as .666 or .667. **A less accurate value such as .66 or .67 will be scored as incorrect.**

Acceptable ways to grid $\frac{2}{3}$ are:

9. Peter goes on a 30-mile bike ride every Sunday. He rides the distance in 3 hours. At this rate, how many miles can he ride in 5 hours and 30 minutes?

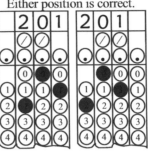

$f(k) = \left|6 - k^2\right|$, $k =$ positive integer

10. If $f(k) = 3$, what is the value of k that satisfies the equation above?

GO ON TO THE NEXT PAGE

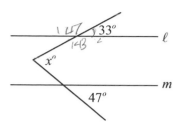

Note: Figure not drawn to scale.

11. In the figure above, line ℓ is parallel to line m. What is the value of x?

12. The average of a set of 8 consecutive odd integers is 18. What is the greatest of these 8 integers?

13. Let the function f be defined by $f(x) = x - p$. If $f(2) = -1$, what is the value of $f(2p)$?

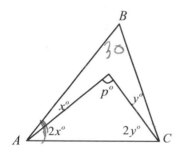

Note: Figure not drawn to scale.

14. If $\angle ABC = 30^o$ in $\triangle ABC$, what is the value of p?

GO ON TO THE NEXT PAGE

15. 20 grams of solution A has 10% of alcohol by mass and 15 grams of solution B has 20% of alcohol by mass. If 10 grams of the solution A is added to 10 grams of the solution B, what is the percent of alcohol in the mixture?

$x^2 - 1$

17. On the number line above, there are 4 equal intervals between 1 and 2. What is the positive value of x?

16. If average of $2a$ and b is equal to 50 percent of $4b$, what is the value of $\dfrac{a}{b}$?

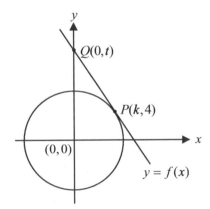

$y = f(x)$

18. In the figure above, a circle is tangent to the line at point P. If the radius of the circle is 5, what is the value of t, the y-intercept of the line?

STOP

If you finish before time is called, you may check your work on this section only.
Do not turn to any other section in the test.

SECTION 7
Time- 20 minutes
16 Questions

Turn to Section 7 (Page 2) of your answer sheet to answer the questions in this section.

Directions: For this section, solve each problem and decide which is the best of the choices given. Fill in the corresponding circle on the answer sheet. You may use any available space for scratchwork.

1. If X is the set of all integers greater than π, and Y is the set of all integers less than π^2, how many integers are in both sets?

   (A) 4
   (B) 5
   (C) 6
   (D) 7
   (E) 8

2. If $\sqrt{k+2} = k$, what is the value of k?

   (A) -1
   (B) 2
   (C) -1 or 0
   (D) -1 or 2
   (E) -2 or 2

GO ON TO THE NEXT PAGE

3. Out of 50 families in a certain town, 20 own a dog and 35 own a cat. If 2 families don't own either of them, what fraction of the families owns both a dog and a cat?

(A) $\dfrac{7}{50}$

(B) $\dfrac{1}{5}$

(C) $\dfrac{11}{50}$

(D) $\dfrac{6}{25}$

(E) $\dfrac{7}{48}$

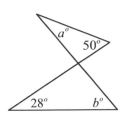

Note: Figure not drawn to scale.

4. In the figure above, what is the value of $|a - b|$?

(A) 10
(B) 17
(C) 20
(D) 22
(E) 25

READING LEVELS

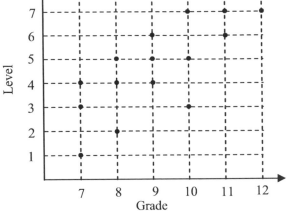

5. The scatterplot above shows the reading levels by grade for 15 students in a certain book-reading club. What is the median reading level for the 15 students?

(A) 3
(B) 5
(C) 5.5
(D) 6
(E) 6.5

6. a and b are different positive integers and a is greater than b. If $a^2 - b^2 = 7$, what is the value of ab?

(A) 6
(B) 8
(C) 10
(D) 12
(E) 14

GO ON TO THE NEXT PAGE

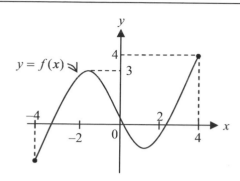

7. In the figure above, the graph of $y = f(x)$ is shown in $-4 \le x \le 4$. If $f(k) = 3$, how many values of k are there in the interval?

(A) 0
(B) 1
(C) 2
(D) 3
(E) 4

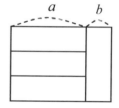

8. The figure shown above is composed of 4 equal rectangles. If the area of each rectangle is 48, what is the value of b?

(A) 4
(B) 5
(C) 6
(D) 7
(E) 8

9. The operation $a \odot b$ is defined by $a \odot b = \dfrac{a+b}{ab}$. If $k^2 \odot 2 = 1$, which of the following could be a possible value of k?

I. 2
II. $\sqrt{2}$
III. $-\sqrt{2}$

(A) I only
(B) II only
(C) III only
(D) I and II only
(E) II and III only

10. The profit p, in dollars, from a car wash is given by the function $p(x) = \dfrac{50x - 200}{x} + k$, where x is the number of cars washed and k is a constant. If 20 cars were washed today for a total profit of $300, what is the value of k?

(A) 100
(B) 150
(C) 200
(D) 260
(E) 300

GO ON TO THE NEXT PAGE

11. In an art class, $\frac{2}{3}$ of the students are girls and $\frac{2}{5}$ of girls are seniors. If $\frac{1}{3}$ of senior girls have passed the final art test, which of the following could be the total number of students in the class?

(A) 15
(B) 20
(C) 30
(D) 45
(E) 60

12. If $a = 2b = 3c$, what is the average (arithmetic mean) of a, b, and c in terms of a ?

(A) $\dfrac{a}{6}$

(B) $\dfrac{2a}{11}$

(C) $\dfrac{6a}{5}$

(D) $\dfrac{11a}{18}$

(E) $\dfrac{2a}{3}$

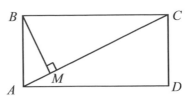

Note: Figure not drawn to scale.

13. In the figure above, $ABCD$ is a rectangle and $\overline{BM}$ is perpendicular to $\overline{AC}$. If the length of $\overline{AM}$ is 2 and the length of $\overline{CM}$ is 18, what is the area of $\triangle ABC$?

(A) 40
(B) 60
(C) 80
(D) 100
(E) 120

14. If $3 < p < 25$ and $12 < q < 16$, which of the following gives the set of all possible values of $\dfrac{p}{q}$?

(A) $0 < \dfrac{p}{q} < \dfrac{1}{4}$

(B) $\dfrac{1}{4} < \dfrac{p}{q} < \dfrac{25}{16}$

(C) $\dfrac{1}{4} < \dfrac{p}{q} < \dfrac{3}{16}$

(D) $\dfrac{3}{16} < \dfrac{p}{q} < \dfrac{25}{16}$

(E) $\dfrac{3}{16} < \dfrac{p}{q} < \dfrac{25}{12}$

GO ON TO THE NEXT PAGE

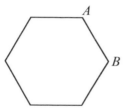

15. The figure above shows a regular hexagon. If the length of $\overline{AB}$ is k, what is the area of the hexagon in terms of k?

(A) $\dfrac{k^2}{3}$

(B) $\dfrac{2k^2}{3}$

(C) $\dfrac{3k^2}{2}$

(D) $\dfrac{3k^2\sqrt{3}}{2}$

(E) $4k^2\sqrt{3}$

16. After the first term, each term in a sequence is k times the preceding term and p is the first term of the sequence. If the ratio of the seventh term to the 3rd term is 16, what is a possible value of k?

(A) p
(B) $2p$
(C) 1
(D) 2
(E) 4

STOP
If you finish before time is called, you may check your work on this section only.
Do not turn to any other section in the test.

#	SECTION 3	SECTION 5	SECTION 7
		TEST 1 **ANSWER KEY**	
1	D	E	C
2	D	C	B
3	B	A	A
4	C	E	D
5	E	B	B
6	E	A	D
7	C	B	C
8	A	D	A
9	C	55	E
10	A	3	D
11	D	80	D
12	B	25	D
13	E	3	B
14	E	80	E
15	D	15	D
16	A	$\frac{3}{2}$ or 1.5	D
17	A	$\frac{3}{2}$ or 1.5	
18	D	$\frac{25}{4}$ or 6.25	
19	D		
20	A		

TEST 1   SECTION 3

1. (D)
$Z = 2$, then $y = 3$,
and $x = 2y + 10 = 2(3) + 10 = 16$

2. (D)
$a_1 = 4$ and $d = 4$. Then
$a_{20} = 4 + 4(20 - 1) = 80$
Or, use pattern.
$4, 8, 12, 16, 20, ... = 4(1, 2, 3, 4......20)$.
Therefore, $4 \cdot 20 = 80$

3. (B)

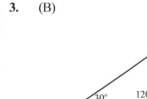

In the figure, the area of
$$\triangle ABC = \frac{4 \times 2\sqrt{3}}{2} = 4\sqrt{3}$$

4. (C)

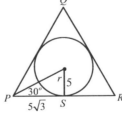

In the figure, $PS = 5\sqrt{3}$, then $PR = 10\sqrt{3}$.
Therefore, Perimeter $= 3 \times 10\sqrt{3} = 30\sqrt{3}$.

5. (E)
$$P(r \text{ and } r) = \frac{{}_3C_2}{{}_9C_2} = \frac{3}{36} = \frac{1}{12}$$

Or, use product of probability. The probability of selecting a red marble first is $\dfrac{3}{9}$ and the probability of selecting a red marble second is $\dfrac{2}{8}$. Then, the probability is $\dfrac{3}{9} \times \dfrac{2}{8} = \dfrac{1}{12}$

6. (E)

Proportion. $\dfrac{{}^\$ k}{1} = \dfrac{{}^\$ p}{x}$. Therefore, $x = \dfrac{p}{k}$

7. (C)

Direction variation is, $y = kx$, where k is a constant. In the xy-plane, y-intercept must be 0.

8. (A)

$rs^2t^3u^3 > 0$ is simplified by dividing $r^2t^2u^2 \, (positive)$.

That is, $rtu > 0$. But $u < 0$. Therefore, rt must be negative.

9. (C)

The speed $= \dfrac{D}{5}$, where D is a distance. And the new speed $= \dfrac{D}{4}$. Therefore, the % of increase

$= \dfrac{\dfrac{D}{4} - \dfrac{D}{5}}{\dfrac{D}{5}} \times 100 = 25\%$.

Or, you can use a convenient number, like 20 (multiple of 4 and 5). The speed $= \dfrac{20}{5} = 4$ and

the new speed $= \dfrac{20}{4} = 5$. Therefore, the % of

increase $= \dfrac{5-4}{4} \times 100 = 25\%$

10. (A)

$\dfrac{(p-3)p}{(p+5)(p-3)} = 0$, where $p \neq -5$ and $p \neq 3$.

Then, sampling as follows. $\dfrac{p}{(p+5)} = 0$.

Therefore, $p = 0$.

11. (D)

Inverse proportion. $xy = k$. $20 \times 4 = 10 \times x$, $x = 8$

12. (B)

The ratio of the areas of the circles $=$ $1^2 : 2^2 = 1 : 4$

You can let the areas as x and $4x$. The area of the shaded region is $3x$. $3x = 12$, $x = 4$. Therefore, the area of the smaller circle is 4.

13. (E)

The linear function has a constant slope.

$m = \dfrac{9-7}{3-2} = 2$.

Between $(2,7)$ and (a,b), the slope $= \dfrac{b-7}{a-2} = 2$.

Therefore, $b = 2a + 3$.

14. (E)

The slope of line $\ell = -\dfrac{1}{3}$. Therefore,

The slope of the perpendicular line $= 3$.

15. (D)

% of increase $= \dfrac{250-150}{150} \times 100 = 66\dfrac{2}{3}\%$.

16. (A)

The volume $= xyz$.

$(xy)(yz)(zx) = (20)(10)(18)$.

Therefore, $xyz = \sqrt{3600} = 60$.

17. (A)

Two of three integers a, b, and c are odd. There are three possible cases.

$(a,b,c) \Rightarrow (O,O,E), (O,E,O), (E,O,O)$.

You can use numbers $(1,1,2), (1,2,1), (2,1,1)$.

Check the choices with the numbers. Choice I is always true.

18. (D)

$\dfrac{a+b}{2} = \dfrac{2ab}{a+b} \Rightarrow (a+b)(a+b) = 4ab$

$a^2 + 2ab + b^2 = 4ab \Rightarrow (a-b)^2 = 0$.

Therefore, $a - b = 0$. (A) and (C) are the answers for " could be true".

19. (D)

The length of the perimeter $= 4 \times \dfrac{360}{45} = 32$

$2\pi r = 32 \Rightarrow r = \dfrac{16}{\pi}$

The area of the sector $= \pi r^2 \times \dfrac{45}{360} = \dfrac{32}{\pi}$

---

20. (A)

Make a table.

	Boys	Girls	
10 ⇑	$9/35$		$2/5$
10 ⇓	$11/35$	$2/7$	$3/5$
	$4/7$	$3/7$	

$\dfrac{2}{3} \times \dfrac{3}{7} = \dfrac{2}{7} \Rightarrow$ Girl who are less the 10 years old.

You can fill the blanks with what you need.

Therefore, $\dfrac{\text{boys and 10} \Uparrow}{\text{boys}} = \dfrac{9/35}{4/7} = \dfrac{9}{20}$

TEST 1 **SECTION 5**

---

1. (E)

$|k - 5| \le 8$. (D) $|-4 - 5| = 9 \le 8$ (False)

---

2. (C)

---

3. (A)

The number must be less then $\dfrac{1}{4}$ of the votes,

that is,

$\dfrac{1}{4} \times 800 = 200$. Only (A) is the possible number.

---

4. (E)

$a = 2b - 5$.

$\dfrac{a}{2b} = \dfrac{2b - 5}{2b} = \dfrac{1}{2} \Rightarrow b = 5$, Therefore,

$2b = 10$

---

5. (B)

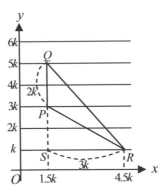

In the figure above, SR is the height. Therefore,

the area of $\triangle PQR = \dfrac{2k \times 3k}{2} = 12 \Rightarrow k = 2$

---

6. (A)

$\dfrac{x^2 y^{-3}}{z^2} = 9y \Rightarrow$

$x^2 = \dfrac{z^2}{y^{-3}}(9y) = 9y^4 z^2 \Rightarrow x = 3y^2 z$

Therefore, $x^{-1} = \dfrac{1}{3y^2 z}$

---

7. (B)

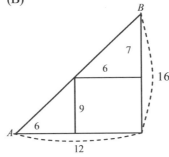

In the figure, $AB = \sqrt{12^2 + 16^2} = 20$

---

8. (D)

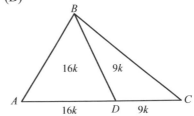

In the figure, the ratio of the areas is equal to the ratio of the lengths of the bases, because the two triangles have the same height.

9. 55

Proportion. $\dfrac{30}{3} = \dfrac{x}{5.5} \Rightarrow x = 55$ miles.

10. 3

$\left|6 - k^2\right| = \left|k^2 - 6\right| = 3 \Rightarrow k^2 - 6 = 3 \Rightarrow k = 3.$

11. 80

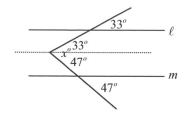

In the figure, the angles are alternate angles.
$x = 33 + 47 = 80$.

12. 25

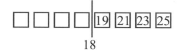

The average is equal to the median.

13. 3

$f(2) = 2 - p = -1 \Rightarrow p = 3$

$f(2p) = f(6) = 6 - 3 = 3$

14. 80

$3x + 3y = 180 - 30 - 150$. Then $x + y = 50$

Therefore,

$2x + 2y = 100 \Rightarrow \angle p = 180 - 100 = 80$

15. 15

	20g solution	10g solution
A	$\dfrac{2g}{20g}$	$\dfrac{1g}{10g}$
B	$\dfrac{3g}{15g}$	$\dfrac{2g}{10g}$

The solute of solution is proportional to the weight.

Therefore, $\dfrac{1+2}{10+10} = \dfrac{3}{20} \Rightarrow \dfrac{3}{20} \times 100 = 15\%$.

16. $\frac{3}{2}$ or 1.5

Translate. $\dfrac{2a+b}{2} = 0.5 \times 4b \Rightarrow 2a + b = 4b$

$a = \dfrac{3b}{2} \Rightarrow \dfrac{a}{b} = \dfrac{3b/2}{b} = \dfrac{3}{2}$

17. 1.5 or $\frac{3}{2}$

The length of each interval $= \dfrac{1}{4} = 0.25$

$x^2 - 1 = 1.25 \Rightarrow x^2 = 2.25 \Rightarrow x = 1.5$

18. $\frac{25}{4}$ or 6.25

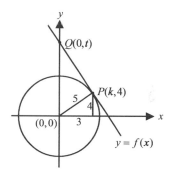

In the figure above, $OP = 5$. Then $k = 3$. The slope of OP is $\dfrac{4}{3}$. Then the slope of

$y = f(x)$ is $-\dfrac{3}{4}$. Therefore, $f(x) = -\dfrac{3}{4}x + b$.

$P(3, 4)$ lies on the line.

$4 = -\dfrac{3}{4}(3) + b \Rightarrow b = \dfrac{25}{4}$

TEST 1   SECTION 7

1. (C)

$\pi < n < \pi^2 \Rightarrow$
$3.14 < n < 9.86 \Rightarrow n = 4, 5, 6, 7, 8, 9$

2. (B)

$\sqrt{k+2} = k$

$k + 2 = k^2 \Rightarrow k^2 - k - 2 = (k-2)(k+1) = 0$.

Therefore, $k = -1, 2$. But k cannot be negative.

3. (A)

n(own dog or cat)=n(dog)+n(cat)-n(own two)

$(50-2)=(20)+(35)-n(\text{own two})$

Therefore, 7 people own both a dog and a cat.

The answer is $\dfrac{7}{50}$.

---

4. (D)

$b+28=a+50 \implies a-b=-22$

$\implies |a-b|=22$

---

5. (B)

Display the reading levels as follows.

1,2,3,3,4,4,4,5,5,5,6,6,7,7,7

The median is 8th number, 5.

---

6. (D)

$a>b$, $a^2-b^2=(a+b)(a-b)=7$. Therefore,

$a+b=7$ and

$a-b=1 \implies 2a=8 \implies a=4$ and $b=3$.

$ab=(3)(4)=12$

---

7. (C)

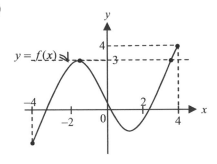

$y=f(x)$

In the graph above, two points have a y-coordinate of 3.

---

8. (A)

In the figure, $a=3b$ and $ab=48$. Therefore,

$ab=(3b)b=48 \implies b^2=16 \implies b=4$.

---

9. (E)

$k^2 \odot 2 = \dfrac{k^2+2}{2k^2}=1 \implies k^2=2 \implies k=\sqrt{2}$ or $-\sqrt{2}$

---

10. (D)

When $x=20$, $p(x)=300$. Therefore, the equation

$300=\dfrac{50(20)-200}{20}+k \implies k=260$

---

11. (D)

The number of girls who passed the test is,

$n=\dfrac{1}{3}\times\dfrac{2}{5}\times\dfrac{2}{3}x$, where x is the number of students.

$n=\dfrac{4}{45}x$. Because n is integral number, x must be a multiple of 45.

---

12. (D)

Average $=\dfrac{a+b+c}{3}$, where $b=\dfrac{a}{2}$ and $c=\dfrac{a}{3}$.

Average $=\dfrac{a+\frac{a}{2}+\frac{a}{3}}{3}=\dfrac{\frac{11a}{6}}{3}=\dfrac{11a}{18}$

---

13. (B)

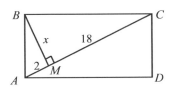

From the similar triangles, $BM^2=AM\times MC$

$x^2=2\times18=36$. Therefore, $x=6$.

Therefore, the area of $\triangle ABC$ is $\dfrac{20\times6}{2}=60$.

---

14. (E)

$3<\ p\ <25$

$12<\ q\ <16$

Therefore, $\dfrac{3}{16}<\dfrac{p}{q}<\dfrac{25}{12}$.

15. (D)

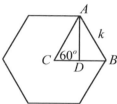

In the figure above, $BD = \dfrac{k}{2}$ and $AD = \dfrac{k}{2}\sqrt{3}$.

The area of $\triangle ABC = \dfrac{k \times \dfrac{k\sqrt{3}}{2}}{2} = \dfrac{k^2\sqrt{3}}{4}$.

Therefore, the area of hexagon is

$6 \times \dfrac{k^2\sqrt{3}}{4} = \dfrac{3k^2\sqrt{3}}{2}$.

16. (D)

The sequence is as follows. $p, pk, pk^2, pk^3 \ldots$

$\dfrac{pk^6}{pk^2} = k^4 = 16 \implies k = 2$

END

NO MATERIAL ON THIS PAGE

Dr. John Chung's SAT Math

TEST
2

ANSWER SHEET　　　TEST #:

SECTION 3

1 Ⓐ Ⓑ Ⓒ Ⓓ Ⓔ	11 Ⓐ Ⓑ Ⓒ Ⓓ Ⓔ	21 Ⓐ Ⓑ Ⓒ Ⓓ Ⓔ	31 Ⓐ Ⓑ Ⓒ Ⓓ Ⓔ
2 Ⓐ Ⓑ Ⓒ Ⓓ Ⓔ	12 Ⓐ Ⓑ Ⓒ Ⓓ Ⓔ	22 Ⓐ Ⓑ Ⓒ Ⓓ Ⓔ	32 Ⓐ Ⓑ Ⓒ Ⓓ Ⓔ
3 Ⓐ Ⓑ Ⓒ Ⓓ Ⓔ	13 Ⓐ Ⓑ Ⓒ Ⓓ Ⓔ	23 Ⓐ Ⓑ Ⓒ Ⓓ Ⓔ	33 Ⓐ Ⓑ Ⓒ Ⓓ Ⓔ
4 Ⓐ Ⓑ Ⓒ Ⓓ Ⓔ	14 Ⓐ Ⓑ Ⓒ Ⓓ Ⓔ	24 Ⓐ Ⓑ Ⓒ Ⓓ Ⓔ	34 Ⓐ Ⓑ Ⓒ Ⓓ Ⓔ
5 Ⓐ Ⓑ Ⓒ Ⓓ Ⓔ	15 Ⓐ Ⓑ Ⓒ Ⓓ Ⓔ	25 Ⓐ Ⓑ Ⓒ Ⓓ Ⓔ	35 Ⓐ Ⓑ Ⓒ Ⓓ Ⓔ
6 Ⓐ Ⓑ Ⓒ Ⓓ Ⓔ	16 Ⓐ Ⓑ Ⓒ Ⓓ Ⓔ	26 Ⓐ Ⓑ Ⓒ Ⓓ Ⓔ	36 Ⓐ Ⓑ Ⓒ Ⓓ Ⓔ
7 Ⓐ Ⓑ Ⓒ Ⓓ Ⓔ	17 Ⓐ Ⓑ Ⓒ Ⓓ Ⓔ	27 Ⓐ Ⓑ Ⓒ Ⓓ Ⓔ	37 Ⓐ Ⓑ Ⓒ Ⓓ Ⓔ
8 Ⓐ Ⓑ Ⓒ Ⓓ Ⓔ	18 Ⓐ Ⓑ Ⓒ Ⓓ Ⓔ	28 Ⓐ Ⓑ Ⓒ Ⓓ Ⓔ	38 Ⓐ Ⓑ Ⓒ Ⓓ Ⓔ
9 Ⓐ Ⓑ Ⓒ Ⓓ Ⓔ	19 Ⓐ Ⓑ Ⓒ Ⓓ Ⓔ	29 Ⓐ Ⓑ Ⓒ Ⓓ Ⓔ	39 Ⓐ Ⓑ Ⓒ Ⓓ Ⓔ
10 Ⓐ Ⓑ Ⓒ Ⓓ Ⓔ	20 Ⓐ Ⓑ Ⓒ Ⓓ Ⓔ	30 Ⓐ Ⓑ Ⓒ Ⓓ Ⓔ	40 Ⓐ Ⓑ Ⓒ Ⓓ Ⓔ

SECTION 5

1 Ⓐ Ⓑ Ⓒ Ⓓ Ⓔ	11 Ⓐ Ⓑ Ⓒ Ⓓ Ⓔ	21 Ⓐ Ⓑ Ⓒ Ⓓ Ⓔ	31 Ⓐ Ⓑ Ⓒ Ⓓ Ⓔ
2 Ⓐ Ⓑ Ⓒ Ⓓ Ⓔ	12 Ⓐ Ⓑ Ⓒ Ⓓ Ⓔ	22 Ⓐ Ⓑ Ⓒ Ⓓ Ⓔ	32 Ⓐ Ⓑ Ⓒ Ⓓ Ⓔ
3 Ⓐ Ⓑ Ⓒ Ⓓ Ⓔ	13 Ⓐ Ⓑ Ⓒ Ⓓ Ⓔ	23 Ⓐ Ⓑ Ⓒ Ⓓ Ⓔ	33 Ⓐ Ⓑ Ⓒ Ⓓ Ⓔ
4 Ⓐ Ⓑ Ⓒ Ⓓ Ⓔ	14 Ⓐ Ⓑ Ⓒ Ⓓ Ⓔ	24 Ⓐ Ⓑ Ⓒ Ⓓ Ⓔ	34 Ⓐ Ⓑ Ⓒ Ⓓ Ⓔ
5 Ⓐ Ⓑ Ⓒ Ⓓ Ⓔ	15 Ⓐ Ⓑ Ⓒ Ⓓ Ⓔ	25 Ⓐ Ⓑ Ⓒ Ⓓ Ⓔ	35 Ⓐ Ⓑ Ⓒ Ⓓ Ⓔ
6 Ⓐ Ⓑ Ⓒ Ⓓ Ⓔ	16 Ⓐ Ⓑ Ⓒ Ⓓ Ⓔ	26 Ⓐ Ⓑ Ⓒ Ⓓ Ⓔ	36 Ⓐ Ⓑ Ⓒ Ⓓ Ⓔ
7 Ⓐ Ⓑ Ⓒ Ⓓ Ⓔ	17 Ⓐ Ⓑ Ⓒ Ⓓ Ⓔ	27 Ⓐ Ⓑ Ⓒ Ⓓ Ⓔ	37 Ⓐ Ⓑ Ⓒ Ⓓ Ⓔ
8 Ⓐ Ⓑ Ⓒ Ⓓ Ⓔ	18 Ⓐ Ⓑ Ⓒ Ⓓ Ⓔ	28 Ⓐ Ⓑ Ⓒ Ⓓ Ⓔ	38 Ⓐ Ⓑ Ⓒ Ⓓ Ⓔ
9 Ⓐ Ⓑ Ⓒ Ⓓ Ⓔ	19 Ⓐ Ⓑ Ⓒ Ⓓ Ⓔ	29 Ⓐ Ⓑ Ⓒ Ⓓ Ⓔ	39 Ⓐ Ⓑ Ⓒ Ⓓ Ⓔ
10 Ⓐ Ⓑ Ⓒ Ⓓ Ⓔ	20 Ⓐ Ⓑ Ⓒ Ⓓ Ⓔ	30 Ⓐ Ⓑ Ⓒ Ⓓ Ⓔ	40 Ⓐ Ⓑ Ⓒ Ⓓ Ⓔ

9　10　11　12　13

14　15　16　17　18

SECTION 7	1 Ⓐ Ⓑ Ⓒ Ⓓ Ⓔ	11 Ⓐ Ⓑ Ⓒ Ⓓ Ⓔ	21 Ⓐ Ⓑ Ⓒ Ⓓ Ⓔ	31 Ⓐ Ⓑ Ⓒ Ⓓ Ⓔ
	2 Ⓐ Ⓑ Ⓒ Ⓓ Ⓔ	12 Ⓐ Ⓑ Ⓒ Ⓓ Ⓔ	22 Ⓐ Ⓑ Ⓒ Ⓓ Ⓔ	32 Ⓐ Ⓑ Ⓒ Ⓓ Ⓔ
	3 Ⓐ Ⓑ Ⓒ Ⓓ Ⓔ	13 Ⓐ Ⓑ Ⓒ Ⓓ Ⓔ	23 Ⓐ Ⓑ Ⓒ Ⓓ Ⓔ	33 Ⓐ Ⓑ Ⓒ Ⓓ Ⓔ
	4 Ⓐ Ⓑ Ⓒ Ⓓ Ⓔ	14 Ⓐ Ⓑ Ⓒ Ⓓ Ⓔ	24 Ⓐ Ⓑ Ⓒ Ⓓ Ⓔ	34 Ⓐ Ⓑ Ⓒ Ⓓ Ⓔ
	5 Ⓐ Ⓑ Ⓒ Ⓓ Ⓔ	15 Ⓐ Ⓑ Ⓒ Ⓓ Ⓔ	25 Ⓐ Ⓑ Ⓒ Ⓓ Ⓔ	35 Ⓐ Ⓑ Ⓒ Ⓓ Ⓔ
	6 Ⓐ Ⓑ Ⓒ Ⓓ Ⓔ	16 Ⓐ Ⓑ Ⓒ Ⓓ Ⓔ	26 Ⓐ Ⓑ Ⓒ Ⓓ Ⓔ	36 Ⓐ Ⓑ Ⓒ Ⓓ Ⓔ
	7 Ⓐ Ⓑ Ⓒ Ⓓ Ⓔ	17 Ⓐ Ⓑ Ⓒ Ⓓ Ⓔ	27 Ⓐ Ⓑ Ⓒ Ⓓ Ⓔ	37 Ⓐ Ⓑ Ⓒ Ⓓ Ⓔ
	8 Ⓐ Ⓑ Ⓒ Ⓓ Ⓔ	18 Ⓐ Ⓑ Ⓒ Ⓓ Ⓔ	28 Ⓐ Ⓑ Ⓒ Ⓓ Ⓔ	38 Ⓐ Ⓑ Ⓒ Ⓓ Ⓔ
	9 Ⓐ Ⓑ Ⓒ Ⓓ Ⓔ	19 Ⓐ Ⓑ Ⓒ Ⓓ Ⓔ	29 Ⓐ Ⓑ Ⓒ Ⓓ Ⓔ	39 Ⓐ Ⓑ Ⓒ Ⓓ Ⓔ
	10 Ⓐ Ⓑ Ⓒ Ⓓ Ⓔ	20 Ⓐ Ⓑ Ⓒ Ⓓ Ⓔ	30 Ⓐ Ⓑ Ⓒ Ⓓ Ⓔ	40 Ⓐ Ⓑ Ⓒ Ⓓ Ⓔ

Math Scoring Worksheet

A. Section 3

_____   _____
numer of correct   number of incorrect

         +   +

B. Section 5 (1-8)

_____   _____
numer of correct   number of incorrect

         +

C. Section 5 (9-18)

_____
numer of correct

         +   +

D. Section 7

_____   _____
numer of correct   number of incorrect

         =   =

E. Total Unrounded Raw Score

_____   −   _____ ÷4 = _____
numer of correct   number of incorrect

F. Total Rounded Raw Score   _____   (See table)

Math Score Range = | _____ |

Math Conversion Table

Raw Score	Scaled Score	Raw Score	Scaled Score
54	800	23	490-550
53	780-800	22	480-540
52	760-800	21	470-530
51	740-800	20	460-520
50	720-780	19	450-510
49	700-760	18	450-510
48	690-750	17	440-500
47	680-740	16	430-490
46	670-730	15	420-480
45	660-720	14	420-480
44	650-710	13	410-470
43	650-710	12	400-460
42	640-700	11	390-450
41	630-690	10	380-440
40	620-680	9	390-430
39	610-670	8	380-420
38	610-670	7	370-410
37	600-660	6	360-400
36	590-650	5	340-380
35	580-640	4	320-370
34	570-630	3	310-360
33	560-620	2	300-350
32	560-620	1	270-320
31	550-610	0	240-300
30	540-600	-1	200-290
29	530-590	-2	200-270
28	530-590	-3	200-260
27	520-580	-4	200-240
26	510-570	-5	200-220
25	500-560	-6 and below	200
24	500-560		

SECTION 3
Time- 25 minutes
20 Questions

Turn to Section 3 (Page 1) of your answer sheet to answer the questions in this section.

Directions: For this section, solve each problem and decide which is the best of the choices given. Fill in the corresponding circle on the answer sheet. You may use any available space for scratchwork.

Notes

1. The use of a calculator is permitted.
2. All numbers used are real numbers.
3. Figures that accompany problems in this test are intended to provide information useful in solving the problems. They are drawn as accurately as possible EXCEPT when it is stated in a specific problem that the figure is not drawn to scale. All figure lie in a plane unless other indicated.
4. Unless otherwise specified, the domain of any function f is assumed to be set of all real numbers x for which $f(x)$ is a real number.

Reference Informatiom

$A = \pi r^2$
$C = 2\pi r$
$A = \ell w$
$A = \dfrac{1}{2}bh$
$V = \ell wh$
$V = \pi r^2 h$
$c^2 = a^2 + b^2$
Special Right Triangles

The numbers of degrees of arc in a circle is $360°$.

The sum of the measures in degrees of the angles is $180°$.

---

1. If $3(a + 2b - c) = 12$, what is the value of $a + 2b$ in terms of c?

(A) $3c - 4$
(B) $c - 12$
(C) $c - 4$
(D) $4 + c$
(E) $12 - c$

$$\begin{array}{r} XY \\ + YX \\ \hline 88 \end{array}$$

2. In the addition problem above, X and Y are distinct positive integers. How many different integer values of X are possible?

(A) 4 (B) 5 (C) 6 (D) 7 (E) 8

GO ON TO THE NEXT PAGE

3. A *clever integer* is defined as an integer that is greater than 20, less than 100, and such that the sum of its digits is 9. What fraction of all clever integers is divisible by 27?

(A) $\dfrac{1}{8}$

(B) $\dfrac{1}{4}$

(C) $\dfrac{3}{8}$

(D) $\dfrac{1}{2}$

(E) $\dfrac{5}{8}$

4. If $8^n \times 4^2 = 2^{10}$, what is the value of n?

(A) 2
(B) 3
(C) 4
(D) 5
(E) 6

5. In the figure above, a player spins the arrow twice. Let the number of the first spin be p where the arrow stops, and the number of the second spin be q where the arrow stops. If each number has an equal probability of being the sector on which the arrow stops, what is the probability that the fraction $\dfrac{p}{q}$ is greater than or equal to 1?

(A) $\dfrac{1}{4}$

(B) $\dfrac{3}{16}$

(C) $\dfrac{5}{16}$

(D) $\dfrac{1}{2}$

(E) $\dfrac{5}{8}$

6. If $n = 4p^2q$, where p and q are distinct prime numbers, and not equal to 2, how many factors does the number n have?

(A) 20
(B) 18
(C) 16
(D) 14
(E) 12

GO ON TO THE NEXT PAGE

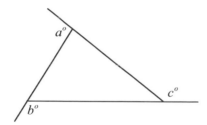

Note: Figure not drawn to scale.

7. In the figure, the angles $a, b,$ and c represent exterior angles of the triangle. What is the value of $a + b + c$?

(A) 180
(B) 240
(C) 300
(D) 360
(E) 540

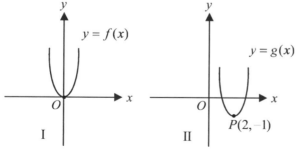

8. The graphs of f and g are shown above. If the equation in figure I is $y = x^2$, which of the following could be the equation of g in figure II?

(A) $g(x) = x^2 - 2$
(B) $g(x) = (x+2)^2 + 1$
(C) $g(x) = (x-2)^2 - 1$
(D) $g(x) = (x+2)^2 - 1$
(E) $g(x) = x^2 - 2x - 1$

9. On a farm, the ratio of cows to goats is 1:2 and the ratio of goats to pigs is 4:5. What is the ratio of cows to pigs?

(A) 1:2
(B) 1:4
(C) 2:5
(D) 1:6
(E) 2:7

10. The area of a sector is directly proportional to the square of the radius. When the value of the radius is 4, the area is 20. If the area is 45, what is the value of the radius?

(A) 4
(B) 6
(C) 8
(D) 10
(E) 12

GO ON TO THE NEXT PAGE

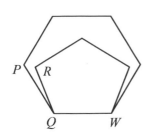

11. If the figure above shows a regular pentagon and a regular hexagon with a common side $\overline{QW}$, what is the value of $\angle PQR$?

(A) 12°
(B) 20°
(C) 22°
(D) 24°
(E) 30°

12. Cathy can do a job in 8 hours while Danny can do the same job in 6 hours. If Cathy and Danny work three hours, what fraction of the job is left to be finished?

(A) $\dfrac{1}{12}$

(B) $\dfrac{1}{8}$

(C) $\dfrac{1}{7}$

(D) $\dfrac{1}{6}$

(E) $\dfrac{1}{4}$

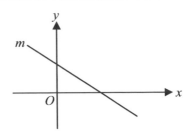

13. The figure above shows the graph of $ax + by + c = 0$. Which of the following could be true?

I. $\dfrac{a}{b} > 0$

II. $\dfrac{c}{b} > 0$

III. $\dfrac{c}{b} < 0$

(A) I only
(B) II only
(C) III only
(D) I and II only
(E) I and III only

14. If X is the set of multiples of 5, and Y is the set of three-digit positive integers which are multiples of 7, how many numbers are common to both sets?

(A) Five
(B) Twenty
(C) Twenty five
(D) Twenty six
(E) Thirty

GO ON TO THE NEXT PAGE

15. What is p percent of q divided by q percent of p?

(A) 1

(B) pq

(C) $\dfrac{p}{q}$

(D) $\dfrac{q}{p}$

(E) $\dfrac{100}{pq}$

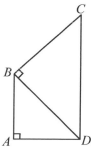

17. The figure above shows two isosceles right triangles. If the length of $\overline{BC}$ is $2a$, what is the perimeter of the quadrilateral $ABCD$ in terms of a?

(A) $2a + 4a\sqrt{2}$

(B) $3a + a\sqrt{2}$

(C) $4a + 4a\sqrt{2}$

(D) $4a + a\sqrt{3}$

(E) $5a$

16. In a plane, the distance between points X and Y is 10, the distance between points X and p is 3, and the distance between points Y and q is 4. Which of the following CANNOT be the length of $\overline{pq}$?

(A) 2

(B) 3

(C) 10

(D) 15

(E) 17

18. If x is the coordinate of the indicated point on the number line above, which of the following points represents the coordinate of $\left|\dfrac{x}{5}\right|$?

(A) A

(B) B

(C) C

(D) D

(E) E

GO ON TO THE NEXT PAGE

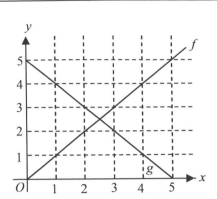

19. The figure above shows the graphs of the functions f and g. If $f\big(g(x)\big) = 4$, what is the value of x?

(A) 1
(B) 2
(C) 3
(D) 4
(E) 5

20. The Maxim Telephone Company charges k cents for the first t minutes of a call and charges for any additional time at the rate of r cents per minute. If a certain customer pays \$10, which of the following could be the length of that phone call in minutes?

(A) $\dfrac{1000}{r} + k$

(B) $\dfrac{1000}{r} + tk$

(C) $\dfrac{1000 - k - t}{r}$

(D) $\dfrac{1000 - k}{r} + t$

(E) $\dfrac{1000 + k}{r} + kt$

STOP

If you finish before time is called, you may check your work on this section only.
Do not turn to any other section in the test.

SECTION 5
Time- 25 minutes
18 Questions

Turn to Section 5 (Page 1) of your answer sheet to answer the questions in this section.

Directions: For this section, solve each problem and decide which is the best of the choices given. Fill in the corresponding circle on the answer sheet. You may use any available space for scratchwork.

Notes

1. The use of a calculator is permitted.
2. All numbers used are real numbers.
3. Figures that accompany problems in this test are intended to provide information useful in solving the problems. They are drawn as accurately as possible EXCEPT when it is stated in a specific problem that the figure is not drawn to scale. All figure lie in a plane unless other indicated.
4. Unless otherwise specified, the domain of any function f is assumed to be set of all real numbers x for which $f(x)$ is a real number.

Reference Informatiom

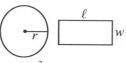   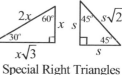

$A = \pi r^2$
$C = 2\pi r$
$A = \ell w$
$A = \frac{1}{2}bh$
$V = \ell wh$
$V = \pi r^2 h$
$c^2 = a^2 + b^2$
Special Right Triangles

The numbers of degrees of arc in a circle is $360°$.

The sum of the measures in degrees of the angles is $180°$.

1. If $4(x+y) - 2(x+y) = 12$, then $x + y =$

(A) 2
(B) 4
(C) 6
(D) 8
(E) 10

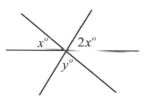

Note: Figure not drawn to scale

2. In the figure above, three lines intersect at a point. If $y = 2x - 30$, what is the value of x?

(A) 30
(B) 37
(C) 42
(D) 50
(E) 60

GO ON TO THE NEXT PAGE

Dr. John Chung's SAT Math Test 2

3. If Sally drives m miles from her house to her office in f hours, and drives back to her house in g hours, what is her average speed of the entire trip, in miles per hour?

(A) $\dfrac{f+g}{2}$

(B) $\dfrac{m}{f+g}$

(C) $\dfrac{2m}{f+g}$

(D) $\dfrac{fg}{f+g}$

(E) $\dfrac{2fg}{f+g}$

4. If $a+b$ is an odd integer, which of the following must be an even integer?

(A) a

(B) b

(C) ab

(D) $ab+1$

(E) $ab+a$

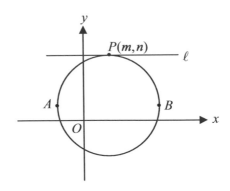

5. In the $xy-$coordinate plane, line ℓ is parallel to $x-$axis, and the points $A(-2,1)$ and $B(4,1)$ lie on a circle. If $\overline{AB}$ (not drawn) is the diameter of the circle, what are the coordinates of point P ?

(A) $(2,4)$
(B) $(1,4)$
(C) $(1,3)$
(D) $(1,5)$
(E) $(2,5)$

6. Morgan took five tests last quarter. If the average of the first three tests was p and the average of the rest of the tests was q , what was his average in the quarter?

(A) $p+q$

(B) $\dfrac{p+q}{2}$

(C) $\dfrac{p+q}{5}$

(D) $\dfrac{3p+2q}{2}$

(E) $\dfrac{3p+2q}{5}$

GO ON TO THE NEXT PAGE

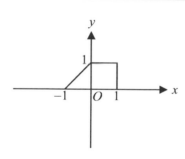

7. The graph of $y = f(x)$ is shown above. Which of the following could be the graph of $y = -f(x+1)$?

(A)

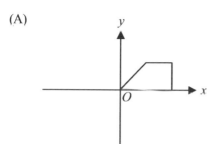

(B)

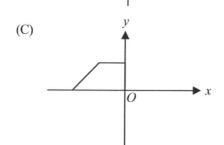

(C)

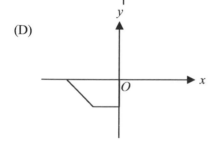

(D)

(E)

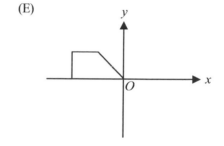

8. For all numbers a and b, let $a \odot b$ be defined as $a \odot b = (a+b)^2 - (a-b)^2$. Which of the following is not the same as the value of $(3 \odot 1) \odot 2$?

(A) $4 \odot 6$

(B) $3 \odot 8$

(C) $2 \odot 12$

(D) $1 \odot 24$

(E) $3 \odot 12$

GO ON TO THE NEXT PAGE

5 □□□ 5 □□□ 5 □□□ 5

Directions: For Students-Produced Response questions 9-18, use the grid at the bottom of the answer sheet page on which you have answered questions 1-8.

Each of the remaining 10 questions requires you to solve the problem and enter your answer by making the circles in the special grid, as shown in the examples below. You may use any available space for scratchwork.

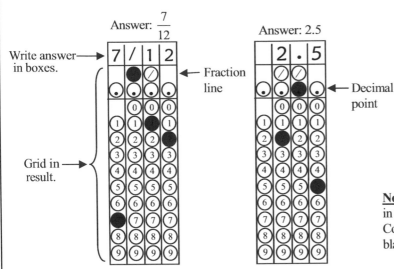

Answer: $\frac{7}{12}$

Write answer in boxes.

Fraction line

Grid in result.

Answer: 2.5

Decimal point

Answer: 201
Either position is correct.

Note: You may start your answers in any column, space permitting. Columns not needed should be left blank.

- Mark no more than one circle in any column.

- Because the answer sheet will be machine-scored, **you will receive credit only if the circles are filled in correctly.**

- Although not required, it is suggested that you write your answer in the boxes at the top of the columns to help you fill in the circles accurately.

- Some problems may have more than one correct answer. In such cases, grid only one answer.

- No question has a negative answer.

- **Mixed numbers** such as $3\frac{1}{2}$ must be gridded as 3.5 or 7/2. (If $\boxed{3\ 1\ /\ 2}$ is gridded, it will be interpreted as $\frac{31}{2}$, not $3\frac{1}{2}$.)

- **Decimal Answers:** If you obtain a decimal answer with more digits than the grid can accommodate, it may be either rounded or truncated, but it must fill the entire grid. For example, if you obtain an answer such as 0.6666..., you should record your result as .666 or .667. **A less accurate value such as .66 or .67 will be scored as incorrect.**

Acceptable ways to grid $\frac{2}{3}$ are:

9. Peter can finish a task in 8 hours, and Sam can finish the same task twice as fast as Peter. If they work together, how many hours would it take to complete the task?

10. Since the beginning of 2002, the number of students in Spring Lake high school has doubled every 3-year period. If there were 800 students in the school at the beginning of 2002, how many students are in the school at the beginning of 2008?

GO ON TO THE NEXT PAGE ▷

- 134 -

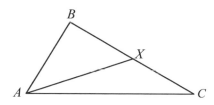

Note: Figure not drawn to scale.

11. In the figure of $\triangle ABC$ above, the area of $\triangle ABX$ is 20 and the area of $\triangle ACX$ is 16.

What is the value of $\dfrac{BX}{CX}$?

$$A = \{0,1,2,3\} \qquad B = \{1,2,3,4\}$$

12. If a is the element from set A, and b is the element from set B, how many distinct possible

values of $\dfrac{a}{b}$ are there?

13. Two marbles at a time are to be drawn out from a jar that contains 5 red marbles and 8 black marbles. What is the probability of drawing one red and one black marble from the jar?

14. If $|10-3x| < 5$, what is one possible integer value of x ?

GO ON TO THE NEXT PAGE

15. If the square of a positive number is same as the number divided by 4, what is the value of the number?

x	$f(x)$		x	$g(x)$
1	7		0	12
3	13		1	14
5	19		2	16
a	b		a	b
Table I			Table II	

17. In the tables above, the points (x, f) and (x, g) represented in the tables lie on straight lines. If the point (a, b) lies on both lines, what is the value of $a + b$?

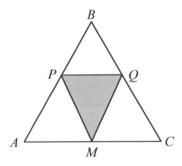

Note: The figure not drawn to scale.

16. The figure above shows an equilateral triangle ABC. If M is the midpoint of $\overline{AC}$, $\overline{BP}$ is $\dfrac{1}{3}$ of $\overline{AB}$, and $\overline{PQ}$ is parallel to $\overline{AC}$, what fraction of the area of $\triangle ABC$ is shaded?

18. If $x^2 + y^2 \leq 25$ and $y \geq 3$, what is the greatest possible value of x ?

STOP

If you finish before time is called, you may check your work on this section only.
Do not turn to any other section in the test.

SECTION 7
Time- 20 minutes
16 Questions

Turn to Section 7 (Page 2) of your answer sheet to answer the questions in this section.

Directions: For this section, solve each problem and decide which is the best of the choices given. Fill in the corresponding circle on the answer sheet. You may use any available space for scratchwork.

Notes

1. The use of a calculator is permitted.
2. All numbers used are real numbers.
3. Figures that accompany problems in this test are intended to provide information useful in solving the problems. They are drawn as accurately as possible EXCEPT when it is stated in a specific problem that the figure is not drawn to scale. All figure lie in a plane unless other indicated.
4. Unless otherwise specified, the domain of any function f is assumed to be set of all real numbers x for which $f(x)$ is a real number.

Reference Informatiom

$A = \pi r^2$
$C = 2\pi r$
$A = \ell w$
$A = \dfrac{1}{2} bh$
$V = \ell wh$
$V = \pi r^2 h$
$c^2 = a^2 + b^2$
Special Right Triangles

The numbers of degrees of arc in a circle is 360°.

The sum of the measures in degrees of the angles is 180°.

1. If $2\sqrt{k}$ is a positive even integer, which of the following MUST be even?

   (A) $\sqrt{k}$
   (B) k
   (C) $\sqrt{2k}$
   (D) $2k$
   (E) $k + 1$

$$S = \{3,\ 6,\ 12\}$$

2. From the given above, how many different values can be created by forming fractions $\dfrac{a}{b}$ such that a and b are distinct elements of the set S ?

   (A) 2
   (B) 3
   (C) 4
   (D) 5
   (E) 6

GO ON TO THE NEXT PAGE

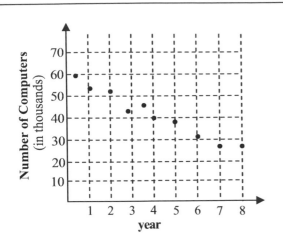

3. The scatterplot above shows the number of computers sold during 8 years of J.C Computer Inc. Which of the following functions best describe the relationship between n, the number of year, and $P(n)$, the number of computers sold per year (in thousands)?

(A) $P(n) = 60n$

(B) $P(n) = 60 - n$

(C) $P(n) = -2n + 60$

(D) $P(n) = 60 - 4n$

(E) $P(n) = 40$

4. A circle and a triangle have equal areas. If the radius of the circle, represented by t, is equal to the base of the triangle, what is the altitude of the triangle in terms of t?

(A) $\dfrac{t}{2}$

(B) $2\pi t$

(C) $\dfrac{\pi t}{2}$

(D) $\dfrac{\pi t^2}{2}$

(E) $\dfrac{2}{\pi t}$

5. If y is directly proportional to x^2, which of the following could be the graph that shows the relationship between x and y?

(A)

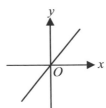

(B)

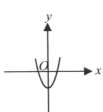

(C)

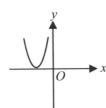

(D)

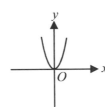

(E)

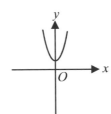

GO ON TO THE NEXT PAGE

Note: Figure not drawn to scale.

6. In the figure above, *ABCD* is an isosceles trapezoid. If the length of $\overline{AD}$ is 24 and the length of $\overline{BC}$ is 16, what is the perimeter of the trapezoid?

(A) 40
(B) 56
(C) $40+4\sqrt{2}$
(D) $40+6\sqrt{2}$
(E) $40+8\sqrt{2}$

7. If $\dfrac{a+b}{x-y}=\dfrac{1}{2}$ and $\dfrac{a-b}{x+y}=\dfrac{2}{3}$, then

$$\frac{2a^2-2b^2}{3x^2-3y^2}=$$

(A) $\dfrac{1}{3}$

(B) $\dfrac{1}{2}$

(C) $\dfrac{2}{7}$

(D) $\dfrac{2}{9}$

(E) $\dfrac{2}{3}$

8. Jimmy drove a car at a constant rate of speed to a shopping mall, spent a while in the mall, and returned home at a constant rate of speed. Which of the following graphs could correctly represent the distance from his house over time?

(A)

(B)

(C)

(D)

(E)

GO ON TO THE NEXT PAGE

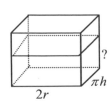

Note: Figure not drawn to scale.

9. In the figures above, the interior dimensions of a cylindrical tank full of water have radius r and height h. All of the water in the tank is poured into a second tank. If the interior dimensions of the second tank are $2r$ feet long and πh feet wide, what is the height of the water in the second tank ?

(A) $\dfrac{1}{2}$

(B) $\dfrac{\pi}{h}$

(C) $\dfrac{r}{2}$

(D) $\dfrac{\pi}{2}$

(E) $\dfrac{\pi r}{2}$

10. If $2^{21} = x + 2^{20}$, what is the value of x ?

(A) 2
(B) 2^2
(C) 3^2
(D) 2^{20}
(E) 2^{21}

NUMBER OF EMPLOYEES		
	Men	Women
Salary over $50,000	10	5
Salary $50,000 or less	40	35

11. The table above shows the number of employees at J.K Computer Technology. If one employee will be picked at random, what is the probability of picking a woman whose salary is over $50,000?

(A) $\dfrac{1}{18}$

(B) $\dfrac{1}{8}$

(C) $\dfrac{1}{5}$

(D) $\dfrac{1}{4}$

(E) $\dfrac{3}{18}$

12. If $5^3 + 5^3 + 5^3 + 5^3 + 5^3 = 5^n$, where n is a positive integer, what is the value of n ?

(A) 3
(B) 4
(C) 5
(D) 6
(E) 7

GO ON TO THE NEXT PAGE

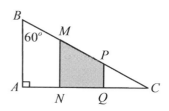

Note: Figure not drawn to scale.

13. In the figure above, $\overline{AB} \parallel \overline{MN} \parallel \overline{PQ}$, $\overline{AN} = \overline{NQ} = \overline{QC}$, and $\overline{AB} = 6$. What is the area of the shaded region?

(A) 8
(B) 10
(C) $6\sqrt{3}$
(D) $12\sqrt{3}$
(E) $18\sqrt{3}$

14. If $\dfrac{1}{p} + \dfrac{1}{q} = t$, and $pq = w$, then $p + q =$

(A) $\dfrac{w}{t}$

(B) $\dfrac{t}{w}$

(C) tw

(D) $\dfrac{tw}{2}$

(E) $t(t + w)$

15. For all numbers a and b , let $a \triangle b$ be defined as $a \triangle b = a^2 - b$. What is the value of $\left(2^{4 \triangle 13}\right) \triangle \left(3^{5 \triangle 23}\right)$?

(A) -17
(B) 17
(C) -55
(D) 55
(E) 70

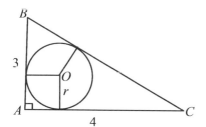

16. In the figure above, a circle O is inscribed in $\triangle ABC$. If $AB = 3$, $AC = 4$, and the radius of the circle is r , what is the value of r ?

(A)   1

(B) $\sqrt{2}$

(C) $\sqrt{3}$

(D) $\dfrac{\sqrt{2}}{3}$

(E) $\sqrt{3} - 1$

STOP

If you finish before time is called, you may check your work on this section only.
Do not turn to any other section in the test.

#	SECTION 3	SECTION 5	SECTION 7
1	D	C	D
2	C	C	C
3	C	C	D
4	A	C	B
5	E	B	D
6	B	E	B
7	D	D	D
8	C	E	E
9	C	$\frac{8}{3}$, 2.66 or 2.67	C
10	B	3200	D
11	A	$5/4$ or 1.25	A
12	B	10	B
13	E	.512 or .513	C
14	D	2, 3, or 4	C
15	A	$1/4$ or 0.25	D
16	A	$2/9$ or .222	A
17	A	36	
18	C	4	
19	A		
20	D		

TEST 2   SECTION 3

1. (D)

$3(a+2b-c)=12 \Rightarrow a+2b-c=4$. Thus, $a+2b=4+c$.

2. (C)

The pairs of (X,Y) are

$(1,7)$ $(2,6)$ $(3,5)$ $(4,4)$ $(5,3)$ $(6,2)$ $(7,1)$

$\Rightarrow$ 7 numbers

3. (C)

The clever integers between 20 and 100 are 27, 36, 45, 54, 63, 72, 81, 90. The numbers divisible by 27 are 27, 54, 81. Thus, the fraction is $\frac{3}{8}$.

4. (A)

$8^n \times 4^2 = 2^{10} \Rightarrow 2^{3n} \times 2^4 = 2^{3n+4} = 2^{10}$. It follows that $3n+4=10$. Therefore, $n=2$.

5. (E)

Since $\frac{p}{q} \geq 1 \Rightarrow p \geq q$. The pairs of (p,q) will be as follows.

$(1,1)$ $(2,1)$ $(2,2)$ $(3,1)$ $(3,2)$ $(3,3)$ $(4,1)$ $(4,2)$ $(4,3)$ $(4,4)$

All possible outcomes are

$4 \times 4 = 16$. Therefore, $P = \frac{10}{16} = \frac{5}{8}$.

6. (B)

$n = 4p^2 q \Rightarrow n = 2^2 p^2 q$.

From prime factorization, the number of factors can be obtained by multiplying the exponents added by 1.

The number of factors

$= (2+1)(2+1)(1+1) = 18$

Or, use $p = 3$ and $q = 5$. Then $n = 180$.

The factors of 180

$\rightarrow 1,2,3,4,5,6,9,10,12,15,18,20,30,45,60,90,$

180. (18 factors)

---

7. (D)

The sum of exterior angles is always 360°.

Or, $(180-a)+(180-b)+(180-c)=180$.

Therefore, $a+b+c=360$.

---

8. (C)

Translate to the right by 2 and down by 1.

Therefore, $g(x)=f(x-2)+1$.

It follows that

$g(x)=(x-2)^2-1$.

---

9. (C)

$\dfrac{C}{G}=\dfrac{1}{2}$ and $\dfrac{G}{P}=\dfrac{4}{5}$. Therefore,

$\dfrac{C}{P}=\dfrac{C}{G}\times\dfrac{G}{P}=\dfrac{1}{2}\times\dfrac{4}{5}=\dfrac{2}{5}$

---

10. (B)

Since $K=\dfrac{A}{r^2}$,

$\dfrac{20}{4^2}=\dfrac{45}{r^2}\Rightarrow r^2=36$. Therefore, $r=6$.

---

11. (A)

$\angle PQW=\dfrac{720}{6}=120$ and

$\angle RQW=\dfrac{540}{5}=108$.

Therefore, $\angle PQR=120-108=12$

---

12. (B)

Cathy's rate per hour$=\dfrac{1}{8}$ and Danny's rate

per hour $=\dfrac{1}{6}$. The sum of the rates is

$\dfrac{1}{8}+\dfrac{1}{6}=\dfrac{7}{24}$.

If they work 3 hours, $\dfrac{7}{24}(3)=\dfrac{7}{8}$ of the job

will be done. $\dfrac{1}{8}$ of the job is left to be

finished.

---

13. (E)

The slope is negative and y-intercept is

positive. Thus, $-\dfrac{a}{b}<0$ and

$-\dfrac{c}{b}>0\Rightarrow\dfrac{a}{b}>0$ and $\dfrac{c}{b}<0$.

---

14. (D)

It is multiple of 35 from 100 to 999. Thus

$\left\lfloor\dfrac{999}{35}\right\rfloor-\left\lfloor\dfrac{99}{35}\right\rfloor=28-2=26$.

---

15. (A)

$\left(\dfrac{p}{100}\times q\right)\div\left(\dfrac{q}{100}\times p\right)=1$

---

16. (A)

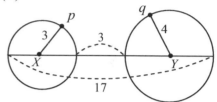

In the figure above, $3\le\overline{pq}\le17$.

---

17. (A)

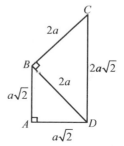

In the figure, the perimeter is $2a+4a\sqrt{2}$.

---

18. (C)

Approximately $x=-2.25$.

$\left|\dfrac{x}{5}\right|=\left|\dfrac{-2.25}{5}\right|=0.45$.

Therefore, point C best represents the coordinate.

---

19. (A)

In the xy-plane,

Since $f(k)=4\Rightarrow k=g(x)=4\Rightarrow x=1$.

---

20. (D)

k cents $\rightarrow$ upto t minutes

For the $(1,000-k)$ cents $\rightarrow$ r cents per minute

It follows that the minutes is $\dfrac{1000-k}{r}$.

Therefore, total length of the phone call is

$\dfrac{1000-k}{r}+t$.

TEST 2   SECTION 5

1. (C)

$4(x+y)-2(x+y)=2(x+y)=2 \Rightarrow x+y=1$.

2. (C)

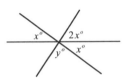

In the figure above,

$3x+y=180 \Rightarrow 3x+(2x-30)=180$.

It follows that

$5x=210 \Rightarrow x=42$.

3. (C)

Average speed$=\dfrac{\text{Entire distance}}{\text{Entire time}}=\dfrac{2m}{f+g}$.

4. (C)

$a+b=odd$. you can use $a=2$ and $b=1$,or $a=1$ and $b=2$.

(C) is always true.

5. (B)

The midpoint of A and

$B=\left(\dfrac{-2+4}{2},\dfrac{1+1}{2}\right)=(1,1)$.

Since the radius of the circle is 3, the coordinates of point P is (1,4).

6. (E)

The sum of the first three tests is $3p$ and the sum of last two tests is $2q$.

The entire average $=\dfrac{3p+2q}{5}$.

7. (D)

Translate to the left by 1 and reflection over the x-axis.

8. (E)

$3\odot1=(3+1)^2-(3-1)^2=12$ and

$(3\odot1)\odot2=12\odot2=14^2-10^2=96$.

Or

$a\odot b=(a+b)^2-(a-b)^2=4ab$

$3\odot1=4(3)(1)=12$ and

$12\odot2=4(12)(2)=96$.

$3\odot12=4(3)(12)=144\neq96$.

9. $\frac{8}{3}$ or 2.66, 2.67

	Hours	Rate	Combined Rate
Peter	8	$\frac{1}{8}$	$\frac{3}{8}$
Sam	4	$\frac{1}{4}$	

In the figure above, $1\div\dfrac{3}{8}=\dfrac{8}{3}$ hours.

10. 3200

$800\rightarrow1600\rightarrow3200$, increased every 3 years.

11. $\frac{5}{4}$ or 1.25

With the same height, The ratio of the areas is equal to the ratio of the lengths. $\dfrac{20}{16}=\dfrac{5}{4}$

12. 10

$\dfrac{a}{b} \Rightarrow \dfrac{0}{1,2,3,4}=0, \dfrac{1}{1},\dfrac{1}{2},\dfrac{1}{3},\dfrac{1}{4},$

$\dfrac{2}{1},\dfrac{\cancel{2}}{\cancel{2}},\dfrac{2}{3},\dfrac{\cancel{2}}{\cancel{4}},\dfrac{3}{1},\dfrac{3}{2},\dfrac{\cancel{3}}{\cancel{3}},\dfrac{3}{4}.$

Therefore, there are 10 distinct values.

13. .512 or .513

There are two possible cases.

(R,B) or (B,R) .

For $(R,B) \Rightarrow \dfrac{5}{13} \times \dfrac{8}{12} = \dfrac{10}{39}$ and

for $(B,R) \Rightarrow \dfrac{8}{13} \times \dfrac{5}{12} = \dfrac{10}{39}$. Therefore,

$\dfrac{10}{39} + \dfrac{10}{39} = \dfrac{20}{39} \Rightarrow .512$ or $.513$.

---

14. 2, 3, or 4

$|10 - 3x| < 5 \Rightarrow |3x - 10| < 5$. It follows that

$-5 < 3x - 10 < 5 \Rightarrow 5 < 3x < 15 \Rightarrow \dfrac{5}{3} < x < 5$.

Therefore, there are three integers, 2, 3, and 4.

---

15. $\frac{1}{4}$ or 0.25

$n^2 = \dfrac{n}{4} \Rightarrow n = \dfrac{1}{4}$

---

16. $\frac{2}{9}$ or .222

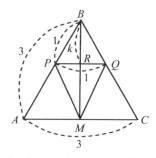

In the figure above, $\triangle RPQ$ and $\triangle BAC$ are

similar. $\dfrac{BR}{BM} = \dfrac{BP}{AB} = \dfrac{1}{3}$.

If $BP = k$, then $BM = 3k$.

Therefore, $RM = 2k$. The area of

$\triangle ABC = \dfrac{1}{2}(3 \times 3k) = 4.5k$ and the area of

$\triangle PQM = \dfrac{1}{2}(1 \times 2k) = k$.

Therefore, $\dfrac{\triangle PQM}{\triangle ABC} = \dfrac{k}{4.5k} = \dfrac{2}{9}$.

---

17. 36

Point (a,b) lies on both of the two lines.

The slope of f is $\dfrac{13 - 7}{3 - 1} = 3 = \dfrac{b - 7}{a - 1}$. Thus,

$b = 3a + 4$ --(1).

The slope of g is $\dfrac{14 - 12}{1 - 0} = 2 = \dfrac{b - 12}{a - 0}$.

Thus $b = 2a + 12$ --(2). From (1) and (2),

$a = 8$ and $b = 28$. Therefore, $a + b = 36$.

---

18. 4

If the value of y is the minimum, the value of x will be the maximum. The minimum of y is 3, then

$x^2 + 9 \le 25 \Rightarrow x^2 \le 16 \Rightarrow -4 \le x \le 4$.

Therefore, the greatest possible value of x is 4. Or, you can solve it geometrically as follows.

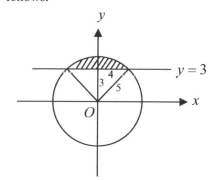

In the figure above, the greatest possible value of x in the shaded region is 4.

TEST 2   **SECTION 7**

1. (D)

$2\sqrt{k} = \text{even} \Rightarrow \sqrt{k}$ is even. $\Rightarrow k$ is even.

        or $\sqrt{k}$ is odd. $\Rightarrow k$ id odd.

Therefore, $2k$ is always even.

Or, use numbers. Let k be 4 or 9.

Only (D) is always even.

---

2. (C)

$\dfrac{a}{b} \Rightarrow \dfrac{3}{6}, \dfrac{3}{12}, \dfrac{6}{3}, \dfrac{\cancel{6}}{\cancel{12}}, \dfrac{12}{3}, \dfrac{\cancel{12}}{\cancel{6}}$.

Therefore, four different values are there.

---

3. (D)

Approximately, the y-intercept is 60, and the

slope is $\dfrac{28 - 60}{8} \cong -4$. Therefore,

$y = -4x + 60$.

---

4. (B)

The areas are equal. $\pi t^2 = \dfrac{t \times h}{2} \Rightarrow h = 2\pi t$.

---

5. (D)

$y = kx^2$ is direct variation. The function passes through the origin.

---

6. (B)

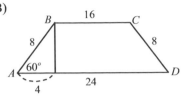

In the figure, Perimeter is 56.

---

7. (D)

$\left(\dfrac{a+b}{x-y}\right)\left(\dfrac{a-b}{x+y}\right) = \dfrac{a^2-b^2}{x^2-y^2} = \left(\dfrac{1}{2}\right)\left(\dfrac{2}{3}\right) = \dfrac{1}{3}$.

$\dfrac{2a^2 - 2b^2}{3x^2 - 3y^2} = \dfrac{2(a^2-b^2)}{3(x^2-y^2)} = \dfrac{2}{3}\left(\dfrac{1}{3}\right) = \dfrac{2}{9}$.

---

8. (E)

---

9. (C)

The volume of the water is equal. Thus,

$\pi r^2 h = 2r \cdot \pi h \cdot H \Rightarrow H = \dfrac{\pi r^2 h}{2\pi rh} = \dfrac{r}{2}$.

---

10. (D)

$2^{21} = 2 \cdot 2^{20} = 2^{20} + 2^{20} = x + 2^{20} \Rightarrow x = 2^{20}$.

---

11. (A)

There are 5 women out of 90 employees.

$P = \dfrac{5}{90} = \dfrac{1}{18}$

---

12. (B)

$5^3 + 5^3 + 5^3 + 5^3 + 5^3 = 5 \cdot 5^3 = 5^4$. Therefore, $x = 4$.

---

13. (C)

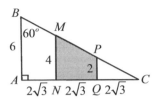

In the figure above, $\triangle CPQ \sim \triangle CMN \sim \triangle CBA$. It follows that the ratio of the length is 1:2:3. $AC = 6\sqrt{3}$. Thus, the area of $\triangle CMN = \dfrac{1}{2}\left(4\sqrt{3} \times 4\right) = 8\sqrt{3}$ and the area of $\triangle CPQ = \dfrac{1}{2}\left(2\sqrt{3} \times 2\right) = 2\sqrt{3}$.

Therefore, $8\sqrt{3} - 2\sqrt{3} = 6\sqrt{3}$.

Or, you can use the ratio of the areas.

$1:4:9 = k, 4k, 9k$

the area of shaded region $= 3k$.

The area of $\triangle ABC$ is

$\dfrac{1}{2}\left(6 \times 6\sqrt{3}\right) = 18\sqrt{3} = 9k$. Thus $k = 2\sqrt{3}$.

Therefore, $3k = 6\sqrt{3}$.

---

14. (C)

$\dfrac{1}{p} + \dfrac{1}{q} = \dfrac{p+q}{pq} = t \Rightarrow \dfrac{p+q}{w} = t$. Therefore, $p + q = tw$.

---

15. (D)

$4 \triangle 13 = 4^2 - 13 = 3$ and $5 \triangle 23 = 5^2 - 23 = 2$.

Therefore,

$2^3 \triangle 3^2 = 8 \triangle 9 = 8^2 - 9 = 64 - 9 = 55$.

---

16. (A)

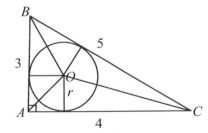

The area of $\triangle ABC$ is $\dfrac{3 \cdot 4}{2} = 6$. The area of

$\triangle ABC$ in terms of $r = \dfrac{1}{2}(5r + 4r + 3r) = 6r$.

Therefore, $6r = 6 \Rightarrow r = 1$.

Or,

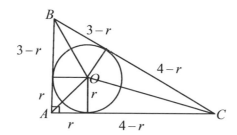

Because the lengths of tangent to a circle is equal,

$BC = (3-r) + (4-r) = 5$.

Therefore, $r = 1$.

$\boxed{END}$

NO MATERIAL ON THIS PAGE

Dr. John Chung's SAT Math

TEST
3

ANSWER SHEET   TEST #:

SECTION 3

1 Ⓐ Ⓑ Ⓒ Ⓓ Ⓔ	11 Ⓐ Ⓑ Ⓒ Ⓓ Ⓔ	21 Ⓐ Ⓑ Ⓒ Ⓓ Ⓔ	31 Ⓐ Ⓑ Ⓒ Ⓓ Ⓔ
2 Ⓐ Ⓑ Ⓒ Ⓓ Ⓔ	12 Ⓐ Ⓑ Ⓒ Ⓓ Ⓔ	22 Ⓐ Ⓑ Ⓒ Ⓓ Ⓔ	32 Ⓐ Ⓑ Ⓒ Ⓓ Ⓔ
3 Ⓐ Ⓑ Ⓒ Ⓓ Ⓔ	13 Ⓐ Ⓑ Ⓒ Ⓓ Ⓔ	23 Ⓐ Ⓑ Ⓒ Ⓓ Ⓔ	33 Ⓐ Ⓑ Ⓒ Ⓓ Ⓔ
4 Ⓐ Ⓑ Ⓒ Ⓓ Ⓔ	14 Ⓐ Ⓑ Ⓒ Ⓓ Ⓔ	24 Ⓐ Ⓑ Ⓒ Ⓓ Ⓔ	34 Ⓐ Ⓑ Ⓒ Ⓓ Ⓔ
5 Ⓐ Ⓑ Ⓒ Ⓓ Ⓔ	15 Ⓐ Ⓑ Ⓒ Ⓓ Ⓔ	25 Ⓐ Ⓑ Ⓒ Ⓓ Ⓔ	35 Ⓐ Ⓑ Ⓒ Ⓓ Ⓔ
6 Ⓐ Ⓑ Ⓒ Ⓓ Ⓔ	16 Ⓐ Ⓑ Ⓒ Ⓓ Ⓔ	26 Ⓐ Ⓑ Ⓒ Ⓓ Ⓔ	36 Ⓐ Ⓑ Ⓒ Ⓓ Ⓔ
7 Ⓐ Ⓑ Ⓒ Ⓓ Ⓔ	17 Ⓐ Ⓑ Ⓒ Ⓓ Ⓔ	27 Ⓐ Ⓑ Ⓒ Ⓓ Ⓔ	37 Ⓐ Ⓑ Ⓒ Ⓓ Ⓔ
8 Ⓐ Ⓑ Ⓒ Ⓓ Ⓔ	18 Ⓐ Ⓑ Ⓒ Ⓓ Ⓔ	28 Ⓐ Ⓑ Ⓒ Ⓓ Ⓔ	38 Ⓐ Ⓑ Ⓒ Ⓓ Ⓔ
9 Ⓐ Ⓑ Ⓒ Ⓓ Ⓔ	19 Ⓐ Ⓑ Ⓒ Ⓓ Ⓔ	29 Ⓐ Ⓑ Ⓒ Ⓓ Ⓔ	39 Ⓐ Ⓑ Ⓒ Ⓓ Ⓔ
10 Ⓐ Ⓑ Ⓒ Ⓓ Ⓔ	20 Ⓐ Ⓑ Ⓒ Ⓓ Ⓔ	30 Ⓐ Ⓑ Ⓒ Ⓓ Ⓔ	40 Ⓐ Ⓑ Ⓒ Ⓓ Ⓔ

SECTION 5

1 Ⓐ Ⓑ Ⓒ Ⓓ Ⓔ	11 Ⓐ Ⓑ Ⓒ Ⓓ Ⓔ	21 Ⓐ Ⓑ Ⓒ Ⓓ Ⓔ	31 Ⓐ Ⓑ Ⓒ Ⓓ Ⓔ
2 Ⓐ Ⓑ Ⓒ Ⓓ Ⓔ	12 Ⓐ Ⓑ Ⓒ Ⓓ Ⓔ	22 Ⓐ Ⓑ Ⓒ Ⓓ Ⓔ	32 Ⓐ Ⓑ Ⓒ Ⓓ Ⓔ
3 Ⓐ Ⓑ Ⓒ Ⓓ Ⓔ	13 Ⓐ Ⓑ Ⓒ Ⓓ Ⓔ	23 Ⓐ Ⓑ Ⓒ Ⓓ Ⓔ	33 Ⓐ Ⓑ Ⓒ Ⓓ Ⓔ
4 Ⓐ Ⓑ Ⓒ Ⓓ Ⓔ	14 Ⓐ Ⓑ Ⓒ Ⓓ Ⓔ	24 Ⓐ Ⓑ Ⓒ Ⓓ Ⓔ	34 Ⓐ Ⓑ Ⓒ Ⓓ Ⓔ
5 Ⓐ Ⓑ Ⓒ Ⓓ Ⓔ	15 Ⓐ Ⓑ Ⓒ Ⓓ Ⓔ	25 Ⓐ Ⓑ Ⓒ Ⓓ Ⓔ	35 Ⓐ Ⓑ Ⓒ Ⓓ Ⓔ
6 Ⓐ Ⓑ Ⓒ Ⓓ Ⓔ	16 Ⓐ Ⓑ Ⓒ Ⓓ Ⓔ	26 Ⓐ Ⓑ Ⓒ Ⓓ Ⓔ	36 Ⓐ Ⓑ Ⓒ Ⓓ Ⓔ
7 Ⓐ Ⓑ Ⓒ Ⓓ Ⓔ	17 Ⓐ Ⓑ Ⓒ Ⓓ Ⓔ	27 Ⓐ Ⓑ Ⓒ Ⓓ Ⓔ	37 Ⓐ Ⓑ Ⓒ Ⓓ Ⓔ
8 Ⓐ Ⓑ Ⓒ Ⓓ Ⓔ	18 Ⓐ Ⓑ Ⓒ Ⓓ Ⓔ	28 Ⓐ Ⓑ Ⓒ Ⓓ Ⓔ	38 Ⓐ Ⓑ Ⓒ Ⓓ Ⓔ
9 Ⓐ Ⓑ Ⓒ Ⓓ Ⓔ	19 Ⓐ Ⓑ Ⓒ Ⓓ Ⓔ	29 Ⓐ Ⓑ Ⓒ Ⓓ Ⓔ	39 Ⓐ Ⓑ Ⓒ Ⓓ Ⓔ
10 Ⓐ Ⓑ Ⓒ Ⓓ Ⓔ	20 Ⓐ Ⓑ Ⓒ Ⓓ Ⓔ	30 Ⓐ Ⓑ Ⓒ Ⓓ Ⓔ	40 Ⓐ Ⓑ Ⓒ Ⓓ Ⓔ

9 10 11 12 13

14 15 16 17 18

Math Scoring Worksheet

A. Section 3

_____   _____
numer of correct   *number of incorrect*

+   +

B. Section 5 (1-8)

_____   _____
numer of correct   *number of incorrect*

+

C. Section 5 (9-18)

_____
numer of correct

+   +

D. Section 7

_____   _____
numer of correct   *number of incorrect*

=   =

E. Total Unrounded Raw Score

_____ − _____ ÷4 = _____
numer of correct   *number of incorrect*

F. Total Rounded Raw Score   _____ (See table)

Math Score Range = | ——— |

Math Conversion Table

Raw Score	Scaled Score	Raw Score	Scaled Score
54	800	23	490-550
53	780-800	22	480-540
52	760-800	21	470-530
51	740-800	20	460-520
50	720-780	19	450-510
49	700-760	18	450-510
48	690-750	17	440-500
47	680-740	16	430-490
46	670-730	15	420-480
45	660-720	14	420-480
44	650-710	13	410-470
43	650-710	12	400-460
42	640-700	11	390-450
41	630-690	10	380-440
40	620-680	9	390-430
39	610-670	8	380-420
38	610-670	7	370-410
37	600-660	6	360-400
36	590-650	5	340-380
35	580-640	4	320-370
34	570-630	3	310-360
33	560-620	2	300-350
32	560-620	1	270-320
31	550-610	0	240-300
30	540-600	-1	200-290
29	530-590	-2	200-270
28	530-590	-3	200-260
27	520-580	-4	200-240
26	510-570	-5	200-220
25	500-560	-6 and below	200
24	500-560		

SECTION 3
Time- 25 minutes
20 Questions

Turn to Section 3 (Page 1) of your answer sheet to answer the questions in this section.

Directions: For this section, solve each problem and decide which is the best of the choices given. Fill in the corresponding circle on the answer sheet. You may use any available space for scratchwork.

Notes

1. The use of a calculator is permitted.
2. All numbers used are real numbers.
3. Figures that accompany problems in this test are intended to provide information useful in solving the problems. They are drawn as accurately as possible EXCEPT when it is stated in a specific problem that the figure is not drawn to scale. All figure lie in a plane unless other indicated.
4. Unless otherwise specified, the domain of any function f is assumed to be set of all real numbers x for which $f(x)$ is a real number.

Reference Informatiom

$A = \pi r^2$
$C = 2\pi r$
$A = \ell w$
$A = \frac{1}{2}bh$
$V = \ell wh$
$V = \pi r^2 h$
$c^2 = a^2 + b^2$
Special Right Triangles

The numbers of degrees of arc in a circle is $360°$.
The sum of the measures in degrees of the angles is $180°$.

1. If $p + q - 2 = 5$, what is the value of $p + q - 7$?

(A) -3
(B) 0
(C) 3
(D) 6
(E) 9

2. Robert earns \$152 in 4 days. At this rate, how many days will it take him to earn \$532?

(A) 10
(B) 12
(C) 14
(D) 16
(E) 18

GO ON TO THE NEXT PAGE

x
↓

−1   0

3. If the tick marks are equally spaced on the number line above, what is the value of $|x+1|$?

(A) $\dfrac{1}{5}$

(B) $\dfrac{2}{5}$

(C) $\dfrac{3}{5}$

(D) $\dfrac{4}{5}$

(E) $\dfrac{7}{5}$

4. If Cathy sells two apple pies and three tacos for $7 and she sells three apple pies and two tacos for $8, what is the cost, in dollars, of three apple pies and three tacos ?

(A) 9
(B) 10
(C) 12
(D) 15
(E) 18

Note: Figure not drawn to scale.

5. The figure above shows that an isosceles triangle of area 100 is cut by $\overline{MN}$ into an isosceles trapezoid and a smaller isosceles triangle. If the area of the trapezoid is 75 and the altitude of $\triangle ABC$ from B is 20, what is the length of $\overline{MN}$?

(A) 3
(B) 5
(C) 10
(D) 12
(E) 15

6. For all values of x, the function is defined by $f(x)=(x-1)(x-5)$. For which of the following intervals does $f(x)$ have negative values ?

(A) $x<-1$
(B) $-1<x<0$
(C) $0<x<1$
(D) $1<x<5$
(E) $x>5$

GO ON TO THE NEXT PAGE

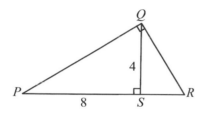

Note: Figure not drawn to scale.

7. In the figure above, $\overline{PS}$ is equal to 8 and $\overline{QS}$ is equal to 4. What is the area of $\triangle PQR$?

    (A) 10
    (B) 20
    (C) 30
    (D) 40
    (E) 50

Row 1			1	2	
Row 2		1	3	2	
Row 3	1	4	5	2	

8. In the array of numbers above, the first number in each row is 1 and the last number is 2. If each of the other entries is the sum of the two numbers nearest it in the row directly above it, what is the sum of all of the numbers in row 5?

    (A) 26
    (B) 48
    (C) 52
    (D) 60
    (E) 70

9. If each of the 8 members from the JFK Hockey team shake hands with all the members in the same team, which of the following is the number of handshakes?

    (A) 28
    (B) 56
    (C) 60
    (D) 120
    (E) 240

10. The figure above shows a dartboard with a radius 9. Each of the concentric circles has a radius 3 less than the next large circle. If 24 darts land randomly on the target, how many darts will be expected to land in the shaded regions?

    (A) 18
    (B) 16
    (C) 14
    (D) 12
    (E) 10

GO ON TO THE NEXT PAGE

11. While driving on an m mile trip from New York City to Washington D.C, Allen drove at the rate of p miles per hour for the first t hours and took w hours to complete the rest of the trip. Which of the following could represent the speed after t hours?

(A) $\dfrac{m-p}{2}$

(B) $\dfrac{pt}{w}$

(C) $\dfrac{m-pt}{w}$

(D) $m - \dfrac{p}{t}$

(E) $m - \dfrac{pt}{w}$

x	y
0	2
k	14
$k+2$	17

12. The table above shows the points (x, y) represented on a straight line. If the point $(16, m)$ lies on the same line, what is the value of m?

(A) 20
(B) 22
(C) 24
(D) 26
(E) 28

13. James spent $\dfrac{3}{4}$ of his allowance on a music CD. He spent $\dfrac{2}{3}$ of what was left on a hamburger. If this left him p dollars, which of the following was his allowance in dollars?

(A) $12p$
(B) $14p$
(C) $16p$
(D) $18p$
(E) $20p$

Note: Figure not drawn to scale.

14. In the figure above, which of the following could not be true?

(A) $x + y = 180$
(B) $x > v$
(C) $w > y$
(D) $x + w > y + v$
(E) $x - y < 0$

GO ON TO THE NEXT PAGE

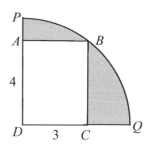

Note: Figure not drawn to scale.

15. In the figure above, a rectangle is inscribed in the part of a circle. If the length of $\overline{AD}$ is 4 and the length of $\overline{DC}$ is 3, what is the area of the shaded region?

(A) $\dfrac{5\pi}{2}$

(B) $25\pi - 12$

(C) $\dfrac{25\pi - 12}{2}$

(D) $\dfrac{25\pi - 24}{2}$

(E) $\dfrac{25\pi - 48}{4}$

16. According to the formula $P = \dfrac{4}{3}K + 81$, if the value of P increases by 16, by how much does the value of K increase?

(A) 12

(B) 14

(C) 25

(D) $\dfrac{64}{3}$

(E) $\dfrac{81}{7}$

17. For all positive values of p and q, let $p \circledast q$ be defined by $\dfrac{pq}{p+q}$. If $p \circledast q = \dfrac{1}{p} \circledast \dfrac{1}{q}$, which of the following must be true?

(A) $p = q + 1$

(B) $p = q$

(C) $p = -q$

(D) $pq = 1$

(E) $pq = p + q$

18. If five lines not parallel to each other are drawn on a sheet of paper, what is the greatest possible number of intersections from the five lines?

(A) 8

(B) 10

(C) 12

(D) 14

(E) 16

GO ON TO THE NEXT PAGE

19. If $2 \le p \le 6$ and $1 \le q \le 12$, what is the smallest value of $(p+q)^2 - (p-q)^2$?

(A) 20
(B) 18
(C) 8
(D) 6
(E) 4

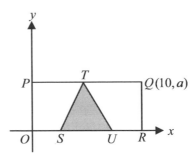

Note: Figure not drawn to scale.

20. In the figure above, $OPQR$ is a rectangle and $\triangle STU$ is an equilateral triangle. A point in the rectangle $OPQR$ is to be chosen at random. If the probability that the point will be in the shaded region is $\dfrac{1}{4}$, what is the value of a ?

(A) $2\sqrt{3}$

(B) $\dfrac{5}{2}$

(C) $\dfrac{3\sqrt{3}}{2}$

(D) $\dfrac{5\sqrt{3}}{2}$

(E) $\dfrac{7\sqrt{3}}{2}$

STOP

If you finish before time is called, you may check your work on this section only.
Do not turn to any other section in the test.

SECTION 5
Time- 25 minutes
18 Questions

Turn to Section 5 (Page 1) of your answer sheet to answer the questions in this section.

Directions: For this section, solve each problem and decide which is the best of the choices given. Fill in the corresponding circle on the answer sheet. You may use any available space for scratchwork.

Notes

1. The use of a calculator is permitted.
2. All numbers used are real numbers.
3. Figures that accompany problems in this test are intended to provide information useful in solving the problems. They are drawn as accurately as possible EXCEPT when it is stated in a specific problem that the figure is not drawn to scale. All figure lie in a plane unless other indicated.
4. Unless otherwise specified, the domain of any function f is assumed to be set of all real numbers x for which $f(x)$ is a real number.

Reference Informatiom

$A = \pi r^2$
$C = 2\pi r$
$A = \ell w$
$A = \frac{1}{2}bh$
$V = \ell wh$
$V = \pi r^2 h$
$c^2 = a^2 + b^2$
Special Right Triangles

The numbers of degrees of arc in a circle is $360°$.

The sum of the measures in degrees of the angles is $180°$.

---

1. If $a = 16 - b$ and $\frac{b}{6} = 3$, what is the value of a?

(A) -4
(B) -2
(C) 2
(D) 4
(E) 6

2. The scale of a map for Bear Mountain National Park is 2 inches = 9 miles. The distance between Discovery Point and Overlook on the map is about $1\frac{1}{2}$ inches. What is the distance between these two places in miles?

(A) 5

(B) 6

(C) $6\frac{1}{3}$

(D) $6\frac{3}{4}$

(E) 8

GO ON TO THE NEXT PAGE

CLASSES STUDENTS ARE TAKING

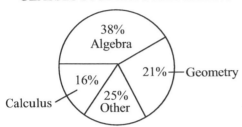

3. The circle graph above shows the percent of which 200 students are taking each subject. How many more students are taking Algebra than Geometry?

(A) 30
(B) 32
(C) 34
(D) 36
(E) 38

4. If k is a positive integer, then $\dfrac{2^{k+5}}{2^4}$ could equal which of the following?

(A) 20
(B) 14
(C) 12
(D) 8
(E) 2

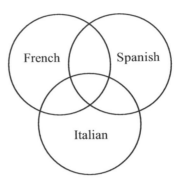

5. In the figure above, of the 40 foreign exchange students attending an International School, 20 speak French, 23 speak Spanish, and 22 speak Italian. Nine students speak French and Spanish, but not Italian. Six students speak French and Italian, but not Spanish. Ten students speak Spanish and Italian but not French. If only 4 of the students speak all three languages, how many exchange students do not speak any of these languages?

(A) 4
(B) 6
(C) 8
(D) 10
(E) 12

GO ON TO THE NEXT PAGE ⇨

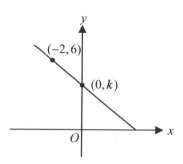

Note: Figure not drawn to scale.

6. In the figure above, the slope of the line is $-\dfrac{2}{3}$.
    What is the value of k ?

    (A) 3

    (B) $\dfrac{10}{3}$

    (C) $\dfrac{11}{3}$

    (D) 4

    (E) $\dfrac{14}{3}$

7. If $\dfrac{a}{b} = 5$ and $\dfrac{p^2}{b^2} = \dfrac{q^2}{a^2}$, where p and q are positive, which of the following must be equal to $\dfrac{p}{q}$?

    (A) 5

    (B) $\dfrac{1}{5}$

    (C) $\dfrac{a}{5}$

    (D) $\dfrac{a}{b}$

    (E) ab

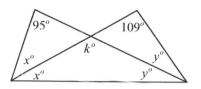

Note: Figure not drawn to scale.

8. In the figure above, what is the value of k ?

    (A) 118
    (B) 120
    (C) 128
    (D) 130
    (E) 132

GO ON TO THE NEXT PAGE

5 ☐☐☐ 5 ☐☐☐ 5 ☐☐☐ 5

Directions: For Students-Produced Response questions 9-18, use the grid at the bottom of the answer sheet page on which you have answered questions 1-8.

Each of the remaining 10 questions requires you to solve the problem and enter your answer by making the circles in the special grid, as shown in the examples below. You may use any available space for scratchwork.

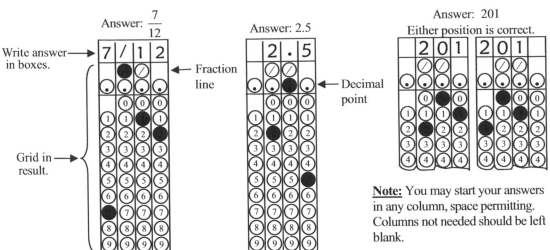

Answer: $\frac{7}{12}$

Write answer in boxes.

Fraction line

Grid in result.

Answer: 2.5

Decimal point

Answer: 201
Either position is correct.

Note: You may start your answers in any column, space permitting. Columns not needed should be left blank.

- Mark no more than one circle in any column.

- Because the answer sheet will be machine-scored, **you will receive credit only if the circles are filled in correctly.**

- Although not required, it is suggested that you write your answer in the boxes at the top of the columns to help you fill in the circles accurately.

- Some problems may have more than one correct answer. In such cases, grid only one answer.

- No question has a negative answer.

- **Mixed numbers** such as $3\frac{1}{2}$ must be gridded as 3.5 or 7/2. (If $\boxed{3\,1\,/\,2}$ is gridded, it will be interpreted as $\frac{31}{2}$, not $3\frac{1}{2}$.)

- **Decimal Answers:** If you obtain a decimal answer with more digits than the grid can accommodate, it may be either rounded or truncated, but it must fill the entire grid. For example, if you obtain an answer such as 0.6666…, you should record your result as .666 or .667. **A less accurate value such as .66 or .67 will be scored as incorrect.**

Acceptable ways to grid $\frac{2}{3}$ are:

9. If $\dfrac{1}{n-1} = \dfrac{2}{n+2}$, what is the value of n ?

10. A recipe for making 20 pepperoni pizza pies requires 32 pounds of flour, 5 pounds of baking powder, and 4 pounds of pepperoni. If the proportions in this recipe are to be used to make 15 pepperoni pizza pies, how many pounds of pepperoni will be needed?

GO ON TO THE NEXT PAGE

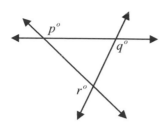

Note: Figure not drawn to scale.

11. In the figure above, three different lines intersect at three points. If $p = 125$, what is the value of $q + r$?

12. The length, width, and height of a rectangular solid have distinct integral values. If the volume of the rectangular solid is 75, what is one possible value for the surface area of the solid?

$$3, \ 3^2, \ 3^3, \ 3^4 \ ...$$

13. The sequence above is formed by multiplying 3 to the preceding term. What is the ones digit of the 100th term of the sequence?

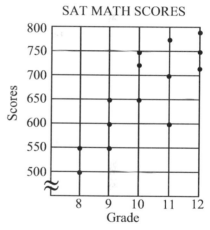

14. The scatterplot above shows SAT math scores for 14 students in a certain group. What is the median math score for the 14 students?

GO ON TO THE NEXT PAGE

15. If $(k-1)x^2 + (k+1)x + 2k = ax^2 + bx + 4$ for all values of x, where k, a, and b are constant, what is the value of $a + b$?

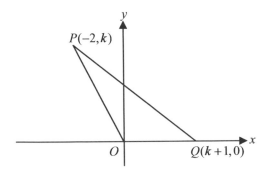

Note: figure not drawn to scale.

16. In the xy – plane above, the area of $\triangle OPQ$ is 3. What is the value of k ?

17. Let the function v be defined so that $v(x)$ is the volume of a cube with side x.
If $v(1) + v(6) + v(8) = v(k)$, what is the value of k ?

18. The total cost of an internet phone-call is the sum of

(1) a basic fixed charge for using the internet

and

(2) an additional charge for each $\dfrac{1}{2}$ of a minute that is used.

If the total cost of a 20 minute-call is \$11.00 and the total cost of a 35 minute-call is \$18.50, what is the total cost, in dollars, of a 40 minute-call?

STOP

If you finish before time is called, you may check your work on this section only.
Do not turn to any other section in the test.

SECTION 7
Time- 20 minutes
16 Questions

Turn to Section 7 (Page 2) of your answer sheet to answer the questions in this section.

Directions: For this section, solve each problem and decide which is the best of the choices given. Fill in the corresponding circle on the answer sheet. You may use any available space for scratchwork.

Notes

1. The use of a calculator is permitted.
2. All numbers used are real numbers.
3. Figures that accompany problems in this test are intended to provide information useful in solving the problems. They are drawn as accurately as possible EXCEPT when it is stated in a specific problem that the figure is not drawn to scale. All figure lie in a plane unless other indicated.
4. Unless otherwise specified, the domain of any function f is assumed to be set of all real numbers x for which $f(x)$ is a real number.

Reference Informatiom

$A = \pi r^2$ $A = \ell w$ $A = \frac{1}{2}bh$ $V = \ell wh$ $V = \pi r^2 h$ $c^2 = a^2 + b^2$ Special Right Triangles
$C = 2\pi r$

The numbers of degrees of arc in a circle is 360°.

The sum of the measures in degrees of the angles is 180°.

---

1. If the price of a jacket was increased by 10% last week and then decreased 10% this week, what is the overall percent change from the original price?

   (A) 0%
   (B) 1%
   (C) 2%
   (D) 20%
   (E) No change

2. If $x = \dfrac{3}{2z}$ and $xy = \dfrac{1}{2z}$ for $z \neq 0$, what is the value of y?

   (A) -1
   (B) $\dfrac{1}{4}$
   (C) $\dfrac{1}{3}$
   (D) $\dfrac{2}{3}$
   (E) 1

GO ON TO THE NEXT PAGE

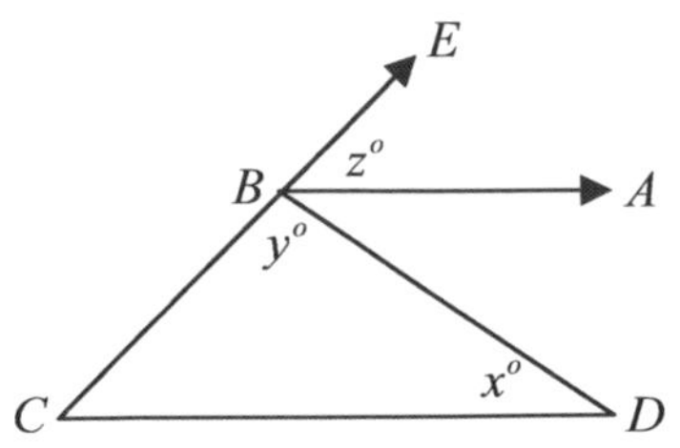

Note: Figure not drawn to scale.

3. In the figure above, $\overline{AB}$ is parallel to $\overline{CD}$. Which of the following MUST be true ?

    (A) $x > z$
    (B) $x = z$
    (C) $z > x$
    (D) $180 = x + y + z$
    (E) $180 = 2x + z$

4. If $p < q < 0$, which of the following MUST be greater than 1 ?

    (A) pq

    (B) $p + q$

    (C) $p - q$

    (D) $\dfrac{p}{q}$

    (E) $\dfrac{q}{p}$

5. If $a + a^{-1} = 3$, what is the value of $a^2 + a^{-2}$?

    (A) 9
    (B) 8
    (C) 7
    (D) 6
    (E) 5

6. If $-12 < w < -4$, which of the following must be true ?

    (A) $|w| = 4$
    (B) $|w| < 4$
    (C) $|w - 8| > 4$
    (D) $|w + 8| > 4$
    (E) $|w + 8| < 4$

GO ON TO THE NEXT PAGE

x	-1	0	1	2
$g(x)$	1	-1	1	7

7. The table above shows values of the function g and values of x. Which of the following defines g ?

(A) $y = 2x - 1$

(B) $y = -2x - 1$

(C) $y = x^2 - 1$

(D) $y = 2x^2 - 1$

(E) $y = -2x^2 - 1$

8. In a chemistry class, some of the students are boys and none of these boys are less then 13 years old. Which of the following MUST be true ?

(A) Most of the students in the class are boys.

(B) Most of the students are older than 13 years old.

(C) None of the students in the class are less than 13 years old.

(D) Some of the students are older than or equal to 13 years old.

(E) Some of the students are less than 13 years old.

9. If $-10 < x < 10$, which of the following must be true ?

    I. $x^2 < 10$

    II. $|x| < 10$

    III. $(x+10)(x-10) < 0$

(A) III only

(B) II and III only

(C) II only

(D) I and II only

(E) I, II, and III

10. The average (arithmetic mean) of a, $2a$, b, and $(b+1)$ is equal to the median of the four numbers, where $a < 2a < b < (b+1)$. What is the value of a ?

(A) $\dfrac{b+1}{2}$

(B) $2b - 3$

(C) 6

(D) 4

(E) 1

GO ON TO THE NEXT PAGE

11. An advertising medium charges d dollars for the basic fixed fee and c cents for every 10 letters for the advertising description. If 300 letters are used for an advertising description, what is the total amount, in dollars, for the advertisement?

(A) $d + 10c$

(B) $d + \dfrac{3c}{10}$

(C) $d + 3c$

(D) $d + 30c$

(E) $d + 300c$

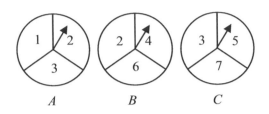

A   B   C

13. The figure above shows three spinners A, B, and C with numbers on it. If each of every number has an equal probability of being the sector on which the arrow stops, what is the probability that the sum of the three numbers is an even number?

(A) $\dfrac{1}{6}$

(B) $\dfrac{1}{4}$

(C) $\dfrac{1}{3}$

(D) $\dfrac{1}{2}$

(E) $\dfrac{2}{3}$

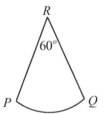

Note: Figure not drawn to scale.

12. The figure above shows a part of a circle with a center at R. If the length of $\overline{PQ}$ (not drawn) is k, what is the perimeter of the figure ?

(A) $3k$

(B) $\dfrac{\pi k}{6}$

(C) $2k + \pi k$

(D) $\dfrac{(\pi + 6)k}{3}$

(E) $\dfrac{(2\pi + 3)k}{3}$

14. A farmer can plow a field in k days. How many days will it take two farmers working at the same rate to plow a field two times larger?

(A) $4k$

(B) $2k$

(C) k

(D) $\dfrac{k}{2}$

(E) $\dfrac{k}{4}$

GO ON TO THE NEXT PAGE

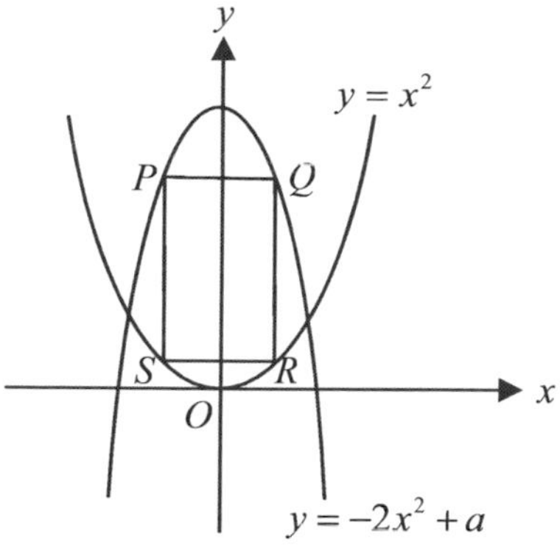

$y = x^2$

$y = -2x^2 + a$

Note: Figure not drawn to scale.

15. In the figure above, $PQRS$ is a rectangle, and points P and Q lie on the graph of $y = -2x^2 + a$ and points S and R lie on the graph of $y = x^2$, where a is a constant. If the area of $PQRS$ is 24 and the length of $\overline{PQ}$ is 4, what is the value of a ?

(A) 8
(B) 12
(C) 14
(D) 16
(E) 18

16. Let the function f be defined as $f(x) = \sqrt{(x-5)(x+5)}$. Which of the following could represent the domain of the function f ?

(A) $x < 5$
(B) $x > 5$
(C) $-5 < x < 5$
(D) $x \geq 5$ or $x \leq -5$
(E) $-5 \leq x \leq 5$

STOP

If you finish before time is called, you may check your work on this section only.
Do not turn to any other section in the test.

#	SECTION 3	SECTION 5	SECTION 7
1	B	B	B
2	C	D	C
3	C	C	D
4	A	D	D
5	B	C	C
6	D	E	E
7	B	B	D
8	B	C	D
9	A	4	B
10	B	3	E
11	C	235	B
12	D	190 or 206	D
13	A	1	E
14	E	675	C
15	E	4	E
16	A	2	D
17	D	9	
18	B	21	
19	C		
20	D		

TEST 3   SECTION 3

1.   (B)

$p+q=7 \Rightarrow p+q-7=7-7=0$.

---

2.   (C)

$\dfrac{152}{4}=\dfrac{532}{x} \Rightarrow x=\dfrac{(532)(4)}{152}=14$.

---

3.   (C)

$x=-0.4$. Thus, $|x+1|=|-0.4+1|=0.6=\dfrac{3}{5}$.

---

4.   (A)

$2a+3t=7$ and $3a+2t=8$. When you add these equations.

$5a+5t=15 \Rightarrow a+t=3$.

Therefore, $3a+3t=9$.

---

5.   (B)

The area of $\triangle ABC=\dfrac{1}{2}(AC\times20)=100$.

$AC=10$.

Since $\triangle ABC \sim \triangle MBN$, the ratio of the areas is $100:25$ and the ratio of the lengths is $\sqrt{100}:\sqrt{25}=10:5=2:1$. Therefore,

$\dfrac{MN}{AC}=\dfrac{1}{2}=\dfrac{x}{10}$.

It follows that $x=5$.

---

6.   (D)

Since $(x-1)(x-5)<0$, the solution is,

$1<x<5$.

Or, from the graph, you can find the solution.

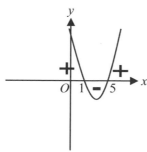

The graph of $f(x) = (x-1)(x-5)$ is shown above. Between 1 and 5, $f < 0$.

---

7. (B)

From the similar triangles,

$QS^2 = PS \times SR \Rightarrow 16 = 8 \times SR$,

$SR = 2$ and $PR = 10$. Therefore, the area of

$\triangle PQR = \dfrac{1}{2}(10 \times 4) = 20$.

---

8. (B)

Row 1			1	2	
Row 2		1	3	2	
Row 3		1	4	5	2
Row 4	1	5	9	7	2
Row 5	1	6	14	16	9 2

Therefore, the sum of Row 5 is 48.

Or, you can use a pattern where sum increases by double.

Sum 1 = 3

Sum 2 = 6

Sum 3 = 12

Sum 4 = 24

Sum 5 = 48

---

9. (A)

Since there are 8 members, the number of handshakes is $7+6+5+4+3+2+1 = 28$

Or, $_8C_2 = \dfrac{8 \cdot 7}{2} = 28$. (When you choose two members, there will be one handshake.)

---

10. (B)

The ratio of the lengths of the three circles $= 9:6:3 = 3:2:1$. Thus, the ratio of the areas$= 9:4:1 = 9k:4k:k$. The area of the shaded region, in terms of k, is ,

$(9k-4k)+k = 6k$. Therefore, the expected

number $= 24 \times \dfrac{6k}{9k} = 16$.

---

11. (C)

The distance left after t hours $= m - pt$. Thus,

the speed after t hours $= \dfrac{m-pt}{w}$.

---

12. (D)

Linear equation has a constant slope.

$\dfrac{14-2}{k-0} = \dfrac{17-2}{k+2-0}$. It follows that $k = 8$ and

the slope is $\dfrac{3}{2}$. Since point $(16, m)$ lies on the

line, the slope between two points must be

$\dfrac{3}{2}$. Therefore, $\dfrac{m-2}{16-0} = \dfrac{3}{2}$. Then, $k = 26$.

---

13. (A)

Let the allowance be x.

The equation is, $\dfrac{3}{4}x + \dfrac{2}{3}\left(\dfrac{1}{4}x\right) + p = x$.

Therefore, $x = 12p$.

Or, $1 - \left(\dfrac{3}{4} + \dfrac{2}{3} \cdot \dfrac{1}{4}\right) = \dfrac{1}{12}$. Therefore,

$\dfrac{1}{12}x = p \rightarrow x = 12p$.

---

14. (E)

In the figure above, $x = v + k$ and $w = y + k$. It follows that $x > v$, $w > y$, and $x + w > y + v$. (E) cannot be determined from the information given.

15. (E)

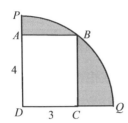

In the figure above, the radius of the circle is 5. Therefore, the area of the shaded region

$$= \frac{\pi(5^2)}{4} - (3 \times 4) = \frac{25\pi - 48}{4}.$$

---

16. (A)

From the formula of $P = \frac{4}{3}K + 81$, $\frac{4}{3}$ is the

slope of P which is defined by $\frac{\Delta P}{\Delta K}$. "When

P increased by 16" $\Rightarrow \Delta P = 16$. Therefore,

$\frac{\Delta P}{\Delta K} = \frac{16}{\Delta K} = \frac{4}{3}$. $\Delta K = 12$.

---

17. (D)

Since $p \circledast q = \frac{1}{p} \circledast \frac{1}{q}$, it follows that

$$\frac{pq}{p+q} = \frac{\left(\frac{1}{p}\right)\left(\frac{1}{q}\right)}{\frac{1}{p} + \frac{1}{q}} \Rightarrow \frac{pq}{p+q} = \frac{1}{p+q}.$$

Therefore, $pq = 1$ is the answer for "Must be true".

---

18. (B)

Like the handshake problem,
$1 + 2 + 3 + 4 = 10$.

Or, $_5C_2 = \frac{5 \times 4}{2} = 10$.

---

19. (C)

$(p+q)^2 - (p-q)^2$
$= p^2 + 2pq + q^2 - p^2 + 2pq - q^2$
$= 4pq = 4(1 \times 2) = 8$

---

20. (D)

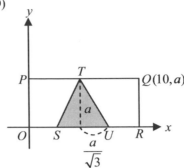

In the figure above, the height of $\triangle STU$ is

a. Thus, $SU = \frac{a}{\sqrt{3}}$ and the area of $\triangle STU$ is,

$\frac{1}{2}\left(\frac{2a}{\sqrt{3}} \times a\right) = \frac{a^2}{\sqrt{3}}$. The area of $OPQR$ is

$10a$. Therefore,

$$P = \frac{\text{area of } \triangle STU}{\text{area of } OPQR} = \frac{a^2/\sqrt{3}}{10a} = \frac{1}{4}. \text{ It follows}$$

that $\frac{4a^2}{\sqrt{3}} = 10a \Rightarrow \frac{4a}{\sqrt{3}} = 10 \Rightarrow a = \frac{5\sqrt{3}}{2}$.

TEST 3   **SECTION 5**

1. (B)
Since $b = 18$, $a = 16 - 18 = -2$.

---

2. (D)

$\frac{2}{9} = \frac{3/2}{x} \Rightarrow 2x = 13.5 \Rightarrow x = \frac{27}{4} = 6\frac{3}{4}$

---

3. (C)
$38 - 21 = 17\% \Rightarrow 17\%$ of $200 = 34$

---

4. (D)

$\frac{2^{k+5}}{2^4} = 2^{k+1}$, where $k =$ integer. Therefore,

2^{k+1} is , 4, 8, ,16, ,32, ... The answer is (D).

---

5. (C)
In the figure below, the number of students who are taking the subjects is,
$(1 + 0 + 2) + (9 + 6 + 10) + (4) = 32$.

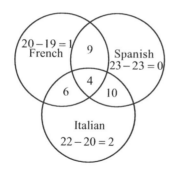

The number of student who speaks any of these languages is 32.
Therefore, $40-32 = 8$.

6. (E)
$$\frac{k-6}{0-(-2)} = -\frac{2}{3} \Rightarrow 3k-18 = -4 \Rightarrow k = \frac{14}{3}$$

7. (B)
Since $a = 5b$, $\dfrac{p^2}{q^2} = \dfrac{b^2}{a^2} = \dfrac{b^2}{(5b)^2} = \dfrac{1}{25}$.

Therefore, $\dfrac{p}{q} = \sqrt{\dfrac{1}{25}} = \dfrac{1}{5}$.

8. (C)
$2x+y = 85$ and $x+2y = 71$. When you add these two equations,
$3x+3y = 156 \Rightarrow x+y = 52$.
It follows that $k = 180-52 = 128$.

9. 4
$$\frac{1}{n-1} = \frac{2}{n+2} \Rightarrow n+2 = 2n-2 \Rightarrow n = 4.$$

10. 3
20 pizzas need 4 pounds of pepperoni.
Proportion:
$$\frac{20}{4} = \frac{15}{x} \Rightarrow x = 3.$$

11. 235
Since $p+q+r = 360$, (Sum of exterior angles of any polygons is $360°$.)
$q+r = 360-125 = 235$.

12. 190 or 206
Since $W \times L \times H = 75$, the pairs of (W, L, H) are $(1,5,15)$, $(1,3,25)$. Therefore, the surface area of the solid is,
$2(1\times5+5\times15+15\times1) = 90$ or
$2(1\times3+3\times25+25\times1) = 206$.

13. 1
The ones digit of the sequence are as follows.
3, 9, 7, 1, 3, 7, 9, 1.......It is repeating every four numbers. Therefore, the ones digit of 100th term is 1.

14. 675
Arrange the scores as follows.
500,550,550,600,600,650,650,700,725,725,750, 750,775,790. Therefore, the median is
$$\frac{650+700}{2} = 675.$$

15. 4
Their coefficients must be equal. $k-1 = a$, $k+1 = b$, and $2k = 4$. Therefore, $k = 2$, $a = 1$, and $b = 3$.
$a+b = 4$.

16. 2
The area of $\triangle OPQ = \dfrac{1}{2}(k+1)(k) = 3$. It follows that $k^2+k-6 = (k+3)(k-2) = 0$.
Therefore, $k = 2$.
k must be positive.

17. 9
$v(1)+v(6)+v(8) = v(k) \Rightarrow$
$1^3+6^3+8^3 = 729 = 9^3$.
Therefore, $k = 9$.

18. 21
Define the basic charge to be a and the additional charge to be b for 0.5 minute.
For a 20 minutes phone call:
20 minutes = 40 of 0.5 minutes. It follows that $a+40b = 11$ ------(1)
For a 35 minutes phone call:
35 minutes = 70 of 0.5 minutes. It follows that $a+70b = 18.50$ ------(2)

(2)-(1) : $30b = 7.5 \Rightarrow b = 0.25$ and $a = 1$.

For the 40 minutes phone call:

$1 + 80 \times 0.25 = 21$

1. (B)

$P = (1 - 0.1)(1 + 0.1)x = 0.99x$.

$\Rightarrow 0.99 = (1 - 0.01)$

Therefore, there is a 1% of change.

2. (C)

Since $x = \dfrac{2}{2z}$, $xy = \left(\dfrac{3}{2z}\right)y = \dfrac{1}{2z}$

$\dfrac{2z}{3} \times \left(\dfrac{3}{2z}\right)y = \dfrac{1}{2z} \times \dfrac{2z}{3}$. It follows that

$y = \dfrac{1}{3}$.

3. (D)

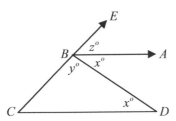

$\angle ABD = x$, because alternate angles are $\cong$.

Therefore, $x + y + z = 180$.

4. (D)

Since $p < q < 0$, when divided by

q (negative number), then $\dfrac{p}{q} > 1$. When

divided by q, $\dfrac{q}{p} < 1$.

Or, use some numbers as $p = -5$ and $q = -1$.

$\dfrac{p}{q} = \dfrac{-5}{-1} = 5 > 1$.

5. (C)

6. (E)

Since $-12 < w < -4$, the midpoint is -8 and the distance between -4 and -8 is 4.

Therefore, $|w - {}^{-}8| < 4 \Rightarrow |w + 8| < 4$.

7. (D)

Check the equations for the values of x.

(D) is true for the values of x.

8. (D)

The only information is " some of them are boys and they are all greater than or equal to 13 years old "

(A) May be girls. (B) May be less than 13 years.

(C) May be less than 13. (D) True (E) maybe they all greater than 13 years old.

9. (B)

$-10 < x < 10$ is equivalent to $|x| < 10$,

$x^2 < 100$, and $(x + 10)(x - 10) < 0$.

I is not true ($x^2 < 100$)

10. (E)

$\dfrac{a + 2a + b + (b + 1)}{4} = \dfrac{2a + b}{2}$. It follows that

$\dfrac{3a + 2b + 1}{4} = \dfrac{2a + b}{2} \Rightarrow a = 1$.

11. (B)

c cents $= \dfrac{c}{100}$ dollars $\Rightarrow$ for 10 letters.

300 letters = 30 of 10 letters. Therefore, the

total amount $= d + \left(\dfrac{c}{100}\right) \times 30 \Rightarrow d + \dfrac{3c}{10}$.

12. (D)

Since $\triangle PQR$ is an equilateral triangle, the

radius of the circle $= k$. $\overset{\frown}{PQ} = 2\pi k \times \dfrac{1}{6} = \dfrac{\pi k}{3}$.

$\left(a + \dfrac{1}{a}\right)^2 = 9 \Rightarrow a^2 + 2 + \dfrac{1}{a^2} = 9$. Therefore,

$a^2 + \dfrac{1}{a^2} = 9 - 2 = 7$.

Therefore, the perimeter of the sector =
$$\frac{\pi k}{3} + 2k = \frac{(\pi+6)k}{3}.$$

---

13. (E)

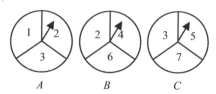

$$A \qquad B \qquad C$$

Since the numbers on spinner B are even, available combinations are as follows:

Spinner A	Spinner B	Spinner C
odd	even	odd
(1,3)	(2,4,6)	(3,5,7)

Therefore, $2 \times 3 \times 3 = 18$. The answer is
$$\frac{18}{27} = \frac{2}{3}.$$

---

14. (C)

Inverse proportion. $1 \times k = 2 \times x \Rightarrow x = \frac{k}{2}$ for two farmers. For double the job, it will take
$$2\left(\frac{k}{2}\right) = k.$$

---

15. (E)

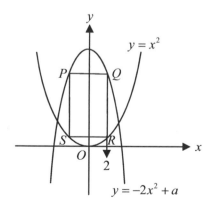

Since $PQ = 4$, $RQ = \frac{24}{4} = 6$. At $x = 2$,

$f(2) = -2(2)^2 + a = -8 + a$ and
$g(2) = 2^2 = 4$. Thus,
the length of $\overline{RQ} = -8 + a - 4 = a - 12$.
Therefore, $a - 12 = 6 \Rightarrow a = 18$.

16. (D)
$(x+5)(x-5) \geq 0 \Rightarrow x \leq -5$ or $x \geq 5$.

$\boxed{END}$

NO MATERIAL ON THIS PAGE

Dr. John Chung's SAT Math

TEST
4

SECTION 3

1	ⒶⒷⒸⒹⒺ	11	ⒶⒷⒸⒹⒺ	21	ⒶⒷⒸⒹⒺ	31	ⒶⒷⒸⒹⒺ
2	ⒶⒷⒸⒹⒺ	12	ⒶⒷⒸⒹⒺ	22	ⒶⒷⒸⒹⒺ	32	ⒶⒷⒸⒹⒺ
3	ⒶⒷⒸⒹⒺ	13	ⒶⒷⒸⒹⒺ	23	ⒶⒷⒸⒹⒺ	33	ⒶⒷⒸⒹⒺ
4	ⒶⒷⒸⒹⒺ	14	ⒶⒷⒸⒹⒺ	24	ⒶⒷⒸⒹⒺ	34	ⒶⒷⒸⒹⒺ
5	ⒶⒷⒸⒹⒺ	15	ⒶⒷⒸⒹⒺ	25	ⒶⒷⒸⒹⒺ	35	ⒶⒷⒸⒹⒺ
6	ⒶⒷⒸⒹⒺ	16	ⒶⒷⒸⒹⒺ	26	ⒶⒷⒸⒹⒺ	36	ⒶⒷⒸⒹⒺ
7	ⒶⒷⒸⒹⒺ	17	ⒶⒷⒸⒹⒺ	27	ⒶⒷⒸⒹⒺ	37	ⒶⒷⒸⒹⒺ
8	ⒶⒷⒸⒹⒺ	18	ⒶⒷⒸⒹⒺ	28	ⒶⒷⒸⒹⒺ	38	ⒶⒷⒸⒹⒺ
9	ⒶⒷⒸⒹⒺ	19	ⒶⒷⒸⒹⒺ	29	ⒶⒷⒸⒹⒺ	39	ⒶⒷⒸⒹⒺ
10	ⒶⒷⒸⒹⒺ	20	ⒶⒷⒸⒹⒺ	30	ⒶⒷⒸⒹⒺ	40	ⒶⒷⒸⒹⒺ

SECTION 5

1	ⒶⒷⒸⒹⒺ	11	ⒶⒷⒸⒹⒺ	21	ⒶⒷⒸⒹⒺ	31	ⒶⒷⒸⒹⒺ
2	ⒶⒷⒸⒹⒺ	12	ⒶⒷⒸⒹⒺ	22	ⒶⒷⒸⒹⒺ	32	ⒶⒷⒸⒹⒺ
3	ⒶⒷⒸⒹⒺ	13	ⒶⒷⒸⒹⒺ	23	ⒶⒷⒸⒹⒺ	33	ⒶⒷⒸⒹⒺ
4	ⒶⒷⒸⒹⒺ	14	ⒶⒷⒸⒹⒺ	24	ⒶⒷⒸⒹⒺ	34	ⒶⒷⒸⒹⒺ
5	ⒶⒷⒸⒹⒺ	15	ⒶⒷⒸⒹⒺ	25	ⒶⒷⒸⒹⒺ	35	ⒶⒷⒸⒹⒺ
6	ⒶⒷⒸⒹⒺ	16	ⒶⒷⒸⒹⒺ	26	ⒶⒷⒸⒹⒺ	36	ⒶⒷⒸⒹⒺ
7	ⒶⒷⒸⒹⒺ	17	ⒶⒷⒸⒹⒺ	27	ⒶⒷⒸⒹⒺ	37	ⒶⒷⒸⒹⒺ
8	ⒶⒷⒸⒹⒺ	18	ⒶⒷⒸⒹⒺ	28	ⒶⒷⒸⒹⒺ	38	ⒶⒷⒸⒹⒺ
9	ⒶⒷⒸⒹⒺ	19	ⒶⒷⒸⒹⒺ	29	ⒶⒷⒸⒹⒺ	39	ⒶⒷⒸⒹⒺ
10	ⒶⒷⒸⒹⒺ	20	ⒶⒷⒸⒹⒺ	30	ⒶⒷⒸⒹⒺ	40	ⒶⒷⒸⒹⒺ

9 10 11 12 13

14 15 16 17 18

SECTION 7	1	Ⓐ Ⓑ Ⓒ Ⓓ Ⓔ	11	Ⓐ Ⓑ Ⓒ Ⓓ Ⓔ	21	Ⓐ Ⓑ Ⓒ Ⓓ Ⓔ	31	Ⓐ Ⓑ Ⓒ Ⓓ Ⓔ
	2	Ⓐ Ⓑ Ⓒ Ⓓ Ⓔ	12	Ⓐ Ⓑ Ⓒ Ⓓ Ⓔ	22	Ⓐ Ⓑ Ⓒ Ⓓ Ⓔ	32	Ⓐ Ⓑ Ⓒ Ⓓ Ⓔ
	3	Ⓐ Ⓑ Ⓒ Ⓓ Ⓕ	13	Ⓐ Ⓑ Ⓒ Ⓓ Ⓕ	23	Ⓐ Ⓑ Ⓒ Ⓓ Ⓔ	33	Ⓐ Ⓑ Ⓒ Ⓓ Ⓔ
	4	Ⓐ Ⓑ Ⓒ Ⓓ Ⓔ	14	Ⓐ Ⓑ Ⓒ Ⓓ Ⓔ	24	Ⓐ Ⓑ Ⓒ Ⓓ Ⓔ	34	Ⓐ Ⓑ Ⓒ Ⓓ Ⓔ
	5	Ⓐ Ⓑ Ⓒ Ⓓ Ⓔ	15	Ⓐ Ⓑ Ⓒ Ⓓ Ⓔ	25	Ⓐ Ⓑ Ⓒ Ⓓ Ⓔ	35	Ⓐ Ⓑ Ⓒ Ⓓ Ⓔ
	6	Ⓐ Ⓑ Ⓒ Ⓓ Ⓔ	16	Ⓐ Ⓑ Ⓒ Ⓓ Ⓔ	26	Ⓐ Ⓑ Ⓒ Ⓓ Ⓔ	36	Ⓐ Ⓑ Ⓒ Ⓓ Ⓔ
	7	Ⓐ Ⓑ Ⓒ Ⓓ Ⓔ	17	Ⓐ Ⓑ Ⓒ Ⓓ Ⓔ	27	Ⓐ Ⓑ Ⓒ Ⓓ Ⓔ	37	Ⓐ Ⓑ Ⓒ Ⓓ Ⓔ
	8	Ⓐ Ⓑ Ⓒ Ⓓ Ⓔ	18	Ⓐ Ⓑ Ⓒ Ⓓ Ⓔ	28	Ⓐ Ⓑ Ⓒ Ⓓ Ⓔ	38	Ⓐ Ⓑ Ⓒ Ⓓ Ⓔ
	9	Ⓐ Ⓑ Ⓒ Ⓓ Ⓔ	19	Ⓐ Ⓑ Ⓒ Ⓓ Ⓔ	29	Ⓐ Ⓑ Ⓒ Ⓓ Ⓔ	39	Ⓐ Ⓑ Ⓒ Ⓓ Ⓔ
	10	Ⓐ Ⓑ Ⓒ Ⓓ Ⓔ	20	Ⓐ Ⓑ Ⓒ Ⓓ Ⓔ	30	Ⓐ Ⓑ Ⓒ Ⓓ Ⓔ	40	Ⓐ Ⓑ Ⓒ Ⓓ Ⓔ

Math Scoring Worksheet

A. Section 3

  _____   _____
  numer of correct   *number of incorrect*

  +   +

B. Section 5 (1-8)

  _____   _____
  numer of correct   *number of incorrect*

  +

C. Section 5 (9-18)

  _____
  numer of correct

  +   +

D. Section 7

  _____   _____
  numer of correct   *number of incorrect*

  =   =

E. Total Unrounded Raw Score

  _____ − _____ ÷4 = _____
  numer of correct   *number of incorrect*

F. Total Rounded Raw Score

  _____ (See table)

  Math Score Range = | _____ |

Math Conversion Table

Raw Score	Scaled Score	Raw Score	Scaled Score
54	800	23	490-550
53	780-800	22	480-540
52	760-800	21	470-530
51	740-800	20	460-520
50	720-780	19	450-510
49	700-760	18	450-510
48	690-750	17	440-500
47	680-740	16	430-490
46	670-730	15	420-480
45	660-720	14	420-480
44	650-710	13	410-470
43	650-710	12	400-460
42	640-700	11	390-450
41	630-690	10	380-440
40	620-680	9	390-430
39	610-670	8	380-420
38	610-670	7	370-410
37	600-660	6	360-400
36	590-650	5	340-380
35	580-640	4	320-370
34	570-630	3	310-360
33	560-620	2	300-350
32	560-620	1	270-320
31	550-610	0	240-300
30	540-600	-1	200-290
29	530-590	-2	200-270
28	530-590	-3	200-260
27	520-580	-4	200-240
26	510-570	-5	200-220
25	500-560	-6 and below	200
24	500-560		

SECTION 3
Time- 25 minutes
20 Questions

Turn to Section 3 (Page 1) of your answer sheet to answer the questions in this section.

Directions: For this section, solve each problem and decide which is the best of the choices given. Fill in the corresponding circle on the answer sheet. You may use any available space for scratchwork.

Notes

1. The use of a calculator is permitted.
2. All numbers used are real numbers.
3. Figures that accompany problems in this test are intended to provide information useful in solving the problems. They are drawn as accurately as possible EXCEPT when it is stated in a specific problem that the figure is not drawn to scale. All figure lie in a plane unless other indicated.
4. Unless otherwise specified, the domain of any function f is assumed to be set of all real numbers x for which $f(x)$ is a real number.

Reference Informatiom

$A = \pi r^2$
$C = 2\pi r$
$A = \ell w$
$A = \frac{1}{2}bh$
$V = \ell w h$
$V = \pi r^2 h$
$c^2 = a^2 + b^2$
Special Right Triangles

The numbers of degrees of arc in a circle is $360°$.

The sum of the measures in degrees of the angles is $180°$.

1. If $3a - b = a + b = 4$, what is the value of $2a - b$?

(A) -2
(B) 0
(C) 2
(D) 4
(E) 6

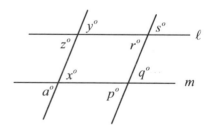

Note: Figure not drawn to scale.

2. In the figure above, line ℓ is parallel to line m. How many angles in the figure are equal to a?

(A) 8
(B) 7
(C) 6
(D) 3
(E) 2

GO ON TO THE NEXT PAGE

3. In a reading group A of 90 students, there are 4 boys for every 5 girls. In the other reading group B, there are 3 boys for every 2 girls. If these two groups are combined, the ratio of boys to girls is $10 : 9$. How many boys are in the combined reading group?

(A) 90
(B) 100
(C) 110
(D) 120
(E) 150

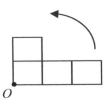

5. The figure above will be rotated counter clockwise 180° about point O. Which of the following could be the rotated figure?

(A)

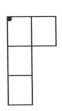

(B)

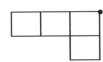

(C)

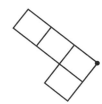

(D)

(E)

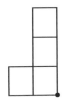

4. If $k^{-3} = 64$, what is the value of k^2?

(A) $\dfrac{1}{16}$

(B) $\dfrac{1}{8}$

(C) $\dfrac{1}{4}$

(D) $\dfrac{1}{2}$

(E) $\dfrac{3}{2}$

GO ON TO THE NEXT PAGE

6. If 10% of 20% of 75 is equal to x percent of 50, what is the value of x ?

(A) 0.03
(B) 0.3
(C) 3
(D) 30
(E) 300

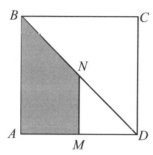

7. In the figure of a square above, the length of $\overline{AD}$ is $2x$, the length of $\overline{AM}$ is x, and $\overline{AB}$ is parallel to $\overline{MN}$. If a point is chosen from the square, what is the probability that the point will lie on the shaded region?

(A) $\dfrac{4}{9}$

(B) $\dfrac{2}{3}$

(C) $\dfrac{3}{8}$

(D) $\dfrac{3}{7}$

(E) $\dfrac{2}{5}$

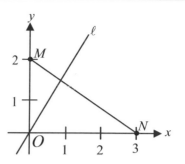

Note: Figure not drawn to scale.

8. In the figure above, line ℓ is perpendicular to $\overline{MN}$. Which of the following points lies on line ℓ ?

(A) $(1, 2)$
(B) $(3, 4)$
(C) $(4, 5)$
(D) $(5, 7)$
(E) $(6, 9)$

9. If $2^{a+b} = 8$ and $3^{a-b} = 81$, what is the value of a ?

(A) 2
(B) 3.5
(C) 6
(D) 8.5
(E) 10

GO ON TO THE NEXT PAGE

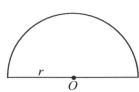

10. The figure above shows a semicircle with a radius r. If the perimeter of the semicircle is p, what is the value of r, in terms of p?

(A) $\dfrac{p}{2}$

(B) $\dfrac{p+2}{\pi}$

(C) $\dfrac{p}{\pi+2}$

(D) $\dfrac{2p}{\pi+2}$

(E) $\dfrac{p}{2+2\pi}$

11. If y is inversely proportional to x, and $y = 25$ when $x = 10$, what is the value of y when $x = 20$?

(A) 50
(B) 30
(C) 25
(D) 12.5
(E) 6.25

x, y, z, w

12. In the sequence above, the first term is x and the difference between any two consecutive terms is d. What is the value of w in terms of x and d?

(A) $4x$
(B) $2x + d$
(C) $3x + d$
(D) $x + 3d$
(E) $x + 4d$

13. Let the function f be defined by $f(x) = 2^{2x+3}$. If $\dfrac{1}{4}f(k) = 16$, what is the value of k?

(A) 1.5
(B) 2
(C) 3
(D) 4
(E) 4.5

GO ON TO THE NEXT PAGE

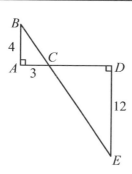

Note: Figure not drawn to scale.

14. In the figure above, $\overline{AB} = 4$, $\overline{AC} = 3$, and $\overline{DE} = 12$. What is the length of $\overline{BE}$?

(A) 5
(B) 10
(C) 15
(D) 20
(E) 25

15. At the beginning of the school year, Lisa's goal was to earn an average of 80% of her 50 quizzes for the year. She earned an average of 78% of the first 30 quizzes. If she wants to achieve her goal, what must be her average of the remaining quizzes?

(A) 83
(B) 84
(C) 85
(D) 86
(E) 87

16. According to the function $p(n) = \dfrac{2}{3}n + 100$, if n increases by 300, by how much does p increase by?

(A) 200
(B) 250
(C) 300
(D) 350
(E) 400

17. For all values of a and b, let $a \otimes b$ be defined by $a \otimes b = a^{\sqrt{b}}$. Which of the following is equal to $(2 \otimes 4) \otimes (8 \otimes 4)$?

(A) 2^4
(B) 2^8
(C) 2^{12}
(D) 2^{16}
(E) 2^{20}

GO ON TO THE NEXT PAGE

18. If $2^a = k$, $k^b = m$, and $m^c = 16$, what is the value of $a \cdot b \cdot c$?

(A) 2
(B) 4
(C) 5
(D) 6
(E) 8

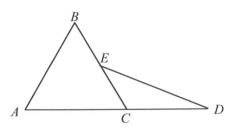

Note: Figure not drawn to scale.

20. In equilateral triangle ABC shown above, $AB = 4$, C is the midpoint of $\overline{AD}$, and E is the midpoint of $\overline{BC}$. What is the area of $\triangle CDE$?

(A) 10
(B) $8\sqrt{2}$
(C) $6\sqrt{2}$
(D) $4\sqrt{3}$
(E) $2\sqrt{3}$

19. The cost of a car rental is the sum of

(1) d dollars per day and

(2) x cents per $\dfrac{1}{4}$ of a mile that is traveled.

If Cathy rents a car for 10 days and drives 200 miles, which of the following could be the cost, in dollars, of the rental?

(A) $2(5d + 4x)$

(B) $10d + 200x$

(C) $10d + 800x$

(D) $\dfrac{10d + 800x}{100}$

(E) $10(d + 8x)$

STOP

If you finish before time is called, you may check your work on this section only.
Do not turn to any other section in the test.

SECTION 5
Time- 25 minutes
18 Questions

Turn to Section 5 (Page 1) of your answer sheet to answer the questions in this section.

Directions: For this section, solve each problem and decide which is the best of the choices given. Fill in the corresponding circle on the answer sheet. You may use any available space for scratchwork.

Notes

1. The use of a calculator is permitted.
2. All numbers used are real numbers.
3. Figures that accompany problems in this test are intended to provide information useful in solving the problems. They are drawn as accurately as possible EXCEPT when it is stated in a specific problem that the figure is not drawn to scale. All figure lie in a plane unless other indicated.
4. Unless otherwise specified, the domain of any function f is assumed to be set of all real numbers x for which $f(x)$ is a real number.

Reference Informatiom

        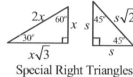

$A = \pi r^2$
$C = 2\pi r$
$\qquad A = \ell w \qquad A = \dfrac{1}{2}bh \qquad V = \ell wh \qquad V = \pi r^2 h \qquad c^2 = a^2 + b^2 \qquad$ Special Right Triangles

The numbers of degrees of arc in a circle is $360°$.

The sum of the measures in degrees of the angles is $180°$.

1. If $\dfrac{6}{k} = 2$ and $3k + h = 15$, what is the value of h?

   (A) 2
   (B) 3
   (C) 4
   (D) 5
   (E) 6

2. There are 600 teachers and students at Hill Side High School. Which of the following COULD NOT be the ratio of teachers to students?

   (A) 1: 5
   (B) 1: 7
   (C) 3: 17
   (D) 3: 20
   (E) 4: 11

GO ON TO THE NEXT PAGE

ALGEBRA TEST RESULTS

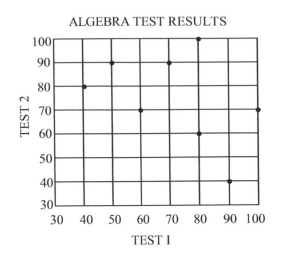

TEST 2

TEST I

3. The scatterplot above shows the results of eight students on their last two algebra tests. Which of the following is the greatest change in scores from test 1 to test 2 ?

(A) 60
(B) 50
(C) 40
(D) 30
(E) 20

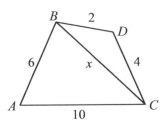

Note: Figure not drawn to scale.

4. In the figure above, x is the length of $\overline{BC}$, the side of $\triangle ABC$ and $\triangle BCD$. Which of the following could be the value of x ?

(A) 8
(B) 7
(C) 6
(D) 5
(E) 4

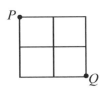

5. The figure above shows the routes from point P to point Q. If a person can only move to the right or down along the segments, what is the greatest possible number of different routes from point P to point Q ?

(A) 2
(B) 3
(C) 4
(D) 6
(E) 8

GO ON TO THE NEXT PAGE

6. If $(m-1)(k-1) = k-1$, which of the following can be true ?

  I. $m = 1$
  II. $k = 1$
  III. $m = 2$

(A) I only
(B) II only
(C) III only
(D) I and II only
(E) II and III only

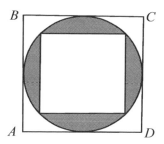

8. In the figure above, a circle is inscribed in a square and a square is inscribed in the circle. If the length of $\overline{AB}$ is $2k$, what is the area of the shaded region?

(A) $k^2(\pi - 2)$
(B) $k^2(\pi - 3)$
(C) $k^2(\pi - 4)$
(D) $k^2(\pi - 6)$
(E) $k^2(\pi - 8)$

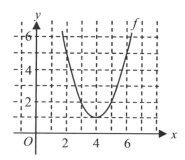

7. The figure above shows the graph of f. Which of the following could be the equation of the graph of function f ?

(A) $f(x) = -x^2 - 8x + 17$
(B) $f(x) = x^2 + 8x + 17$
(C) $f(x) = x^2 - 8x + 17$
(D) $f(x) = x^2 - 4x + 1$
(E) $f(x) = x^2 - 6x + 17$

GO ON TO THE NEXT PAGE

5 □□□ 5 □□□ 5 □□□ 5

Directions: For Students-Produced Response questions 9-18, use the grid at the bottom of the answer sheet page on which you have answered questions 1-8.

Each of the remaining 10 questions requires you to solve the problem and enter your answer by making the circles in the special grid, as shown in the examples below. You may use any available space for scratchwork.

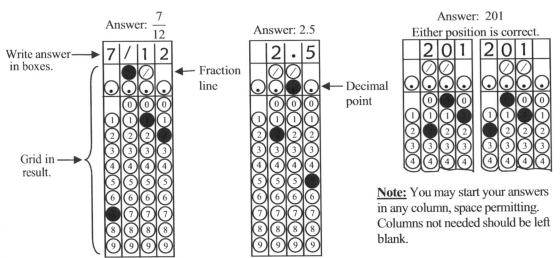

Answer: $\frac{7}{12}$

Write answer in boxes. →
← Fraction line

Grid in result. →

Answer: 2.5
← Decimal point

Answer: 201
Either position is correct.

Note: You may start your answers in any column, space permitting. Columns not needed should be left blank.

- Mark no more than one circle in any column.

- Because the answer sheet will be machine-scored, **you will receive credit only if the circles are filled in correctly.**

- Although not required, it is suggested that you write your answer in the boxes at the top of the columns to help you fill in the circles accurately.

- Some problems may have more than one correct answer. In such cases, grid only one answer.

- No question has a negative answer.

- **Mixed numbers** such as $3\frac{1}{2}$ must be gridded as 3.5 or 7/2. (If $\boxed{3\,1\,/\,2}$ is gridded, it will be interpreted as $\frac{31}{2}$, not $3\frac{1}{2}$.)

- **Decimal Answers:** If you obtain a decimal answer with more digits than the grid can accommodate, it may be either rounded or truncated, but it must fill the entire grid. For example, if you obtain an answer such as 0.6666…, you should record your result as .666 or .667. **A less accurate value such as .66 or .67 will be scored as incorrect.**

Acceptable ways to grid $\frac{2}{3}$ are:

9. Robert earns $152 in 4 days. At this rate, how many days will it take him to earn $532?

10. If $\dfrac{m}{n} = \dfrac{3}{4}$ and $3m = 18$, what is the value of $3n$?

GO ON TO THE NEXT PAGE ⇒

11. In isosceles triangle ABC, the measure of angle A is $k°$. If the measure of another angle is $2k°$, what is one possible value of k?

12. The volume V of a certain sphere is directly proportional to r^3. If $V = 2$ when $r = \dfrac{1}{2}$, what is the value of r when $V = 128$?

HONORS CLASSES

English	25
Chemistry	20
Physics	18

13. The table above shows the enrollments for the 11th grade honors program at Spring Lake high school. Of these students, 8 take English and Chemistry, 10 take Chemistry and Physics, and 6 take English and Physics. Three of the students take all three classes. If all of the students take at least one class, what is the number of students in the honors program?

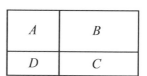

14. In the figure above, A, B, C, and D are the areas of the rectangles respectively. If $A = 9$, $B = 15$, and $D = 4$, what is the value of C?

GO ON TO THE NEXT PAGE

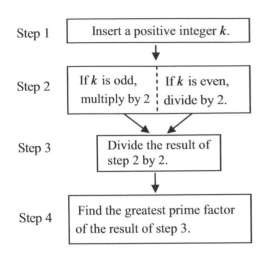

Step 1 — Insert a positive integer k.

Step 2 — If k is odd, multiply by 2 | If k is even, divide by 2.

Step 3 — Divide the result of step 2 by 2.

Step 4 — Find the greatest prime factor of the result of step 3.

15. In the chart above, if the number 88 is inserted in step 1, what number will be the result of step 4?

16. If pipe F can fill a certain water-tank in 3 hours and pipe E can empty it in 4 hours, how long will it take to fill the tank when both pipes are open?

17. If $a^2 - b^2 = 20$, where a and b are positive integers, and $a > b$, what is the value of a?

18. How many positive integers less than 1000 are multiples of 3 or multiples of 5?

STOP

If you finish before time is called, you may check your work on this section only.
Do not turn to any other section in the test.

SECTION 7
Time- 20 minutes
16 Questions

Turn to Section 7 (Page 2) of your answer sheet to answer the questions in this section.

Directions: For this section, solve each problem and decide which is the best of the choices given. Fill in the corresponding circle on the answer sheet. You may use any available space for scratchwork.

1. If $\dfrac{3}{2x-4} = \dfrac{3}{x}$, what is the value of x?

    (A) 2
    (B) 4
    (C) 6
    (D) 8
    (E) 10

A ———— P ———— B
—8 ························ 2

Note: Figure not drawn to scale.

2. If the ratio of AP to PB is 3:2, what is the coordinate of the point P on the number line above?

    (A) -4
    (B) -2
    (C) -1
    (D) 0
    (E) 1

GO ON TO THE NEXT PAGE

1, 2, 2, 2, 3, 3, 4, 5, 8, 10

3. For the list of 10 numbers above, which of the following must be true?

(A) mode < mean < median
(B) median < mode < mean
(C) median < mean < mode
(D) mean < median < mean
(E) mode < median < mean

5. If a number is increased by 45 and the result is multiplied by 2, the result is 18 less than the original number. In which of the following equations does x represent the number?

(A) $x + 2 \times 45 = 18 - x$
(B) $2x + 45 = x - 18$
(C) $2(x + 45) = x + 18$
(D) $2(x - 45) = x - 18$
(E) $2(x + 45) = x - 18$

$k + 5$

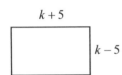

$k - 5$

Note: Figure not drawn to scale.

4. If the area of the rectangle above is 24, what is the perimeter of the rectangle?

(A) 28
(B) 26
(C) 24
(D) 22
(E) 20

$$f(x) = 2x^3$$

6. For the function above, which of the following is true?

I. $f(k) = f(-k)$
II. $f(k) = -f(-k)$
III. $f(k) = -f(k)$

(A) I only
(B) II only
(C) III only
(D) I and II only
(E) II and III only

GO ON TO THE NEXT PAGE ▷

7. If $k^6 = m^{-2}n^4$, where k, m, and n are positive numbers, which of the following is the value of m?

(A) $k^3 n^2$

(B) $\dfrac{k^3}{n^2}$

(C) $\dfrac{1}{k^3 n^2}$

(D) $\dfrac{k^6}{n^3}$

(E) $\dfrac{n^2}{k^3}$

8. Which of the following must be true for all positive numbers a, b, and c?

I. $a - b = -(b - a)$

II. $(a - b) + c = a - (b - c)$

III. $a(b + c) = b(a + c)$

(A) I only
(B) II only
(C) I and II only
(D) I and III only
(E) I, II, and III

9. A cake recipe calls for $\dfrac{1}{3}$ teaspoon of baking powder for $\dfrac{3}{4}$ cup of the flour. If the recipe is changed to include c cups of flour, how much teaspoon of baking powder, in terms of c, is needed?

(A) $2c$

(B) $\dfrac{8c}{9}$

(C) $\dfrac{2c}{3}$

(D) $\dfrac{4c}{9}$

(E) $\dfrac{c}{3}$

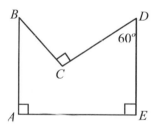

10. In the figure, $AB = AE = DE = 2$ and $\angle CDE = 60^\circ$, what is the perimeter of the figure?

(A) $6 + \sqrt{3}$
(B) $7 + \sqrt{3}$
(C) $6 + 2\sqrt{3}$
(D) $7 + 2\sqrt{3}$
(E) $8 + 2\sqrt{2}$

GO ON TO THE NEXT PAGE

11. Ten members registered as a team for a math contest. If one hand shake is made between every two members, which of the following could be the total number of hand shakes in the team?

(A) 45
(B) 50
(C) 75
(D) 100
(E) 120

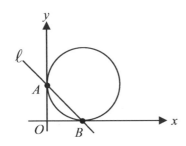

12. In the figure above, a circle is tangent to the two axes at point A and B. If the equation of the line ℓ is $f(x) = 5 - x$, what is the area of the circle?

(A) 5π
(B) 10π
(C) 20π
(D) 25π
(E) 30π

13. If the function f is defined by $f(x) = ax^2 + bx + c$, where constants a and b are positive, and constant c is negative, which of the following could be the graph of f?

(A)

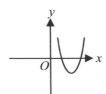

(B)

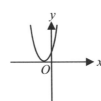

(C)

(D)

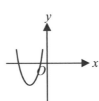

(E)

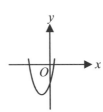

GO ON TO THE NEXT PAGE

| A | B | C | D | E |

14. If the five cards shown above are placed in a row so that cars B and C must be next to each other, how many different arrangements are possible?

(A) 24
(B) 36
(C) 48
(D) 72
(E) 120

$$h(t) = 36t - 6t^2$$

16. The function h above shows the height, in feet, of an object thrown upward after t seconds. What is the maximum height, in feet, the object can reach?

(A) 36
(B) 48
(C) 54
(D) 72
(E) 90

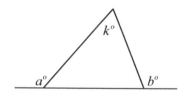

Note: Figure not drawn to scale.

15. In the figure above, what is k in terms of a and b?

(A) $a + b + 90$
(B) $a + b - 90$
(C) $a + b + 180$
(D) $a + b - 180$
(E) $180 - a - b$

STOP

If you finish before time is called, you may check your work on this section only.
Do not turn to any other section in the test.

TEST 4   ANSWER KEY

#	SECTION 3	SECTION 5	SECTION 7
1	C	E	B
2	D	D	B
3	B	B	E
4	A	D	A
5	B	D	E
6	C	E	B
7	C	C	E
8	E	A	C
9	B	14	D
10	C	24	B
11	D	36 or 45	A
12	D	2	D
13	A	42	E
14	D	$20/3$, 6.66 or 6.67	C
15	A	11	D
16	A	12	C
17	D	6	
18	B	466	
19	A		
20	E		

TEST 4   SECTION 3

1.   (C)

$$3a - b = 4$$
$$\frac{a+b=4}{4a \quad = 8} \Rightarrow a = 2 \text{ and } b = 2. \text{ thus,}$$
$$2a - b = 2(2) - 2 = 2.$$

---

2.   (D)

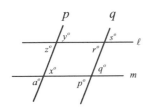

Since $\ell \parallel m$, $a = x = z = y$. But line p could not be parallel to line q.

---

3.   (B)

Boys: Girls = $4k : 5k$ $\Rightarrow 9k = 90 \Rightarrow k = 10$.
Group $B \rightarrow$ boys:girls = 3:2 = $3a : 2a$. Thus,
$$\frac{Boys}{Girls} = \frac{40 + 3a}{50 + 2a} = \frac{10}{9} \Rightarrow a = 20 .$$
The number of boys in group B is $3(20) = 60$.
Therefore, the number of boys of the combined classes = $40 + 60 = 100$.

---

4.   (A)

$$k^{-3} = 64 \Rightarrow k^3 = \frac{1}{64} \Rightarrow k = \frac{1}{4} \Rightarrow k^2 = \frac{1}{16} .$$

---

5.   (B)

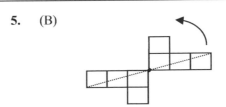

It must be symmetrical with the center.

6. (C)

$\left(\dfrac{10}{100}\right)\left(\dfrac{20}{100}\right)(75)=\dfrac{x}{100}(50)\ \Rightarrow x=3.$

---

7. (C)

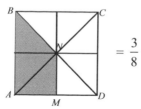

$=\dfrac{3}{8}$

---

8. (E)

The slope of $\overline{MN}=-\dfrac{2}{3}$ and the slope of

line $\ell=\dfrac{3}{2}$.

Therefore, the equation of line $\ell \Rightarrow y=\dfrac{3}{2}x.$

Point $(6,9) \Rightarrow 9=\dfrac{3}{2}(6)$, lies on the line.

---

9. (B)

$2^{a+b}=2^3$ and $3^{a-b}=3^4\ \Rightarrow a+b=3$ and

$a-b=4$. Use addition. (Eliminate b)

Therefore, $a=\dfrac{7}{2}$.

---

10. (C)

Since $p=\dfrac{2\pi r}{2}+2r=r(\pi+2)$, $r=\dfrac{p}{\pi+2}$.

---

11. (D)

$xy=k(\text{constant}) \Rightarrow (10)(25)=(20)(y)$

$\Rightarrow y=12.5$

---

12. (D)

$w=x+d+d+d=x+3d$. Or use

$a_n=a_1+(n-1)d.$

---

13. (A)

$\dfrac{1}{4}\left(2^{2k+3}\right)=16 \Rightarrow 2^{2k+3}=64=2^6$. Therefore,

$2k+3=6 \Rightarrow k=1.5$

---

14. (D)

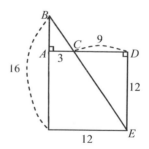

Since

$\triangle ABC \sim \triangle DEC,$

$\dfrac{AB}{AC}=\dfrac{DE}{CD} \Rightarrow \dfrac{4}{3}=\dfrac{12}{CD} \Rightarrow CD=9$.

Thus, $BC=5$ and $CE=15$. Therefore,

$BE=20$. Or

$BE=\sqrt{16^2+12^2}=20$.

---

15. (A)

Let x be the average of the remaining

quizzes, then $\dfrac{(30\times 78)+(20x)}{50}=80$.

$2340+20x=4000 \rightarrow 20x=1660.$

Therefore, $x=83.$

---

16. (A)

Since $p(n)=\dfrac{2}{3}n+100$, $\dfrac{\Delta p}{\Delta n}=\dfrac{2}{3}$ = slope.

Therefore, $\dfrac{\Delta p}{\Delta n}=\dfrac{\Delta p}{300}=\dfrac{2}{3} \Rightarrow \Delta p=200.$

---

17. (D)

$(2\otimes 4)\otimes(8\otimes 4)=$

$\left(2^{\sqrt 4}\right)\otimes\left(8^{\sqrt 4}\right)=(4)\otimes(64)$

Therefore, $4\otimes 64=4^{\sqrt{64}}=4^8=2^{16}.$

---

18. (B)

Since $2^a=k$, $k^b=m$, and $m^c=16$,

$\left(\left(2^a\right)^b\right)^c=16$. Therefore,

$2^{abc}=2^4 \Rightarrow abc=4.$

---

19. (A)

10 days $=10d$. 200 miles $=800$ of $\dfrac{1}{4}$ mile.

Therefore, cost in dollars

$$= 10d + \frac{800x}{100} = 10d + 8x = 2(5d + 4x).$$

20. (E)

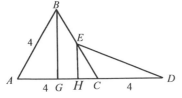

Since $BG = 2\sqrt{3}$,

$EH = \frac{1}{2}BG = \sqrt{3}$. Therefore, the area of

$\triangle CDE = \frac{1}{2} \times 4 \times \sqrt{3} = 2\sqrt{3}.$

TEST 4   SECTION 5

1. (E)
Since $k = 3$, $3(3) + h = 15. \Rightarrow h = 6.$

2. (D)
If $\dfrac{\text{teachers}}{\text{students}} = \dfrac{a}{b} = \dfrac{ak}{bk}$, the sum is
$ak + bk = (a + b)k$.
Therefore, $(a + b)k = 600$, where, k must be an integer. (A) $1 : 5 = k : 5k$, $k + 5k = 600$.
$k = 100$. But
(D) $(3 + 20)k = 600$, k cannot be an integer.

3. (B)
In the figure below, 50 is the largest difference.

ALGEBRA TEST RESULTS

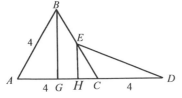

4. (D)
From the triangular inequality, $2 < x < 2 + 4$ and $4 < x < 6 + 10$. Therefore, $4 < x < 6$.

5. (D)

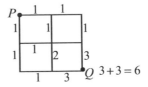

6. (E)
$(m - 1)(k - 1) = k - 1 \Rightarrow (m - 2)(k - 1) = 0$.
Therefore, $m = 2$ or $k = 1$. The answer is (E).

7. (C)
The equation will be $y = (x - 4)^2 + 1$.
Therefore, $y = x^2 - 8x + 17$.

8. (A)

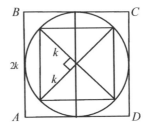

Since $AB = 2k$, the radius of the circle is $2k$ and the length of the diagonal of the square is also $2k$. There are 4 congruent triangles. Therefore, the shaded region

is, $\pi(k)^2 - 4\left(\dfrac{k \times k}{2}\right) = \pi k^2 - 2k^2 = k^2(\pi - 2)$.

9. 14
Proportion. $\dfrac{152}{4} = \dfrac{532}{x}$. $x = \dfrac{532 \times 4}{152} = 14$.

10. 24
Since $m = 6$, $\dfrac{6}{n} = \dfrac{3}{4}$. Thus, $n = 8$ and
$3n = 24$.

11. 36 or 45
There are two possible triangles as follows.

$(k,k,2k)$ or $(k,2k,2k)$. Thus,

$4k=180$ or $5k=180$. Therefore, $k=45$ or $k=36$.

---

12. 2

Direct proportion: $\dfrac{V}{r^3}=k$ (constant)

$$\dfrac{2}{\left(\dfrac{1}{2}\right)^3}=\dfrac{128}{r^3}$$

$\Rightarrow 2y^3=\dfrac{1}{8}\times128=16 \Rightarrow r^3=8$. Therefore,

$r=2$.

---

13. 42

The number of students
$n(E\cup C\cup P)=$

$n(E)+n(C)+n(P)-n(E\cap C)-n(C\cap P)$

$-n(P\cap E)+n(E\cap P\cap P)$

$=25+20+18-8-10-6+3=42$

Or, you can use a Venn diagram.

---

14. $\dfrac{20}{3}$ or 6.66, 6.67

$9k$	$15k$
A	B
D	C

In the figure above, $\dfrac{A}{B}=\dfrac{D}{C}=\dfrac{9}{15}$. Therefore,

$\dfrac{4}{C}=\dfrac{9}{15} \Rightarrow C=\dfrac{60}{9}=\dfrac{20}{3}$.

---

15. 11

$88 \Rightarrow \dfrac{88}{2}=44 \Rightarrow \dfrac{44}{2}=22 \Rightarrow 2\times11$

---

16. 12

	Hours	Rate
Pipe A	3	$\frac{1}{3}$
Pipe B	4	$\frac{1}{4}$

$\dfrac{1}{3}-\dfrac{1}{4}=\dfrac{1}{12}$

In the above, combined rate is $\dfrac{1}{12}$. Therefore,

it will take 12 hours.

---

17. 6

$(a+b)(a-b)=20$

$\quad 10 \times \ 2$

$a+b=10$

$a-b=2$ Therefore $2a=12 \Rightarrow a=6$.

---

18. 466

The number of integers $=$

$\left\lfloor\dfrac{999}{3}\right\rfloor+\left\lfloor\dfrac{999}{5}\right\rfloor-\left\lfloor\dfrac{999}{15}\right\rfloor=333+199-66$

$=466$

TEST 4   SECTION 7

1. (B)

$3x=6x-12 \Rightarrow 3x=12 \Rightarrow x=4$.

---

2. (B)

Since $AP=3k$ and $PB=2k$,

$5k=10 \Rightarrow k=2$. $AP=6$. Therefore,

$P=-8+6=-2$.

---

3. (E)

Mode $=2$, Mean $=4$, and Median $=3$.

---

4. (A)

$(k+5)(k-5)=k^2-25=24$. Thus

$k^2=49 \Rightarrow k=7$.

Therefore, $P=2(k+5+k-5)=4k=28$.

---

5. (E)

---

6. (B)

Odd function. $f(k)=k^3$ and $-f(-k)=k^3$.

---

7. (E)

Since $m^2=\dfrac{n^4}{k^6}$, $m=\left(\dfrac{n^4}{k^6}\right)^{\frac{1}{2}}=\dfrac{n^2}{k^3}$.

---

8. (C)

III. $a(b+c)=ab+ac$, $b(a+c)=ab+bc$

$\quad a(b+c)\neq b(a+c)$

I and II are correct.

---

9. (D)

Since $\dfrac{\frac{1}{3}}{\frac{3}{4}} = \dfrac{x}{c}$, $\dfrac{3}{4}x = \dfrac{1}{3}c$. Therefore,

$x = \dfrac{4c}{9}$.

10. (B)

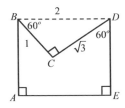

In the figure, $\triangle BCD$ is a special triangle. Therefore,

the perimeter $= 2 + 2 + 2 + 1 + \sqrt{3} = 7 + \sqrt{3}$.

11. (A)

$9 + 8 + 7 + 6 + 5 + 4 + 3 + 2 + 1 = 45$

Or, $_{10}C_2 = \dfrac{10 \times 9}{2} = 45$.

12. (D)

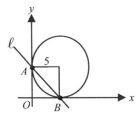

In the figure, the radius of the circle is 5. Therefore, the area $= \pi (5)^2 = 25\pi$.

13. (E)

Axis of symmetry $= -\dfrac{b}{2a} < 0$ and

y-intercept $c < 0$.

Therefore, (E) is correct.

14. (C)

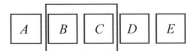

The number to arrange four card is $4 \times 3 \times 2 \times 1 = 24..$

The cards B and C have two arrangements of BC or CB. Therefore $24 \times 2 = 48$.

15. (D)

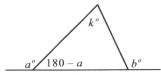

Exterior angle is equal to the sum of the two non-adjacent interior angles.

Since $180 - a + k = b$, $k = a + b - 180$.

16. (C)

$h(t) = -6t^2 + 36t$

Let $a = -6$ and $b = 36$.

Axis of symmetry $t = \dfrac{-b}{2a}$

$= \dfrac{-36}{-12} = 3$. Therefore, The maximum height

$h(3) = 36(3) - 6(3)^2 = 54$.

$\boxed{END}$

Dr. John Chung's SAT Math

TEST

5

ANSWER SHEET   TEST #:

SECTION 3

1 Ⓐ Ⓑ Ⓒ Ⓓ Ⓔ	11 Ⓐ Ⓑ Ⓒ Ⓓ Ⓔ	21 Ⓐ Ⓑ Ⓒ Ⓓ Ⓔ	31 Ⓐ Ⓑ Ⓒ Ⓓ Ⓔ
2 Ⓐ Ⓑ Ⓒ Ⓓ Ⓔ	12 Ⓐ Ⓑ Ⓒ Ⓓ Ⓔ	22 Ⓐ Ⓑ Ⓒ Ⓓ Ⓔ	32 Ⓐ Ⓑ Ⓒ Ⓓ Ⓔ
3 Ⓐ Ⓑ Ⓒ Ⓓ Ⓔ	13 Ⓐ Ⓑ Ⓒ Ⓓ Ⓔ	23 Ⓐ Ⓑ Ⓒ Ⓓ Ⓔ	33 Ⓐ Ⓑ Ⓒ Ⓓ Ⓔ
4 Ⓐ Ⓑ Ⓒ Ⓓ Ⓔ	14 Ⓐ Ⓑ Ⓒ Ⓓ Ⓔ	24 Ⓐ Ⓑ Ⓒ Ⓓ Ⓔ	34 Ⓐ Ⓑ Ⓒ Ⓓ Ⓔ
5 Ⓐ Ⓑ Ⓒ Ⓓ Ⓔ	15 Ⓐ Ⓑ Ⓒ Ⓓ Ⓔ	25 Ⓐ Ⓑ Ⓒ Ⓓ Ⓔ	35 Ⓐ Ⓑ Ⓒ Ⓓ Ⓔ
6 Ⓐ Ⓑ Ⓒ Ⓓ Ⓔ	16 Ⓐ Ⓑ Ⓒ Ⓓ Ⓔ	26 Ⓐ Ⓑ Ⓒ Ⓓ Ⓔ	36 Ⓐ Ⓑ Ⓒ Ⓓ Ⓔ
7 Ⓐ Ⓑ Ⓒ Ⓓ Ⓔ	17 Ⓐ Ⓑ Ⓒ Ⓓ Ⓔ	27 Ⓐ Ⓑ Ⓒ Ⓓ Ⓔ	37 Ⓐ Ⓑ Ⓒ Ⓓ Ⓔ
8 Ⓐ Ⓑ Ⓒ Ⓓ Ⓔ	18 Ⓐ Ⓑ Ⓒ Ⓓ Ⓔ	28 Ⓐ Ⓑ Ⓒ Ⓓ Ⓔ	38 Ⓐ Ⓑ Ⓒ Ⓓ Ⓔ
9 Ⓐ Ⓑ Ⓒ Ⓓ Ⓔ	19 Ⓐ Ⓑ Ⓒ Ⓓ Ⓔ	29 Ⓐ Ⓑ Ⓒ Ⓓ Ⓔ	39 Ⓐ Ⓑ Ⓒ Ⓓ Ⓔ
10 Ⓐ Ⓑ Ⓒ Ⓓ Ⓔ	20 Ⓐ Ⓑ Ⓒ Ⓓ Ⓔ	30 Ⓐ Ⓑ Ⓒ Ⓓ Ⓔ	40 Ⓐ Ⓑ Ⓒ Ⓓ Ⓔ

SECTION 5

1 Ⓐ Ⓑ Ⓒ Ⓓ Ⓔ	11 Ⓐ Ⓑ Ⓒ Ⓓ Ⓔ	21 Ⓐ Ⓑ Ⓒ Ⓓ Ⓔ	31 Ⓐ Ⓑ Ⓒ Ⓓ Ⓔ
2 Ⓐ Ⓑ Ⓒ Ⓓ Ⓔ	12 Ⓐ Ⓑ Ⓒ Ⓓ Ⓔ	22 Ⓐ Ⓑ Ⓒ Ⓓ Ⓔ	32 Ⓐ Ⓑ Ⓒ Ⓓ Ⓔ
3 Ⓐ Ⓑ Ⓒ Ⓓ Ⓔ	13 Ⓐ Ⓑ Ⓒ Ⓓ Ⓔ	23 Ⓐ Ⓑ Ⓒ Ⓓ Ⓔ	33 Ⓐ Ⓑ Ⓒ Ⓓ Ⓔ
4 Ⓐ Ⓑ Ⓒ Ⓓ Ⓔ	14 Ⓐ Ⓑ Ⓒ Ⓓ Ⓔ	24 Ⓐ Ⓑ Ⓒ Ⓓ Ⓔ	34 Ⓐ Ⓑ Ⓒ Ⓓ Ⓔ
5 Ⓐ Ⓑ Ⓒ Ⓓ Ⓔ	15 Ⓐ Ⓑ Ⓒ Ⓓ Ⓔ	25 Ⓐ Ⓑ Ⓒ Ⓓ Ⓔ	35 Ⓐ Ⓑ Ⓒ Ⓓ Ⓔ
6 Ⓐ Ⓑ Ⓒ Ⓓ Ⓔ	16 Ⓐ Ⓑ Ⓒ Ⓓ Ⓔ	26 Ⓐ Ⓑ Ⓒ Ⓓ Ⓔ	36 Ⓐ Ⓑ Ⓒ Ⓓ Ⓔ
7 Ⓐ Ⓑ Ⓒ Ⓓ Ⓔ	17 Ⓐ Ⓑ Ⓒ Ⓓ Ⓔ	27 Ⓐ Ⓑ Ⓒ Ⓓ Ⓔ	37 Ⓐ Ⓑ Ⓒ Ⓓ Ⓔ
8 Ⓐ Ⓑ Ⓒ Ⓓ Ⓔ	18 Ⓐ Ⓑ Ⓒ Ⓓ Ⓔ	28 Ⓐ Ⓑ Ⓒ Ⓓ Ⓔ	38 Ⓐ Ⓑ Ⓒ Ⓓ Ⓔ
9 Ⓐ Ⓑ Ⓒ Ⓓ Ⓔ	19 Ⓐ Ⓑ Ⓒ Ⓓ Ⓔ	29 Ⓐ Ⓑ Ⓒ Ⓓ Ⓔ	39 Ⓐ Ⓑ Ⓒ Ⓓ Ⓔ
10 Ⓐ Ⓑ Ⓒ Ⓓ Ⓔ	20 Ⓐ Ⓑ Ⓒ Ⓓ Ⓔ	30 Ⓐ Ⓑ Ⓒ Ⓓ Ⓔ	40 Ⓐ Ⓑ Ⓒ Ⓓ Ⓔ

Grid-in questions: 9, 10, 11, 12, 13, 14, 15, 16, 17, 18

Each grid-in contains four columns with bubbles for digits 0–9, plus fraction bar (/) and decimal point (.) positions.

SECTION 7				
1 Ⓐ Ⓑ Ⓒ Ⓓ Ⓔ	11 Ⓐ Ⓑ Ⓒ Ⓓ Ⓔ	21 Ⓐ Ⓑ Ⓒ Ⓓ Ⓔ	31 Ⓐ Ⓑ Ⓒ Ⓓ Ⓔ
2 Ⓐ Ⓑ Ⓒ Ⓓ Ⓔ	12 Ⓐ Ⓑ Ⓒ Ⓓ Ⓔ	22 Ⓐ Ⓑ Ⓒ Ⓓ Ⓔ	32 Ⓐ Ⓑ Ⓒ Ⓓ Ⓔ
3 Ⓐ Ⓑ Ⓒ Ⓓ Ⓔ	13 Ⓐ Ⓑ Ⓒ Ⓓ Ⓔ	23 Ⓐ Ⓑ Ⓒ Ⓓ Ⓔ	33 Ⓐ Ⓑ Ⓒ Ⓓ Ⓔ
4 Ⓐ Ⓑ Ⓒ Ⓓ Ⓔ	14 Ⓐ Ⓑ Ⓒ Ⓓ Ⓔ	24 Ⓐ Ⓑ Ⓒ Ⓓ Ⓔ	34 Ⓐ Ⓑ Ⓒ Ⓓ Ⓔ
5 Ⓐ Ⓑ Ⓒ Ⓓ Ⓔ	15 Ⓐ Ⓑ Ⓒ Ⓓ Ⓔ	25 Ⓐ Ⓑ Ⓒ Ⓓ Ⓔ	35 Ⓐ Ⓑ Ⓒ Ⓓ Ⓔ
6 Ⓐ Ⓑ Ⓒ Ⓓ Ⓔ	16 Ⓐ Ⓑ Ⓒ Ⓓ Ⓔ	26 Ⓐ Ⓑ Ⓒ Ⓓ Ⓔ	36 Ⓐ Ⓑ Ⓒ Ⓓ Ⓔ
7 Ⓐ Ⓑ Ⓒ Ⓓ Ⓔ	17 Ⓐ Ⓑ Ⓒ Ⓓ Ⓔ	27 Ⓐ Ⓑ Ⓒ Ⓓ Ⓔ	37 Ⓐ Ⓑ Ⓒ Ⓓ Ⓔ
8 Ⓐ Ⓑ Ⓒ Ⓓ Ⓔ	18 Ⓐ Ⓑ Ⓒ Ⓓ Ⓔ	28 Ⓐ Ⓑ Ⓒ Ⓓ Ⓔ	38 Ⓐ Ⓑ Ⓒ Ⓓ Ⓔ
9 Ⓐ Ⓑ Ⓒ Ⓓ Ⓔ	19 Ⓐ Ⓑ Ⓒ Ⓓ Ⓔ	29 Ⓐ Ⓑ Ⓒ Ⓓ Ⓔ	39 Ⓐ Ⓑ Ⓒ Ⓓ Ⓔ
10 Ⓐ Ⓑ Ⓒ Ⓓ Ⓔ	20 Ⓐ Ⓑ Ⓒ Ⓓ Ⓔ	30 Ⓐ Ⓑ Ⓒ Ⓓ Ⓔ	40 Ⓐ Ⓑ Ⓒ Ⓓ Ⓔ

Math Scoring Worksheet

A. Section 3

_____   _____
numer of correct   number of incorrect

+   +

B. Section 5 (1-8)

_____   _____
numer of correct   number of incorrect

+

C. Section 5 (9-18)

_____
numer of correct

+   +

D. Section 7

_____   _____
numer of correct   number of incorrect

=   =

E. Total Unrounded Raw Score

_____ − _____ ÷4 = _____
numer of correct   number of incorrect

F. Total Rounded Raw Score   _____ (See table)

Math Score Range = | _____ |

Math Conversion Table

Raw Score	Scaled Score	Raw Score	Scaled Score
54	800	23	490-550
53	780-800	22	480-540
52	760-800	21	470-530
51	740-800	20	460-520
50	720-780	19	450-510
49	700-760	18	450-510
48	690-750	17	440-500
47	680-740	16	430-490
46	670-730	15	420-480
45	660-720	14	420-480
44	650-710	13	410-470
43	650-710	12	400-460
42	640-700	11	390-450
41	630-690	10	380-440
40	620-680	9	390-430
39	610-670	8	380-420
38	610-670	7	370-410
37	600-660	6	360-400
36	590-650	5	340-380
35	580-640	4	320-370
34	570-630	3	310-360
33	560-620	2	300-350
32	560-620	1	270-320
31	550-610	0	240-300
30	540-600	-1	200-290
29	530-590	-2	200-270
28	530-590	-3	200-260
27	520-580	-4	200-240
26	510-570	-5	200-220
25	500-560	-6 and below	200
24	500-560		

SECTION 3
Time- 25 minutes
20 Questions

Turn to Section 3 (Page 1) of your answer sheet to answer the questions in this section.

Directions: For this section, solve each problem and decide which is the best of the choices given. Fill in the corresponding circle on the answer sheet. You may use any available space for scratchwork.

Notes

1. The use of a calculator is permitted.
2. All numbers used are real numbers.
3. Figures that accompany problems in this test are intended to provide information useful in solving the problems. They are drawn as accurately as possible EXCEPT when it is stated in a specific problem that the figure is not drawn to scale. All figure lie in a plane unless other indicated.
4. Unless otherwise specified, the domain of any function f is assumed to be set of all real numbers x for which $f(x)$ is a real number.

Reference Informatiom

$A = \pi r^2$
$C = 2\pi r$ $\qquad A = \ell w \qquad A = \frac{1}{2}bh \qquad V = \ell wh \qquad V = \pi r^2 h \qquad c^2 = a^2 + b^2 \qquad$ Special Right Triangles

The numbers of degrees of arc in a circle is $360°$.

The sum of the measures in degrees of the angles is $180°$.

---

1. If $\dfrac{a}{b} = \dfrac{1}{3}$ and $\dfrac{b}{c} = \dfrac{3}{2}$, what is the value of $\dfrac{a}{c}$?

(A) 2

(B) 1

(C) $\dfrac{1}{2}$

(D) $\dfrac{2}{9}$

(E) $\dfrac{1}{3}$

2. If $(x-2)^2 = 16$, which of the following could be a value of x?

(A) -6
(B) -4
(C) -2
(D) 2
(E) 8

GO ON TO THE NEXT PAGE

3. Two players in a baseball team are to be selected as a pitcher and catcher. If there are 15 players in the team, how many different outcomes are possible?

(A) 5
(B) 15
(C) 30
(D) 210
(E) 225

5. If $-3 \le m \le -1$ and $0 \le n \le 2$, what is the greatest possible value of mn?

(A) -6
(B) -2
(C) 0
(D) 2
(E) 6

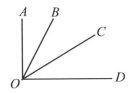

Note: Figure not drawn to scale.

4. In the figure above, $\overline{AO}$ is perpendicular to $\overline{DO}$. If $\angle AOC = 65^o$ and $\angle BOD = 78^o$, what is the value of $\angle BOC$?

(A) 12
(B) 25
(C) 35
(D) 47
(E) 53

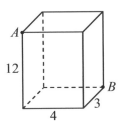

Note: Figure not drawn to scale.

6. In the figure above, point A and B are on the vertices of the rectangular solid. What is the length of $\overline{AB}$ (not shown)?

(A) 13
(B) 14
(C) 15
(D) 16
(E) 17

GO ON TO THE NEXT PAGE

7. In a sequence of numbers, the first number is k and each number after the first is 1 more than 2 times the preceding number. If the fourth term is 47, what is the value of k ?

   (A) 4
   (B) 5
   (C) 6
   (D) 7
   (E) 8

x	$f(x)$
0	2
a	b
1	0

8. The table above shows values of the linear function f for selected values of x. Which of the following defines b in terms of a ?

   (A) $b = 2a + 2$
   (B) $b = 3a + 2$
   (C) $b = -2a + 2$
   (D) $b = 2a - 2$
   (E) $b = a - 1$

9. Which of the following could be the graph in the xy-plane of the function $f(x) = x^2 - 2x - 3$?

(A)

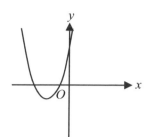

(B)

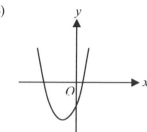

(C)

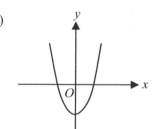

(D)

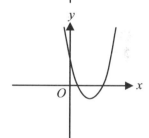

(E)

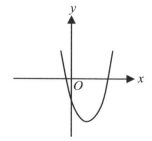

GO ON TO THE NEXT PAGE

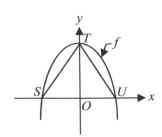

Note: Figure not drawn to scale.

10. The figure above shows the graph of the function f. If the function f is defined by $f(x) = ax^2 + 18$ and the area of $\triangle STU$ is 54, what is the value of a?

(A) 9
(B) 3
(C) −2
(D) −3
(E) −9

11. If k is a positive integer, and the number n is defined by $n = k(k+1)(k+2)$, which of the following could be equal to n?

(A) 12
(B) 18
(C) 24
(D) 30
(E) 42

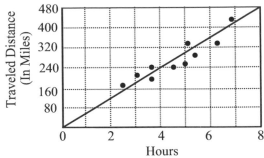

12. The scatterplot above shows the distance traveled for hours for 10 taxi drivers and the line of best fit for the data is shown above. Which of the following is closest to the average speed, in miles per hour, for the 10 taxi drivers?

(A) 50
(B) 54
(C) 59
(D) 65
(E) 68

13. A square and a circle drawn in a plane can intersect in at most how many points?

(A) 8
(B) 6
(C) 4
(D) 3
(E) 2

GO ON TO THE NEXT PAGE

14. If a and b are positive integers and

$8(2^{4a}) = \dfrac{2^{2b}}{2^3}$, what is b in terms of a ?

(A) $a+1$
(B) $a+2$
(C) $2a+1$
(D) $2a+2$
(E) $2a+3$

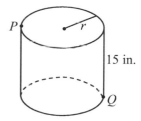

Note: Figure not drawn to scale.

15. The figure above is a cylinder that is 15 inches high and has a base of radius r. If the length of $\overline{PQ}$ (not shown) is 17, what is the value of r ?

(A) 2
(B) 4
(C) 8
(D) 10
(E) 12

16. When the number m is divided by 4, the remainder is 3. If $20m$ is divided by 8, what is the remainder?

(A) 7
(B) 6
(C) 5
(D) 4
(E) 3

17. If $\dfrac{a+b}{2} = \dfrac{b+c}{3} = \dfrac{c+a}{4} = k$, what is the value of $(a+b+c)$ in terms of k ?

(A) $3k$
(B) $4k$
(C) $4.5k$
(D) $5k$
(E) $5.5k$

GO ON TO THE NEXT PAGE

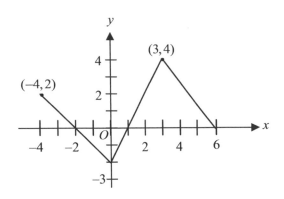

18. The figure above shows the graph of the function f defined for $-4 \le x \le 6$. For which of the following values of x is $f(x) < |f(x)|$?

(A) -3
(B) -2
(C) -0.5
(D) 2
(E) 3

19. If a and b are positive integers and $a^2 - b^2 = 24$, which of the following could be the smallest value of a?

(A) 1
(B) 2
(C) 4
(D) 5
(E) 6

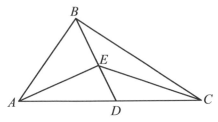

Note: Figure not drawn to scale.

20. In the figure above, the ratio of the length of $\overline{AD}$ to the length of $\overline{DC}$ is 3:2 and the area of $\triangle AED$ is 9. If the area of $\triangle ABE$ is 7, what is the area of $\triangle BEC$?

(A) 6
(B) $5\dfrac{2}{3}$
(C) 5
(D) $4\dfrac{2}{3}$
(E) $4\dfrac{1}{3}$

STOP

If you finish before time is called, you may check your work on this section only.
Do not turn to any other section in the test.

SECTION 5
Time- 25 minutes
18 Questions

Turn to Section 5 (Page 1) of your answer sheet to answer the questions in this section.

Directions: For this section, solve each problem and decide which is the best of the choices given. Fill in the corresponding circle on the answer sheet. You may use any available space for scratchwork.

1. Cathy had d dollars. After she bought a apples for c cents each, there are k dollars left. What is k in terms of a, c, and d ?

   (A) $d - ac$

   (B) $d + ac$

   (C) $100d - ac$

   (D) $\dfrac{d - ac}{100}$

   (E) $\dfrac{100d - ac}{100}$

All numbers that are divisible by 3 and 4 are also divisible by 8

2. Which of the following numbers can be used to show that the statement above is FALSE?

   (A) 24
   (B) 28
   (C) 40
   (D) 48
   (E) 60

GO ON TO THE NEXT PAGE

COMPUTER SALES AT COMPANY X

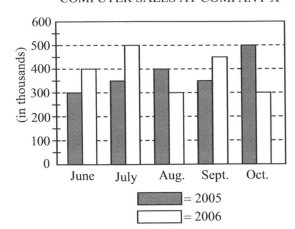

= 2005
= 2006

3. According to the graph above, for which month was the greatest percent of change from 2005 to 2006?

(A) June
(B) July
(C) August
(D) September
(E) October

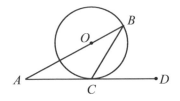

Note: Figure not drawn to scale.

4. In the figure above, O is the center of the circle and $\overline{AD}$ is tangent to the circle at point C. If the measure of $\angle BAD$ is 40°, what is the measure of $\angle ACB$?

(A) 100°
(B) 115°
(C) 118°
(D) 120°
(E) 130°

5. For O.K theater tickets, a ticket for an adult is $5 more than a ticket for a child. If a group of 6 adults and 10 children pay a total of $142, what is the cost of a ticket for one adult and one child?

(A) $19
(B) $18
(C) $17
(D) $16
(E) $15

GO ON TO THE NEXT PAGE

Figure A Figure B

Note: Figures not drawn to scale

6. Figure A and Figure B shows two squares with the length of diagonal a and b. If the ratio of a to b is 5:2 and the area of figure A is 50, what is the area of figure B?

(A) 25
(B) 20
(C) 10
(D) 8
(E) 4

8. The smallest integer of a set of even consecutive integers is -20. If the sum of these integers is 72, how many integers are in this set?

(A) 20
(B) 21
(C) 22
(D) 23
(E) 24

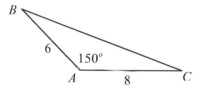

Note: Figure not drawn to scale.

7. In $\triangle ABC$ above, $AB = 6$, $AC = 8$, and $\angle BAC = 150^{o}$. What is the area of $\triangle ABC$?

(A) 12
(B) 18
(C) 20
(D) 24
(E) 48

GO ON TO THE NEXT PAGE

Directions: For Students-Produced Response questions 9-18, use the grid at the bottom of the answer sheet page on which you have answered questions 1-8.

Each of the remaining 10 questions requires you to solve the problem and enter your answer by making the circles in the special grid, as shown in the examples below. You may use any available space for scratchwork.

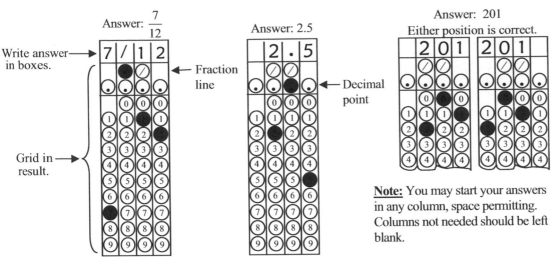

Answer: $\frac{7}{12}$

Write answer in boxes.

← Fraction line

Grid in result.

Answer: 2.5

← Decimal point

Answer: 201
Either position is correct.

Note: You may start your answers in any column, space permitting. Columns not needed should be left blank.

- Mark no more than one circle in any column.

- Because the answer sheet will be machine-scored, **you will receive credit only if the circles are filled in correctly.**

- Although not required, it is suggested that you write your answer in the boxes at the top of the columns to help you fill in the circles accurately.

- Some problems may have more than one correct answer. In such cases, grid only one answer.

- No question has a negative answer.

- **Mixed numbers** such as $3\frac{1}{2}$ must be gridded as 3.5 or 7/2. (If [3 1 / 2] is gridded, it will be interpreted as $\frac{31}{2}$, not $3\frac{1}{2}$.)

- **Decimal Answers:** If you obtain a decimal answer with more digits than the grid can accommodate, it may be either rounded or truncated, but it must fill the entire grid. For example, if you obtain an answer such as 0.6666..., you should record your result as .666 or .667. **A less accurate value such as .66 or .67 will be scored as incorrect.**

Acceptable ways to grid $\frac{2}{3}$ are:

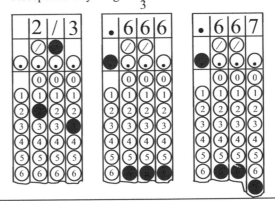

9. According to a formula, the surface area of a sphere is directly proportional to the square of the radius of the sphere. If the surface area is 36 when the radius is 3, what is the surface area when the radius is 4?

10. What is the measure of the smaller angle, in degrees, formed by the hands of a clock at 3:30 pm?

GO ON TO THE NEXT PAGE ⟹

11. If $x + \dfrac{1}{x} = 2$, what is the value of $\left(x - \dfrac{1}{x} \right)$?

$$c(x) = \frac{400x - 200}{x} + k$$

12. The cost, in dollars, of framing x pictures is given by the function $c(x)$ above. If 20 pictures are to be framed at a cost of $520, what is the value of k?

13. For all x, the function f is defined by $f(x) = 3^x$ and the function g is defined by $g(x) = 3f(x) + 1$. If $g(k) = 28$, what is the value of k?

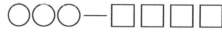

Set $F = \{A, B, C\}$ Set $R = \{1, 2, 3, 4\}$

14. A plate number of a car is arranged as shown above. A selection is made by choosing the first three letters from the set F and the next four numbers from the set R. If each arrangement must start with a letter, and no letter or number appears more than once in an arrangement, how many arrangements can be formed?

GO ON TO THE NEXT PAGE

15. Let the a_n, nth term, be defined by $a_n = 2n^2 + k$, where k is a constant. How much larger is the 15th term of the sequence than the 14th term?

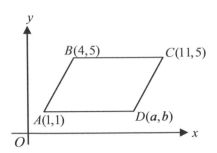

Note: Figure not drawn to scale.

17. In the figure above, $ABCD$ is a parallelogram. What is the perimeter of the figure $ABCD$?

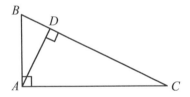

Note: Figure not drawn to scale.

16. In the figure above, $AB = 6$ and $BD = 2$. What is the length of $\overline{CD}$?

18. If a number $12k$ can be formed into a cube of a positive integer, that is $12k = m^3$, what could be the least possible value of k?

STOP

If you finish before time is called, you may check your work on this section only.
Do not turn to any other section in the test.

SECTION 7
Time- 20 minutes
16 Questions

Turn to Section 7 (Page 2) of your answer sheet to answer the questions in this section.

Directions: For this section, solve each problem and decide which is the best of the choices given. Fill in the corresponding circle on the answer sheet. You may use any available space for scratchwork.

Notes

1. The use of a calculator is permitted.
2. All numbers used are real numbers.
3. Figures that accompany problems in this test are intended to provide information useful in solving the problems. They are drawn as accurately as possible EXCEPT when it is stated in a specific problem that the figure is not drawn to scale. All figure lie in a plane unless other indicated.
4. Unless otherwise specified, the domain of any function f is assumed to be set of all real numbers x for which $f(x)$ is a real number.

Reference Information

$A = \pi r^2$
$C = 2\pi r$

$A = \ell w$

$A = \dfrac{1}{2}bh$

$V = \ell wh$

$V = \pi r^2 h$

$c^2 = a^2 + b^2$

Special Right Triangles

The numbers of degrees of arc in a circle is $360°$.

The sum of the measures in degrees of the angles is $180°$.

---

1. Which of the following represents the total cost, in dollars, of the ticket price for a adults that cost \$10 each and c children that cost \$5 each?

(A) $5a + 10c$
(B) $5(a + c)$
(C) $10(c + a)$
(D) $5(2a + c)$
(E) $5(a + 2c)$

2. If $ax^2 + bx + c = 0$ for all values of x, what is the value of $a + b + c$?

(A) -2
(B) -1
(C) 0
(D) 1
(E) 2

GO ON TO THE NEXT PAGE

3. In an xy-coordinate plane, the distance between points A $(1, 3)$ and B $(5, t)$ is 5. Which of the following could be the value of t ?

(A) -3
(B) -2
(C) -1
(D) 0
(E) 2

4. If $p - q = 30$ and $p > 12$, which of the following MUST be true?

(A) $q < 0$
(B) $q < -10$
(C) $q < -18$
(D) $q > -16$
(E) $q > -18$

PEOPLE AT X HIGH SCHOOL

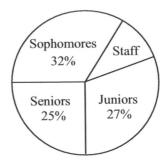

Note: Figure not drawn to scale.

5. The graph above shows the number of students at X High School. If the number of sophomores is 25 more than the number of juniors, how many students are there in the school?

(A) 500
(B) 450
(C) 400
(D) 350
(E) 325

$$a^2 - b^2 = 12$$
$$a + b > 4$$

6. If a and b are positive integers in the inequalities above and $a > b$, what is the value of a ?

(A) 1
(B) 2
(C) 3
(D) 4
(E) 5

GO ON TO THE NEXT PAGE

7. If $(x-1)(x+2) < 0$, which of the following must be true?

(A) $x > 2$
(B) $-2 < x < 1$
(C) $-1 < x < 0$
(D) $x < -2$
(E) $x = -5$

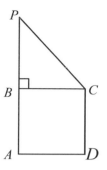

Note: Figure not drawn to scale.

9. The figure above, $ABCD$ is a square and $\triangle PBC$ is an isosceles triangle. If $AB = 4$, what is the length of $\overline{PD}$ (not shown)?

(A) $4\sqrt{2}$ (approximately 5.66)
(B) $4\sqrt{5}$ (approximately 8.94)
(C) $5\sqrt{2}$ (approximately 7.07)
(D) $5\sqrt{3}$ (approximately 8.66)
(E) 8

8. A 100-gallon tank is 10% full of water. If the water is poured into a 30-gallon empty tank, what percent of the 30-gallon tank has been filled?

(A) $\dfrac{1}{3}$%

(B) 30%

(C) 33%

(D) $33\dfrac{1}{2}$%

(E) $33\dfrac{1}{3}$%

10. What is the equation of a line which has an x-intercept of 5 and a slope parallel to $f(x) = -2x + 3$?

(A) $f(x) = -2x + 5$
(B) $f(x) = -2x - 5$
(C) $f(x) = 2x + 10$
(D) $f(x) = -2x + 10$
(E) $f(x) = -2x - 10$

GO ON TO THE NEXT PAGE

PROFIT FROM BOOKS SOLD

Number of books sold	10	15	20	25	30
Profit in dollars	40	45	55	70	93

11. Which of the following graphs best represents the information in the table above?

(A)

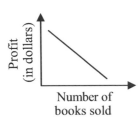

(B)

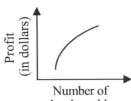

(C)

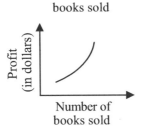

(D)

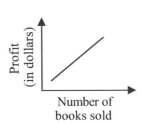

(E)

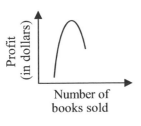

Set $A = \{0, 1, 2, 3\}$

Set $B = \{4, 5, 6\}$

12. If a is any element from set A and b is any element from set B, how many different values are possible for $\dfrac{a}{b}$?

(A) 8
(B) 9
(C) 11
(D) 12
(E) 13

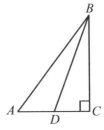

Note: Figure not drawn to scale.

13. In the figure above, $AB = 15$, $AD = 4$, and $CD = 5$. What is the area of $\triangle ABD$?

(A) 12
(B) 24
(C) 48
(D) 54
(E) 108

GO ON TO THE NEXT PAGE

14. If a, b, c, and d are positive integers, and $\dfrac{a}{b} = \dfrac{c}{d} = k$, where k is a constant, which of the following must be true?

I. $ad = bc$

II. $\dfrac{a+b}{b} = \dfrac{c+d}{d}$

III. $\dfrac{a+c}{b+d} = k$

(A) I only
(B) I and II only
(C) I and III only
(D) II and III only
(E) I, II, and III

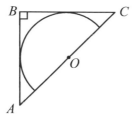

15. In the figure above, a semicircle is inscribed in $\triangle ABC$. If the area of the semicircle is 4π and $AB = BC$, what is the area of $\triangle ABC$?

(A) 16
(B) 18
(C) 20
(D) 22
(E) 24

16. During a sale last week, the original price of a hamburger was discounted by p percent. But this week the price of the hamburger has increased by q percent. If the price of the hamburger this week is x dollars, what is the original price, in dollars?

(A) $x(1 - \dfrac{p}{100})(1 - \dfrac{q}{100})$

(B) $x(1 - \dfrac{p}{100})(1 + \dfrac{q}{100})$

(C) $x\left(\dfrac{1-p}{100}\right)\left(\dfrac{1+q}{100}\right)$

(D) $x\left(\dfrac{(100-p)(100+q)}{10000}\right)$

(E) $\dfrac{10000x}{(100-p)(100+q)}$

STOP

If you finish before time is called, you may check your work on this section only.
Do not turn to any other section in the test.

TEST 5   ANSWER KEY

#	SECTION 3	SECTION 5	SECTION 7
1	C	E	D
2	C	E	C
3	D	B	D
4	E	B	E
5	C	A	A
6	A	D	D
7	B	A	B
8	C	E	E
9	E	64	B
10	C	75	D
11	C	0	C
12	C	130	B
13	A	2	B
14	E	144	E
15	B	58	A
16	D	16	E
17	C	24	
18	C	18	
19	D		
20	D		

TEST 5   SECTION 3

1. (C)
$$\frac{a}{c} = \frac{a}{b} \times \frac{b}{a} = \frac{1}{3} \times \frac{3}{2} = \frac{1}{2}$$

2. (C)
Since $x - 2 = 4, -4 \Rightarrow x = 6, -2$

3. (D)
Pitcher   Catcher
   15   ×   14   =210

4. (E)
Since $\angle BOC = \angle AOC + \angle BOC - 90$,
$\angle BOC = 65 + 78 - 90 = 53$.

5. (C)
Since $mn \le 0$, the greatest value is 0.
$n = 0$.

6. (A)
$$AB = \sqrt{12^2 + 4^2 + 3^2} = 13$$

7. (B)
The fourth term = $8k + 7 = 47$, $k = 5$.

8. (C)
Linear function has a constant slope. Thus,
$\frac{2-0}{0-1} = \frac{b-0}{a-1}$. Therefore,
$2a - 2 = -b \Rightarrow b = -2a + 2$.

9. (E)
The axis of symmetry= $\frac{-(-2)}{2} = 1$ and the
y-intercept = $f(0) = -3$. (E) is correct.

10. (C)

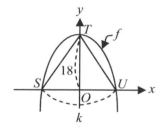

The area of $\triangle STU = \dfrac{18 \times k}{2} = 9k = 54$. Thus,

$k = 6$. The coordinates of point U is $(3,0)$.

Therefore, $0 = a(3)^2 + 18 \Rightarrow a = -2$.

---

11. (C)

$$24 = 2 \times 3 \times 4$$

---

12. (C)

Average speed = The slope of the line.

Since $m = \dfrac{480 - 0}{8} \simeq 60$, (C) is the best

answer.

---

13. (A)

In the figure above, there are 8 intersections.

---

14. (E)

$8\left(2^{4a}\right) = 2^3 2^{4a} = 2^{3+4a}$. $2^{3+4a} = 2^{2b-3}$. Thus,

$3 + 4a = 2b - 3$. Therefore, $b = 2a + 3$.

---

15. (B)

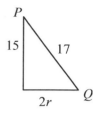

In the figure above,

$(2r)^2 = 17^2 - 15^2 = 64$. Thus,

$2r = 8 \Rightarrow r = 4$.

---

16. (D)

Choose the number, 7.

$20m = 20(7) = 140$. When 140 is divided by

8, the remainder is 4.

Or, algebraically,

let m be $m = 4q + 3$. Then $20m = 80q + 60$.

When it is divided by 8,

$80q + 60 = 8(10q + 7) + 4$. The remainder is 4.

---

17. (C)

$a + b = 2k$, $b + c = 3k$, and $c + a = 4k$.

Therefore,

$$2(a + b + c) = 9k \Rightarrow a + b + c = \dfrac{9}{2}k.$$

Or,

$$\dfrac{a+b}{2} = \dfrac{b+c}{3} = \dfrac{c+a}{4} = \dfrac{2a + 2b + 2c}{2 + 3 + 4} = k$$

$$\dfrac{2(a + b + c)}{9} = k.$$

Therefore, $a + b + c = 4.5k$.

---

18. (C)

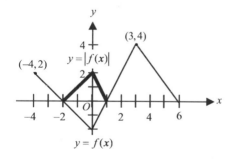

In the graph above, $f(x) < |f(x)|$ for

$-2 < x < 2$.

Therefore, -0.5 is a possible answer.

---

19. (D)

Since $(a + b)(a - b) = 24$, there are two

possible solutions.

$a + b = 12$ and $a - b = 2$ ---(1) or

$a + b = 6$ and $a - b = 4$ -----(2). From (1)

$a = 7$ and from (2) $a = 5$. Therefore, the

answer is (D)

20. (D)

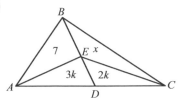

Since $\triangle AED$ and $\triangle CED$ have the same height. Thus the ratio of the lengths is equal to the ratio of the areas that is 3:2, or $3k:2k$. In the figure above, $\triangle ABD$ and $\triangle BDC$ have the same height, then

$\dfrac{\text{area of } \triangle ABD}{\text{area of } \triangle CBD} = \dfrac{7+3k}{x+2k} = \dfrac{3}{2}$. Therefore,

$14+6k = 3x+6k \Rightarrow x = 4\dfrac{2}{3}$.

Or, by the formula,

$\dfrac{7}{3} = \dfrac{x}{2} \Rightarrow x = \dfrac{14}{3} = 4\dfrac{2}{3}$.

TEST 5   SECTION 5

---

1. (E)

$d - \dfrac{c}{100} \times a = k$. Thus, $k = \dfrac{100d - ac}{100}$.

---

2. (E)

The false statement is " All the numbers that are divisible by 3 and 4 are not divisible by 8."

First, the numbers divisible by 3 and 4 are 24, 48, 60. But 60 is not divisible by 8.

---

3. (B)

June: $\dfrac{100}{300} \simeq 0.33$ July: $\dfrac{150}{350} \simeq 0.43$

Aug: $\dfrac{100}{400} \simeq 0.25$ Sept.: $\dfrac{100}{350} \simeq 0.29$

Oct.: $\dfrac{200}{500} = 0.40$.

Therefore, the greatest % change was in July.

---

4. (B)

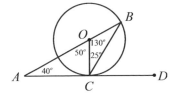

In the figure above, $\triangle OBC$ is an isosceles triangle. Thus, $\angle OCB = 25^o$. Therefore, $\angle ACB = 90 + 25 = 115^o$

---

5. (A)

If a child's ticket is $\$k$, an adult's ticket is $\$(k+5)$. Thus, $6(k+5)+10k = 142 \Rightarrow k = 7$, and an adult ticket is $12. Therefore, the answer is 7+12 = $19.

---

6. (D)

The ratio of the areas is $5^2:2^2 = 25k:4k$.

Therefore, $25k = 50$, or $k = 2$.

The area of figure $B = 4k = 4\times 2 = 8$.

Or, use proportion. $\dfrac{25}{50} = \dfrac{4}{x} \Rightarrow x = 8$.

---

7. (A)

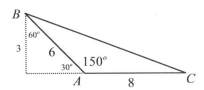

In the figure above, the height of $\triangle ABC$ is 3.

Therefore, the area is, $\dfrac{1}{2}(3\times 8) = 12$.

---

8. (E)

$-20, -18, -16, ...0,...2, 4,....20, \underbrace{22, 24, 26}_{72}$.

Therefore, there are 24 numbers.

---

9. 64

Direct proportion: $\dfrac{A}{r^2} = K$ (constant)

$\dfrac{36}{3^2} = \dfrac{A}{4^2}$. Therefore, $A = 64$.

---

10. 75

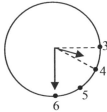

In the figure above, the hour hand moves

15° during 30 minutes. Therefore, $60° + 15° = 75°$.

---

11. 0

Since $\left(x - \dfrac{1}{x}\right)^2 = x^2 + \dfrac{1}{x^2} - 2$, and

$\left(x + \dfrac{1}{x}\right)^2 = x^2 + \dfrac{1}{x^2} + 2 = 2$. $x^2 + \dfrac{1}{x^2} = 2$.

Therefore, by substitution: $2 - 2 = 0$

Or, when $x = 1$, $x + \dfrac{1}{x} = 2$, then $x - \dfrac{1}{x} = 0$.

---

12. 130

From the equation $c = \dfrac{400x - 200}{x} + k$.

$520 = \dfrac{400(20) - 200}{20} + k = 390 + k$.

Therefore, $k = 130$.

---

13. 2

Since $g(k) = 3f(k) + 1 = 28$, $f(k) = 9$.

$f(k) = 3^k = 9 = 3^2 \Rightarrow k = 2$.

---

14. 144

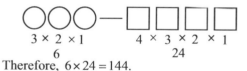

$3 \times 2 \times 1$ $4 \times 3 \times 2 \times 1$
6   24

Therefore, $6 \times 24 = 144$.

---

15. 58

Since $a_n = 2n^2 + k$,

$a_{15} = 2(15^2) + k = 450 + k$ and

$a_{14} = 2(14^2) + k = 392 + k$. Therefore,

$a_{15} - a_{14} = 58$.

---

16. 16

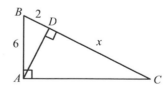

In the figure, $AB^2 = BD \times CD$.

Therefore, $36 = 2(2 + x)$. $x = 16$.

---

17. 24

$AB = \sqrt{(4-1)^2 + (5-1)^2} = 5$ and
$BC = 11 - 4 = 7$. Therefore, the perimeter is
$2(5 + 7) = 24$.

---

18. 18

Since $12k = 2 \times 2 \times 3 \times k = (\quad)^3$, the least value of k must be $2 \times 3 \times 3 = 18$.

TEST 5   **SECTION 7**

1. (D)
$10a + 5c = 2(5a + c)$

---

2. (C)
If $ax^2 + bx + c = 0$ for all values of x, then $a = b = c = 0$. Therefore, $a + b + c = 0$.

---

3. (D)
$\sqrt{(t-3)^2 + (5-1)^2} = 5 \Rightarrow (t-3)^2 = 9$. Thus, $t - 3 = 3, -3$. Therefore, $t = 6$, or 0.

---

4. (E)
Since $p = q + 30 > 12$, $q > -18$.

---

5. (A)
5% of the students is 25. Thus, $0.05x = 25$ or $x = 500$.

---

6. (D)
$(a+b)(a-b) = 12$ and $(a+b) > 4$.
$(a+b)$ and $(a-b)$ are integers.
If $a + b = 6$, then $a - b = 2$.

Therefore, $\dfrac{\begin{array}{c} a+b = 6 \\ a-b = 2 \end{array}}{2a \;=\; 8} \rightarrow a = 4$.

---

7. (B)
The solution of $(x-1)(x+2) < 0$ is $-2 < x < 1$.

---

8. (E)
$10\% \times 100 = 0.1(100) = 10G$. Therefore,
$\dfrac{10G}{30G} = \dfrac{1}{3}$.

---

9. (B)

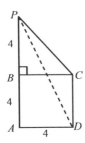

In the figure above,
$$PD = \sqrt{8^2 + 4^2} = \sqrt{80} = 4\sqrt{5}.$$

---

10. (D)

Since the slope is equal to -2,

$f(x) = -2x + b$.

$0 = -2(5) + b \Rightarrow b = 10$.

Therefore, $f(x) = -2x + 10$.

---

11. (C)

The data is increasing exponentially.

---

12. (B)

Set $A = \{0, 1, 2, 3\}$ Set $B = \{4, 5, 6\}$

$\dfrac{a}{b} = \dfrac{0}{4,5,6} = 0$ $\dfrac{1}{4,5,6} = \dfrac{1}{4}, \dfrac{1}{5}, \dfrac{1}{6}$

$\dfrac{2}{4,5,6} = \dfrac{2}{4}, \dfrac{2}{5}, \dfrac{2}{6}$

$\dfrac{3}{4,5,6} = \dfrac{3}{4}, \dfrac{3}{5}, \dfrac{3}{6}$

Therefore,

$0, \dfrac{1}{4}, \dfrac{1}{5}, \dfrac{1}{6}, \dfrac{1}{2}, \dfrac{2}{5}, \dfrac{1}{3}, \dfrac{3}{4}, \dfrac{3}{5} \Rightarrow 9$ different values.

---

13. (B)

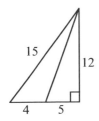

The area of the triangle $= \dfrac{4 \times 12}{2} = 24$.

---

14. (E)

$\dfrac{a}{b} = \dfrac{c}{d} = k \Rightarrow a = bk$ and $c = dk$

I. $ad = bc$ (cross multiplication)

II. $\dfrac{a+b}{b} = \dfrac{c+d}{d} \Rightarrow \dfrac{a}{b} + 1 = \dfrac{c}{d} + 1$ (True)

III.

$\dfrac{a+c}{b+d} = k \Rightarrow \dfrac{bk+dk}{b+d} = \dfrac{(b+d)k}{b+d} = k$ (True)

---

15. (A)

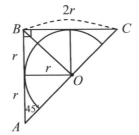

In the figure above, the area of the semicircle

is $\dfrac{\pi r^2}{2} = 4\pi$. Thus, $r = \sqrt{8}$. Therefore,

the area of $\triangle ABC = \dfrac{1}{2}\left(2\sqrt{8} \times 2\sqrt{8}\right) = 16$.

---

16. (E)

The first discounted price $= \left(1 - \dfrac{p}{100}\right)y$

The final price $= \left(1 + \dfrac{q}{100}\right)\left(1 - \dfrac{p}{100}\right)y = x$

Therefore,

$y = \dfrac{x}{\left(1 + \dfrac{q}{100}\right)\left(1 - \dfrac{p}{100}\right)} = \dfrac{10000x}{(100-p)(100+q)}$

$\boxed{END}$

Dr. John Chung's SAT Math

TEST
6

ANSWER SHEET   TEST #:

SECTION 3

1 Ⓐ Ⓑ Ⓒ Ⓓ Ⓔ	11 Ⓐ Ⓑ Ⓒ Ⓓ Ⓔ	21 Ⓐ Ⓑ Ⓒ Ⓓ Ⓔ	31 Ⓐ Ⓑ Ⓒ Ⓓ Ⓔ
2 Ⓐ Ⓑ Ⓒ Ⓓ Ⓔ	12 Ⓐ Ⓑ Ⓒ Ⓓ Ⓔ	22 Ⓐ Ⓑ Ⓒ Ⓓ Ⓔ	32 Ⓐ Ⓑ Ⓒ Ⓓ Ⓔ
3 Ⓐ Ⓑ Ⓒ Ⓓ Ⓔ	13 Ⓐ Ⓑ Ⓒ Ⓓ Ⓔ	23 Ⓐ Ⓑ Ⓒ Ⓓ Ⓔ	33 Ⓐ Ⓑ Ⓒ Ⓓ Ⓔ
4 Ⓐ Ⓑ Ⓒ Ⓓ Ⓔ	14 Ⓐ Ⓑ Ⓒ Ⓓ Ⓔ	24 Ⓐ Ⓑ Ⓒ Ⓓ Ⓔ	34 Ⓐ Ⓑ Ⓒ Ⓓ Ⓔ
5 Ⓐ Ⓑ Ⓒ Ⓓ Ⓔ	15 Ⓐ Ⓑ Ⓒ Ⓓ Ⓔ	25 Ⓐ Ⓑ Ⓒ Ⓓ Ⓔ	35 Ⓐ Ⓑ Ⓒ Ⓓ Ⓔ
6 Ⓐ Ⓑ Ⓒ Ⓓ Ⓔ	16 Ⓐ Ⓑ Ⓒ Ⓓ Ⓔ	26 Ⓐ Ⓑ Ⓒ Ⓓ Ⓔ	36 Ⓐ Ⓑ Ⓒ Ⓓ Ⓔ
7 Ⓐ Ⓑ Ⓒ Ⓓ Ⓔ	17 Ⓐ Ⓑ Ⓒ Ⓓ Ⓔ	27 Ⓐ Ⓑ Ⓒ Ⓓ Ⓔ	37 Ⓐ Ⓑ Ⓒ Ⓓ Ⓔ
8 Ⓐ Ⓑ Ⓒ Ⓓ Ⓔ	18 Ⓐ Ⓑ Ⓒ Ⓓ Ⓔ	28 Ⓐ Ⓑ Ⓒ Ⓓ Ⓔ	38 Ⓐ Ⓑ Ⓒ Ⓓ Ⓔ
9 Ⓐ Ⓑ Ⓒ Ⓓ Ⓔ	19 Ⓐ Ⓑ Ⓒ Ⓓ Ⓔ	29 Ⓐ Ⓑ Ⓒ Ⓓ Ⓔ	39 Ⓐ Ⓑ Ⓒ Ⓓ Ⓔ
10 Ⓐ Ⓑ Ⓒ Ⓓ Ⓔ	20 Ⓐ Ⓑ Ⓒ Ⓓ Ⓔ	30 Ⓐ Ⓑ Ⓒ Ⓓ Ⓔ	40 Ⓐ Ⓑ Ⓒ Ⓓ Ⓔ

SECTION 5

1 Ⓐ Ⓑ Ⓒ Ⓓ Ⓔ	11 Ⓐ Ⓑ Ⓒ Ⓓ Ⓔ	21 Ⓐ Ⓑ Ⓒ Ⓓ Ⓔ	31 Ⓐ Ⓑ Ⓒ Ⓓ Ⓔ
2 Ⓐ Ⓑ Ⓒ Ⓓ Ⓔ	12 Ⓐ Ⓑ Ⓒ Ⓓ Ⓔ	22 Ⓐ Ⓑ Ⓒ Ⓓ Ⓔ	32 Ⓐ Ⓑ Ⓒ Ⓓ Ⓔ
3 Ⓐ Ⓑ Ⓒ Ⓓ Ⓔ	13 Ⓐ Ⓑ Ⓒ Ⓓ Ⓔ	23 Ⓐ Ⓑ Ⓒ Ⓓ Ⓔ	33 Ⓐ Ⓑ Ⓒ Ⓓ Ⓔ
4 Ⓐ Ⓑ Ⓒ Ⓓ Ⓔ	14 Ⓐ Ⓑ Ⓒ Ⓓ Ⓔ	24 Ⓐ Ⓑ Ⓒ Ⓓ Ⓔ	34 Ⓐ Ⓑ Ⓒ Ⓓ Ⓔ
5 Ⓐ Ⓑ Ⓒ Ⓓ Ⓔ	15 Ⓐ Ⓑ Ⓒ Ⓓ Ⓔ	25 Ⓐ Ⓑ Ⓒ Ⓓ Ⓔ	35 Ⓐ Ⓑ Ⓒ Ⓓ Ⓔ
6 Ⓐ Ⓑ Ⓒ Ⓓ Ⓔ	16 Ⓐ Ⓑ Ⓒ Ⓓ Ⓔ	26 Ⓐ Ⓑ Ⓒ Ⓓ Ⓔ	36 Ⓐ Ⓑ Ⓒ Ⓓ Ⓔ
7 Ⓐ Ⓑ Ⓒ Ⓓ Ⓔ	17 Ⓐ Ⓑ Ⓒ Ⓓ Ⓔ	27 Ⓐ Ⓑ Ⓒ Ⓓ Ⓔ	37 Ⓐ Ⓑ Ⓒ Ⓓ Ⓔ
8 Ⓐ Ⓑ Ⓒ Ⓓ Ⓔ	18 Ⓐ Ⓑ Ⓒ Ⓓ Ⓔ	28 Ⓐ Ⓑ Ⓒ Ⓓ Ⓔ	38 Ⓐ Ⓑ Ⓒ Ⓓ Ⓔ
9 Ⓐ Ⓑ Ⓒ Ⓓ Ⓔ	19 Ⓐ Ⓑ Ⓒ Ⓓ Ⓔ	29 Ⓐ Ⓑ Ⓒ Ⓓ Ⓔ	39 Ⓐ Ⓑ Ⓒ Ⓓ Ⓔ
10 Ⓐ Ⓑ Ⓒ Ⓓ Ⓔ	20 Ⓐ Ⓑ Ⓒ Ⓓ Ⓔ	30 Ⓐ Ⓑ Ⓒ Ⓓ Ⓔ	40 Ⓐ Ⓑ Ⓒ Ⓓ Ⓔ

9 10 11 12 13

14 15 16 17 18

1	Ⓐ Ⓑ Ⓒ Ⓓ Ⓔ	11	Ⓐ Ⓑ Ⓒ Ⓓ Ⓔ	21	Ⓐ Ⓑ Ⓒ Ⓓ Ⓔ	31	Ⓐ Ⓑ Ⓒ Ⓓ Ⓔ
2	Ⓐ Ⓑ Ⓒ Ⓓ Ⓔ	12	Ⓐ Ⓑ Ⓒ Ⓓ Ⓔ	22	Ⓐ Ⓑ Ⓒ Ⓓ Ⓔ	32	Ⓐ Ⓑ Ⓒ Ⓓ Ⓔ
3	Ⓐ Ⓑ Ⓒ Ⓓ Ⓔ	13	Ⓐ Ⓑ Ⓒ Ⓓ Ⓔ	23	Ⓐ Ⓑ Ⓒ Ⓓ Ⓕ	33	Ⓐ Ⓑ Ⓒ Ⓓ Ⓔ
4	Ⓐ Ⓑ Ⓒ Ⓓ Ⓔ	14	Ⓐ Ⓑ Ⓒ Ⓓ Ⓔ	24	Ⓐ Ⓑ Ⓒ Ⓓ Ⓔ	34	Ⓐ Ⓑ Ⓒ Ⓓ Ⓔ
5	Ⓐ Ⓑ Ⓒ Ⓓ Ⓔ	15	Ⓐ Ⓑ Ⓒ Ⓓ Ⓔ	25	Ⓐ Ⓑ Ⓒ Ⓓ Ⓔ	35	Ⓐ Ⓑ Ⓒ Ⓓ Ⓔ
6	Ⓐ Ⓑ Ⓒ Ⓓ Ⓔ	16	Ⓐ Ⓑ Ⓒ Ⓓ Ⓔ	26	Ⓐ Ⓑ Ⓒ Ⓓ Ⓔ	36	Ⓐ Ⓑ Ⓒ Ⓓ Ⓔ
7	Ⓐ Ⓑ Ⓒ Ⓓ Ⓔ	17	Ⓐ Ⓑ Ⓒ Ⓓ Ⓔ	27	Ⓐ Ⓑ Ⓒ Ⓓ Ⓔ	37	Ⓐ Ⓑ Ⓒ Ⓓ Ⓔ
8	Ⓐ Ⓑ Ⓒ Ⓓ Ⓔ	18	Ⓐ Ⓑ Ⓒ Ⓓ Ⓔ	28	Ⓐ Ⓑ Ⓒ Ⓓ Ⓔ	38	Ⓐ Ⓑ Ⓒ Ⓓ Ⓔ
9	Ⓐ Ⓑ Ⓒ Ⓓ Ⓔ	19	Ⓐ Ⓑ Ⓒ Ⓓ Ⓔ	29	Ⓐ Ⓑ Ⓒ Ⓓ Ⓔ	39	Ⓐ Ⓑ Ⓒ Ⓓ Ⓔ
10	Ⓐ Ⓑ Ⓒ Ⓓ Ⓔ	20	Ⓐ Ⓑ Ⓒ Ⓓ Ⓔ	30	Ⓐ Ⓑ Ⓒ Ⓓ Ⓔ	40	Ⓐ Ⓑ Ⓒ Ⓓ Ⓔ

Math Scoring Worksheet

A. Section 3

    _____   _____
    numer of correct   *number of incorrect*

    +   +

B. Section 5 (1-8)

    _____   _____
    numer of correct   *number of incorrect*

    +

C. Section 5 (9-18)

    _____
    numer of correct

    +   +

D. Section 7

    _____   _____
    numer of correct   *number of incorrect*

    =   =

E. Total Unrounded Raw Score

    _____ − _____ ÷4 = _____
    numer of correct   *number of incorrect*

F. Total Rounded Raw Score   _____ (See table)

           Math Score Range = | _____ — _____ |

Math Conversion Table

Raw Score	Scaled Score	Raw Score	Scaled Score
54	800	23	490-550
53	780-800	22	480-540
52	760-800	21	470-530
51	740-800	20	460-520
50	720-780	19	450-510
49	700-760	18	450-510
48	690-750	17	440-500
47	680-740	16	430-490
46	670-730	15	420-480
45	660-720	14	420-480
44	650-710	13	410-470
43	650-710	12	400-460
42	640-700	11	390-450
41	630-690	10	380-440
40	620-680	9	390-430
39	610-670	8	380-420
38	610-670	7	370-410
37	600-660	6	360-400
36	590-650	5	340-380
35	580-640	4	320-370
34	570-630	3	310-360
33	560-620	2	300-350
32	560-620	1	270-320
31	550-610	0	240-300
30	540-600	-1	200-290
29	530-590	-2	200-270
28	530-590	-3	200-260
27	520-580	-4	200-240
26	510-570	-5	200-220
25	500-560	-6 and below	200
24	500-560		

SECTION 3
Time- 25 minutes
20 Questions

Turn to Section 3 (Page 1) of your answer sheet to answer the questions in this section.

Directions: For this section, solve each problem and decide which is the best of the choices given. Fill in the corresponding circle on the answer sheet. You may use any available space for scratchwork.

Notes

1. The use of a calculator is permitted.
2. All numbers used are real numbers.
3. Figures that accompany problems in this test are intended to provide information useful in solving the problems. They are drawn as accurately as possible EXCEPT when it is stated in a specific problem that the figure is not drawn to scale. All figure lie in a plane unless other indicated.
4. Unless otherwise specified, the domain of any function f is assumed to be set of all real numbers x for which $f(x)$ is a real number.

Reference Informatiom

$A = \pi r^2$
$C = 2\pi r$ $A = \ell w$ $A = \frac{1}{2}bh$ $V = \ell wh$ $V = \pi r^2 h$ $c^2 = a^2 + b^2$ Special Right Triangles

The numbers of degrees of arc in a circle is $360°$.
The sum of the measures in degrees of the angles is $180°$.

---

1. Which of the following points (x, y) does NOT satisfy $|5 - x| > y - 5$?

   (A) $(12, 10)$
   (B) $(11, 10)$
   (C) $(6, 5)$
   (D) $(2, 3)$
   (E) $(4, 7)$

2. If x percent of x is equal to a^4, what is the value of x in terms of a?

   (A) $10a^2$
   (B) $8a^2$
   (C) $6a^2$
   (D) $8a$
   (E) $10a$

GO ON TO THE NEXT PAGE

3. If $(k-1)^2 < 225$, where k is a positive integer, what is the greatest possible value of k ?

(A) 12
(B) 13
(C) 14
(D) 15
(E) 16

PROFIT FROM BOOK SALE

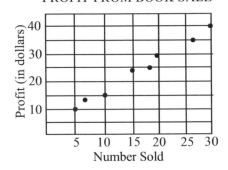

4. Based on the scatter plot above, which function best models the relationship between n , the number of books, and p , the profit realized from sales?

(A) $p(n) = 20$

(B) $p(n) = 2n + 10$

(C) $p(n) = 2n + 20$

(D) $p(n) = 1.2n + 5$

(E) $p(n) = 40 - 2n$

5. Five years ago the ratio of Alice's to Bernard's age was 1:4. 10 years from now, the ratio of Alice's to Bernard's age will be 2:3. How old is Bernard?

(A) 15
(B) 17
(C) 25
(D) 27
(E) 30

Set $A= \{x | x = \text{prime number}\}$

Set $B= \{x | x = 2n + 1, n = \text{positive integer}\}$

6. Which of the following is an element of both set A and set B shown above?

(A) 2
(B) 9
(C) 13
(D) 15
(E) 33

GO ON TO THE NEXT PAGE

7. Two farmers, working together, can plow a field in 3 days. One farmer working alone can plow the field in 4 days. How many days will it take the second farmer, working alone, to plow the entire field?

(A) 4
(B) 6
(C) 8
(D) 10
(E) 12

Note: Figure not drawn to scale.

8. In the figure above, if

$$AB = \frac{1}{2}BC = \frac{1}{3}CD = \frac{1}{4}DE$$, what is the ratio

of the length of $\overline{DE}$ to the length of $\overline{BD}$?

(A) $\frac{1}{2}$

(B) $\frac{2}{5}$

(C) $\frac{3}{4}$

(D) $\frac{4}{5}$

(E) $\frac{5}{6}$

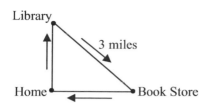

9. Bernard began to ride a bicycle slowly to the town library, and he spent 30 minutes in the library, and then rode quickly to the book store to buy a novel. If after 10 minutes, he began to ride home slowly again, which of the following could possibly represent the trip?

(A)

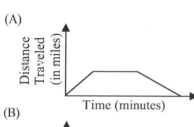

(B)

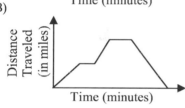

(C)

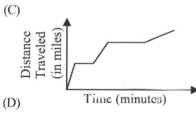

(D)

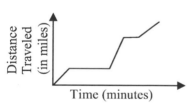

(E)

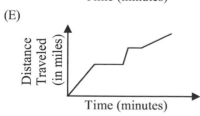

GO ON TO THE NEXT PAGE ⟩

10. If each of 8 boys played a game of chess with each of 6 girls, and then each girl played a game with each of the other girls, which of the following could be the total number of games played?

(A) 63
(B) 65
(C) 69
(D) 75
(E) 78

12. Let the function f be defined by

$f(x) = x^2 - 2x + 1$ for all numbers x. If $f(a) = 4$, which of the following could be the value of a?

(A) -1
(B) 0
(C) 1
(D) 2
(E) 4

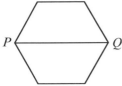

11. In the regular hexagon above, the length of a side of the hexagon is 10. What is the length of $\overline{PQ}$?

(A) 15
(B) 20
(C) $20\sqrt{2}$
(D) $20\sqrt{3}$
(E) 25

13. In the figure above, the part of a circle has an area of 30π and its center at point O. If the measure of $\angle AOB$ is 60°, what is the perimeter of the figure?

(A) $12 + 2\pi$
(B) $12 + 5\pi$
(C) $12 + 10\pi$
(D) $24 - 5\pi$
(E) $24 - 10\pi$

GO ON TO THE NEXT PAGE

14. If a and b are positive integers, and $(a-1)(b+1) = 0$, which of the following MUST be true?

(A) $a > b$
(B) $a < b$
(C) $a = b$
(D) $a = 1$
(E) $b = 1$

Note: Figure not drawn to scale.

16. In the figure above, $x > 75$ and $z = y - 30$. If y is an integer, which of the following is the greatest possible value of y ?

(A) 114
(B) 72
(C) 67
(D) 61
(E) 60

15. If $-3 < k < 3$, which of the following must be true?

(A) $k^2 < 3$
(B) $k^2 > 3$
(C) $|k| > 3$
(D) $k^2 > 9$
(E) $k^2 < 9$

17. For all integers p and q, let $\boxtimes$ be defined by $p \boxtimes q = p^2 + q^2 + 2pq$. Which of the following CANNOT be the value of $p \boxtimes q$?

(A) 0
(B) 1
(C) 2
(D) 4
(E) 9

GO ON TO THE NEXT PAGE

18. In the *xy*-plane, the equation of line ℓ is

$y = -\dfrac{3}{5}x + 4$. If line *m* is a reflection of line ℓ

over the *x*-axis, what is the equation of line *m* ?

(A) $y = \dfrac{3}{5}x + 4$

(B) $y = \dfrac{5}{3}x + 4$

(C) $y = -\dfrac{3}{5}x - 4$

(D) $y = \dfrac{3}{5}x - 4$

(E) $y = \dfrac{5}{3}x - 4$

19. If $x - 1 = kx$, for which of the following values of k will the equation have NO solution?

(A) -1
(B) 0
(C) 1
(D) 2
(E) 3

MATH TEST RESULTS

Test Scores	Number of Students
100	3
95	10
90	20
85	30
70	17
less than 70	20

20. The table above shows the results for 100 students on a Math test. Which of the following can be determined from the information in the table?

   I. The average (arithmetic mean) score of the test.
  II. The median score of the test.
 III. The mode of the test

(A) I only
(B) II only
(C) I and II only
(D) II and III only
(E) I, II, and III

STOP

If you finish before time is called, you may check your work on this section only.
Do not turn to any other section in the test.

SECTION 5
Time- 25 minutes
18 Questions

Turn to Section 5 (Page 1) of your answer sheet to answer the questions in this section.

Directions: For this section, solve each problem and decide which is the best of the choices given. Fill in the corresponding circle on the answer sheet. You may use any available space for scratchwork.

Notes

1. The use of a calculator is permitted.
2. All numbers used are real numbers.
3. Figures that accompany problems in this test are intended to provide information useful in solving the problems. They are drawn as accurately as possible EXCEPT when it is stated in a specific problem that the figure is not drawn to scale. All figure lie in a plane unless other indicated.
4. Unless otherwise specified, the domain of any function f is assumed to be set of all real numbers x for which $f(x)$ is a real number.

Reference Informatiom

$A = \pi r^2$
$C = 2\pi r$
$A = \ell w$
$A = \dfrac{1}{2}bh$
$V = \ell wh$
$V - \pi r^2 h$
$c^2 = a^2 + b^2$
Special Right Triangles

The numbers of degrees of arc in a circle is $360°$.

The sum of the measures in degrees of the angles is $180°$.

---

1. If $xyz = 15$ and $\dfrac{x}{y} = 3$, what is the value of z ?

(A) $5y$

(B) $\dfrac{5y}{x}$

(C) $\dfrac{5}{y^2}$

(D) $\dfrac{15}{x^2}$

(E) $\dfrac{5}{xy}$

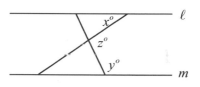

Note: Figure not drawn to scale

2. In the figure above, lines ℓ and m are parallel. If $x = 30$ and $y = 120$, what is the value of z ?

(A) 90
(B) 95
(C) 100
(D) 110
(E) 120

GO ON TO THE NEXT PAGE

TELEPHONE CHARGE

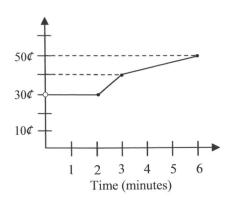

Time (minutes)

3. The figure above shows the graph of telephone charge as a function of the length of a call, in minutes. Which of the following is the most consistent with the information in the graph?

(A) The company charges the same rate for any length of a call.
(B) The company charges 15 ¢ per minute for the first minute.
(C) The company charges a minutely rate for the first minute.
(D) The company charges a fixed amount for the first two minutes or less. The company charges at a rate of 10 ¢ per minute between 2 and 3 minutes
(E) The company charges a fixed amount for the first two minutes or less. The company charges at a rate of 5 ¢ per minute beyond 3 minutes.

$$\left| k^2 - 2 \right| = 2$$

4. For how many integer values of k is the equation above true?

(A) One
(B) Two
(C) Three
(D) Four
(E) More than four

5. If k is a number between 0 and 1, which of the following must be true?

    I. $k > k^3$

    II. $k > \dfrac{1}{\sqrt{k}}$

    III. $k > \dfrac{1}{k}$

(A) I only
(B) II only
(C) III only
(D) I and II only
(E) I and III only

6. There are s students in an art class, and x percent of the students are boys. If y percent of the boys are qualified to advance toward the U.S Math Olympiad, which of the following could be the number of boys who does not advance to the USMO?

(A) $\dfrac{xys}{100}$

(B) $\dfrac{(x+y)s}{100}$

(C) $\dfrac{x(1-y)s}{10000}$

(D) $\dfrac{x(y-1)s}{10000}$

(E) $\dfrac{x(100-y)s}{10000}$

GO ON TO THE NEXT PAGE

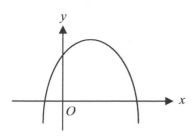

7. The quadratic function graphed above has an equation of $f(x) = ax^2 + bx + c$. Which of the following must be true?

(A) $a > 0$
(B) $b > 0$
(C) $b < 0$
(D) $c < 0$
(E) c is even.

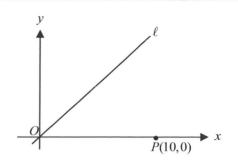

Note: Figure not drawn to scale.

8. The equation of line ℓ in the figure above is $y = 2x$. If line m (not shown) is perpendicular to line ℓ and contains point P, at what point does m intersect the y-axis?

(A) 5
(B) 7
(C) 7.5
(D) 8
(E) 10

GO ON TO THE NEXT PAGE

5 ▢▢▢ 5 ▢▢▢ 5 ▢▢ 5

Directions: For Students-Produced Response questions 9-18, use the grid at the bottom of the answer sheet page on which you have answered questions 1-8.

Each of the remaining 10 questions requires you to solve the problem and enter your answer by making the circles in the special grid, as shown in the examples below. You may use any available space for scratchwork.

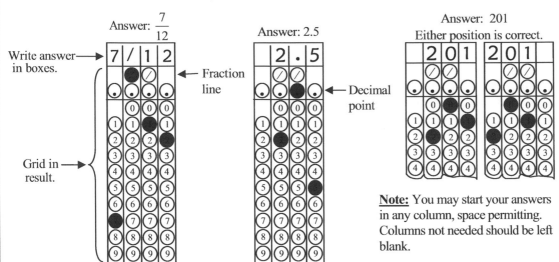

Answer: $\frac{7}{12}$

Write answer in boxes.

← Fraction line

Grid in result.

Answer: 2.5

← Decimal point

Answer: 201
Either position is correct.

Note: You may start your answers in any column, space permitting. Columns not needed should be left blank.

- Mark no more than one circle in any column.

- Because the answer sheet will be machine-scored, **you will receive credit only if the circles are filled in correctly.**

- Although not required, it is suggested that you write your answer in the boxes at the top of the columns to help you fill in the circles accurately.

- Some problems may have more than one correct answer. In such cases, grid only one answer.

- No question has a negative answer.

- **Mixed numbers** such as $3\frac{1}{2}$ must be gridded as 3.5 or 7/2. (If 3|1|/|2 is gridded, it will be interpreted as $\frac{31}{2}$, not $3\frac{1}{2}$.)

- **Decimal Answers:** If you obtain a decimal answer with more digits than the grid can accommodate, it may be either rounded or truncated, but it must fill the entire grid. For example, if you obtain an answer such as 0.6666..., you should record your result as .666 or .667. **A less accurate value such as .66 or .67 will be scored as incorrect.**

Acceptable ways to grid $\frac{2}{3}$ are:

9. The scale of a map for Bear Mountain National Park is 1.5 inches = 5 miles. The distance between Discovery point and Overlook on the map is about 5.1 inches. What is the distance between these two places in miles?

10. If $(x-5)y = x^2 - y^2 + 11$, what is the positive value of y when $x = 5$?

GO ON TO THE NEXT PAGE

11. A man can wear any combination of one of 5 shirts, one of 7 ties, and one of 3 pants. How many different outfits can he wear?

Ms. Evan's Class

English	30
Chemstry	25
Algebra	20

Figure 1

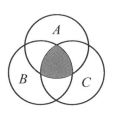

Figure 2

Note: Figure 2 not drawn to scale.

13. The figures above show the enrollment for three subjects in Ms. Evan's class in different ways. Of these students, 15 students take English and Chemistry, 10 students take Chemistry and Algebra, and 8 students take Algebra and English. If there are 47 students in the class, what is the number of students represented by the shaded region in figure 2?

12. The function f is defined by $f(x) = kx + 5$. If $f(3) = 15$, what is the value of k?

GO ON TO THE NEXT PAGE

14. If k is a positive integer, and $4\sqrt{k^2} - 3 = 13$, what is the value of k?

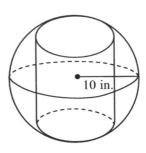

16. In the figure above, a cylinder is inscribed in a sphere with a radius of 10 inches. If the circular base of the cylinder has a radius of 6 inches, what is the length of the height of the cylinder?

15. If $(0.213 \times 10^k) + (9 \times 10^7) = 921.3 \times 10^5$, what is the value of k?

GO ON TO THE NEXT PAGE ⟩

17. If the value of k^{-5} is twice the value of $4k^{-2}$, what is the value of k ?

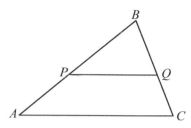

18. In the figure above, P is the midpoint of $\overline{AB}$ and Q is the midpoint of $\overline{BC}$. If the area of trapezoid $APQC$ is 20, what is the area of $\triangle ABC$?

STOP
If you finish before time is called, you may check your work on this section only.
Do not turn to any other section in the test.

SECTION 7
Time- 20 minutes
16 Questions

Turn to Section 7 (Page 2) of your answer sheet to answer the questions in this section.

Directions: For this section, solve each problem and decide which is the best of the choices given. Fill in the corresponding circle on the answer sheet. You may use any available space for scratchwork.

Notes

1. The use of a calculator is permitted.
2. All numbers used are real numbers.
3. Figures that accompany problems in this test are intended to provide information useful in solving the problems. They are drawn as accurately as possible EXCEPT when it is stated in a specific problem that the figure is not drawn to scale. All figure lie in a plane unless other indicated.
4. Unless otherwise specified, the domain of any function f is assumed to be set of all real numbers x for which $f(x)$ is a real number.

Reference Informatiom

$A = \pi r^2$
$C = 2\pi r$ $A = \ell w$ $A = \frac{1}{2}bh$ $V = \ell wh$ $V = \pi r^2 h$ $c^2 = a^2 + b^2$ Special Right Triangles

The numbers of degrees of arc in a circle is 360°.

The sum of the measures in degrees of the angles is 180°.

---

1. If $|10 - x^2| \geq 5$, which of the following is NOT a possible value of x?

(A) -2
(B) -1
(C) 0
(D) 1
(E) 3

2. In a class of 20 boys and 15 girls, 10 boys and 8 girls speak more than one language.
If a representative for student council is to be chosen at random from the class, what is the probability that the representative will be a boy who speaks more than one language?

(A) $\frac{10}{18}$

(B) $\frac{10}{20}$

(C) $\frac{2}{7}$

(D) $\frac{8}{35}$

(E) $\frac{10}{30}$

GO ON TO THE NEXT PAGE ⟩

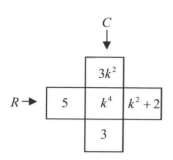

3. In the figure above, the sum of the numbers in row R is the same as the sum of the numbers in column C. What is the sum of the numbers in row R ?

(A) 15
(B) 14
(C) 13
(D) 12
(E) 10

$$y = -3x + 2$$

4. According to the equation above, if the value of x increases by 10, which of the following statements is true?

(A) The value of y will increase by 30.
(B) The value of y will increase by 28
(C) The value of y will decrease by 28
(D) The value of y will decrease by 30
(E) The value of y will decrease by 2

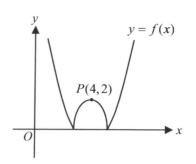

5. Which of the following functions best represents the graph shown above?

(A) $y = x^2 - 8x + 14$
(B) $y = x^2 + 8x + 14$
(C) $y = -x^2 + 8x + 14$
(D) $y = \left| -x^2 + 8x + 14 \right|$
(E) $y = \left| x^2 - 8x + 14 \right|$

6. If the average (arithmetic mean) of p, q, and 10 is 12, and the average of $2p$, q, and 7 is 13, what is the value of p ?

(A) 3
(B) 6
(C) 9
(D) 12
(E) 15

GO ON TO THE NEXT PAGE

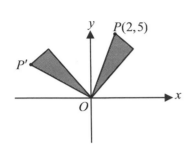

Note: Figure not drawn to scale.

7. In the figure above, the areas of the two shaded triangles are equal. If $\angle POP' = 90^o$, which of the following could be the coordinates of point P' ?

(A) $(2,-5)$
(B) $(-2,5)$
(C) $(-5,-2)$
(D) $(-5,2)$
(E) $(5,-2)$

8. The necklace above contains one bead of each color : red, yellow, green, white, blue, and purple. If yellow and red beads must be placed at either end of the necklace, how many different arrangements are possible?

(A) 48
(B) 36
(C) 24
(D) 16
(E) 12

9. Mary and Ned each have a bag that contains one marble of each of the colors blue, green, red, and white. Mary randomly selects one marble from her bag and puts it into Ned's bag, and then Ned randomly selects one marble from his bag and puts it into Mary's bag again. After this process, what is the probability that the contents of the two bags will be the same?

(A) $\dfrac{1}{25}$

(B) $\dfrac{1}{16}$

(C) $\dfrac{1}{5}$

(D) $\dfrac{2}{5}$

(E) $\dfrac{1}{4}$

10. The first term is 5, and each term after the first is 3 more than the preceding term, which of the following is an expression for nth term of the sequence for any positive integer n ?

(A) $5n$
(B) $4n+1$
(C) $3n+2$
(D) $2n+3$
(E) $n+4$

GO ON TO THE NEXT PAGE

$$x + 2y = 6$$
$$2x - ky = 5$$

11. For which of the following values for k will the two lines be perpendicular?

(A) -1
(B) 0
(C) 1
(D) 2
(E) 3

13. If $(k-1)x + 3k = ax + 6$ for all value of x, where k and a are constant, what is the value of a?

(A) 2
(B) 1
(C) 0
(D) -1
(E) -2

12. In the figure above, $ABCD$ is a square, and a circle is inscribed in the square. If a point is chosen from $ABCD$, what is the probability that the point will be from the shaded region?

(A) $\dfrac{\pi}{4}$

(B) $\dfrac{2}{\pi}$

(C) $\dfrac{2-\pi}{4}$

(D) $\dfrac{4-\pi}{4}$

(E) $\dfrac{8-\pi}{8}$

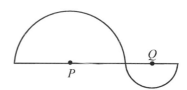

Note: Figure not drawn to scale.

14. In the figure above, the diameter of semicircle P is twice the diameter of semicircle Q. If the area of semicircle P is 12π, what is the area of semicircle Q?

(A) 8π
(B) 6π
(C) 3π
(D) 6
(E) 4

GO ON TO THE NEXT PAGE

15. A carton contains k boxes of paper cups, and each box contains 100 paper cups. If the carton cost x dollars, what is the cost per paper cup, in cents?

(A) kx

(B) $\dfrac{x}{k}$

(C) $\dfrac{k}{x}$

(D) $\dfrac{100k}{x}$

(E) $\dfrac{x}{100k}$

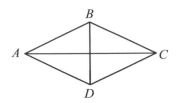

Note: Figure not drawn to scale.

16. In the figure above, $\triangle ABC$ and $\triangle ADC$ are isosceles triangles. If $AB = AD = BD = 10$, what is the ratio of BD to AC?

(A) $\dfrac{1}{\sqrt{2}}$

(B) $\dfrac{2}{\sqrt{2}}$

(C) $\dfrac{1}{\sqrt{3}}$

(D) $\dfrac{2}{\sqrt{3}}$

(E) $\dfrac{\sqrt{2}}{\sqrt{3}}$

STOP

If you finish before time is called, you may check your work on this section only.
Do not turn to any other section in the test.

NO MATERIAL ON THIS PAGE

TEST 6 ANSWER KEY

#	SECTION 3	SECTION 5	SECTION 7
1	E	C	E
2	A	A	C
3	D	D	C
4	D	C	D
5	B	A	E
6	C	E	B
7	E	B	D
8	D	A	A
9	D	17	D
10	A	6	C
11	B	105	C
12	A	$10/3$ or 3.33	D
13	C	5	B
14	D	4	C
15	E	7	B
16	C	16	C
17	C	$1/2$ or 0.5	
18	D	$80/3$ or 26.7	
19	C		
20	D		

TEST 6   SECTION 3

1. (E)
Substitution: For (4,7),
$|5-4| > 7-5 \Rightarrow 1 > 2$ (False)

2. (A)
Translation:
$\dfrac{x}{100} \times x = a^4 \Rightarrow x^2 = 100a^4 \Rightarrow x = 10a^2$

3. (D)
Since $k-1 < 15$, $k < 16$. Therefore, the greatest integer of k is 15.

4. (D)
Slope $\simeq \dfrac{40-10}{30-5} = 1.2$ and the y-intercept is around 5.

5. (B)

	←5	Now	→10
A	x		$x+15$
B	$4x$	?	$4x+15$

$\dfrac{x+15}{4x+15} = \dfrac{2}{3} \Rightarrow x = 3$. Therefore, Bernard is $4(3)+5 = 17$ years old.

6. (C)
13 is a prime number and also an odd integer.

7. (E)

	Days	Rate	Combined Rate
Farmer A	4	$1/4$	$\dfrac{1}{4} + \dfrac{1}{k} = \dfrac{1}{3}$
Farmer B	k	$1/k$	

- 252 -

Therefore, $k = 12$.

8. (D)

If $AB = k$, then $BC = 2k$, $CD = 3k$, and $DE = 4k$.

Therefore, $\dfrac{DE}{BD} = \dfrac{4k}{5k} = \dfrac{4}{5}$.

9. (D)

Graph D best represents the trip.

10. (A)

There are 48 games between 8 boys and 6 girls, and 15 games between the girls.
$8 \times 6 = 48$ and $5+4+3+2+1 = 15$ games.
Therefore $48 + 15 = 63$ games.

11. (B)

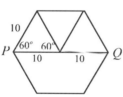

The triangles in the figure are equilateral triangles.
Therefore, $PQ = 20$.

12. (A)

Since $f(a) = a^2 - 2a + 1 = 4$, $a^2 - 2a - 3 = 0$.
Thus, $(a-3)(a+1) = 0$. Therefore, $a = 3$ or $a = -1$.

13. (C)

The area of the circle $= 30\pi \times \dfrac{360}{300} = 36\pi$

$\pi r^2 = 36\pi \Rightarrow r = 6$. Thus the length of

$\overset{\frown}{AB} = 12\pi \times \dfrac{300}{360} = 10\pi$. Therefore, the

perimeter is $10\pi + 12$.

14. (D)

Since $b + 1 > 0$, a must be -1.

15. (E)

$-3 < k < 3$ is equivalent to $|k| < 3$ or $k^2 < 9$

16. (C)

$x = 180 - y - z = 180 - y - (y-30) = 210 - 2y$

Thus,
$x = 210 - 2y > 75 \Rightarrow 2y < 135 \Rightarrow y < 67.5$.
Therefore, the greatest integer is 67.

Or, you know that all three angles are integers. To get the greatest of y, choose the smallest of $x(= 76)$. Then
$y + z = y + (y - 30) = 104 \rightarrow 2y = 134$
Therefore, the greatest $y = 67$.

17. (C)

$p \boxtimes q = p^2 + q^2 + 2pq = (p+q)^2 =$ squares of a number. 2 is not a square number.

18. (D)

The slope of line $m = \dfrac{3}{5}$ and y-intercept is

-4. Therefore, $y = \dfrac{3}{5}x - 4$.

19. (C)

$x - 1 = kx \Rightarrow x(1-k) = 1 \Rightarrow x = \dfrac{1}{1-k}$

If $k = 1$, it doesn't have a solution (undefined).

20. (D)

I. In order to have the average, accurate test scores are needed ("less than 70" is not accurate)
II. The median is 85.
III. The mode is 85.

TEST 6   **SECTION 5**

1. (C)

Since $z = \dfrac{15}{xy}$ and $x = 3y$, then

$z = \dfrac{15}{3y \cdot y} = \dfrac{5}{y^2}$.

2. (A)

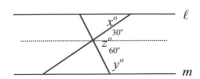

In the figure above, $z = 30 + 60 = 90$.

---

3. (D)

---

4. (C)

Since $|k^2 - 2| = 2$, $k^2 - 2 = 2$, -2. Thus $k^2 = 4, 0$. Therefore, $k = 2, -2$, and 0.

---

5. (A)

Since $0 < k < 1$, choose $k = \dfrac{1}{2}$.

$k > k^3$ (True), $k > \dfrac{1}{\sqrt{k}}$ (False),

$k > \dfrac{1}{k}$ (False)

---

6. (E)

Boys $= \dfrac{x}{100} \times s$ and Boys advanced USMO $=$

$\dfrac{y}{100}\left(\dfrac{x}{100} \times s\right)$. Therefore, boys not advanced

USMO $=$

$\dfrac{xs}{100} - \dfrac{xys}{10000} = \dfrac{100xs - xys}{10000} = \dfrac{xs(100-y)}{10000}$.

---

7. (B)

Axis of symmetry $= -\dfrac{b}{a} > 0$, where $a < 0$.

$f(0) = c > 0$. Therefore, $b > 0$.

---

8. (A)

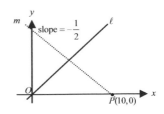

In the plane, the equation of line m is,

$y = -\dfrac{1}{2}x + b$, where b is y-intercept. Point

P lies on the line, $0 = -\dfrac{1}{2}(10) + b$.

Therefore, $b = 5$.

Or, the slope between $(0, b)$ and $(10, 0)$ is

$-\dfrac{1}{2}$. Therefore, $\dfrac{b-0}{0-10} = \dfrac{-1}{2}$.

$2b = 10 \rightarrow b = 5$

---

9. 17

Proportion: $\dfrac{1.5}{5} = \dfrac{5.1}{x} \Rightarrow x = 17$.

---

10. 6

When $x = 5$, $25 - y^2 + 11 = 0$.

$y^2 = 36 \Rightarrow y = 6$.

---

11. 105

$5 \times 7 \times 3 = 105$.

---

12. $10/3$ or 3.33

$f(3) = 3k + 5 = 15 \Rightarrow k = \dfrac{10}{3}$.

---

13. 5

$n(A \cup B \cup C) = n(A) + n(B) + n(C) - n(A \cap B)$
$- n(B \cap C) - n(C \cap A) + n(A \cap B \cap C)$
$30 + 25 + 20 - (15 + 8 + 10) + x = 47. \quad x = 5.$

---

14. 4

$\sqrt{k^2} = 4 \Rightarrow k^2 = 16 \Rightarrow k = 4$.

---

15. 7

$0.213 \times 10^k + 900 \times 10^5 = 21.3 \times 10^5$.

Therefore $0.213 \times 10^k = 21.3 \times 10^5 \rightarrow k = 7$.

---

16. 16

From the view of the front,
$h = \sqrt{20^2 - 12^2} = \sqrt{256} = 16$

17. $\dfrac{1}{2}$ or 0.5

$k^{-5} = 2\left(4k^{-2}\right) = 8k^{-2}. \Rightarrow \dfrac{k^{-5}}{k^{-2}} = 8 \Rightarrow k^{-3} = 8$.

Therefore, $k^3 = \dfrac{1}{8} \Rightarrow k = \dfrac{1}{2}$.

18. $\dfrac{80}{3}$ or 26.7

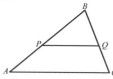

Since $\dfrac{\text{Area of } \triangle PBQ}{\text{Area of } \triangle ABC} = \dfrac{1}{4} = \dfrac{k}{4k}$,

the area of the trapezoid is $3k$. Thus,

$3k = 20 \Rightarrow k = \dfrac{20}{3}$. Therefore, the area of

$\triangle ABC = 4k = 4\left(\dfrac{20}{3}\right) = \dfrac{80}{3}$.

TEST 6   SECTION 7

1. (E)
When $x = 3$, $\left|10 - 3^2\right| = 1 \geq 5$ (False)

2. (C)
$P = \dfrac{10 \text{ boys}}{35} = \dfrac{2}{7}$.

3. (C)
In the figure, $5 + k^2 + 2 = 3k^2 + 3$. That is,
$k^2 = 2$. Therefore,
$R \to 5 + k^4 + k^2 + 2 = 5 + 4 + 2 + 2 = 13$.

4. (D)
Slope $= \dfrac{\Delta y}{\Delta x} = \dfrac{\Delta y}{10} = -3$. Therefore,
$\Delta y = -30$.

5. (E)
Axis of symmetry $= -4$. (E) is correct.

Or, check which one has $y = 2$ when $x = 4$.
(E) $\left|4^2 - 8(4) + 14\right| = 2$

6. (B)
$\dfrac{p + q + 10}{3} = 12$ and $\dfrac{2p + q + 7}{3} = 13$.
$p + q = 26 --(1)$ and $2p + q = 32 --(2)$
From (1) and (2), $p = 6$.

7. (D)
Counter clockwise rotation:
$(x, y) \xrightarrow{\;90^o\;} (-y, x)$. Therefore,
$(2, 5) \to (-5, 2)$
Or,

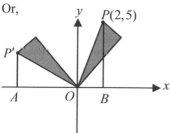

Hence $\triangle POB \cong \triangle P'OA$, $OA = 5$ and
$AP' = 2$. Therefore the coordinates are
$P'(-5, 2)$

8. (A)
$YOOOOR \Rightarrow 4 \times 3 \times 2 \times 1 = 24$
$ROOOOY \Rightarrow 4 \times 3 \times 2 \times 1 = 24$
Therefore, $24 + 24 = 48$.

9. (D)

Mary	Ned
b	b
g	g
r	r
	w
	w

From the figure above, in order to be the
same, Ned must choose white. Therefore,
$P = \dfrac{2}{5}$.

Or, if you choose a white, then you have to
choose a white again from the Ned's bag.

$\frac{1}{4} \times \frac{2}{5} = \frac{1}{10}$. And it is same probability for

other colors. Therefore $\frac{1}{10} \times 4 = \frac{2}{5}$.

---

10. (C)

Since $a_1 = 5$ and $d = 3$, $a_n = a_1 + (n-1)d$.

Therefore, $a_n = 5 + (n-1)3 = 3n + 2$.

Or, check all choices using the numbers.

---

11. (C)

Slope (1) $= -\frac{1}{2}$ $\Rightarrow$ $y = -\frac{1}{2}x + 3$

Slope (2) $= \frac{2}{k}$ $\Rightarrow$ $y = \frac{2}{k}x - \frac{5}{k}$

Perpendicular: $\frac{2}{k} = 2$. Therefore, $k = 1$.

---

12. (D)

Geometric Probability:

If $AB = 2$, the area of the circle $= \pi$ and the area of the square $= 4$. Therefore,

$P = \dfrac{\text{Area of the shaded region}}{\text{Entire area}} = \dfrac{4 - \pi}{4}$.

---

13. (B)

Identical equation: $k - 1 = a$ and

$3k = 6. \Rightarrow k = 2$.

Therefore, $2 - 1 = a$. $a = 1$.

---

14. (C)

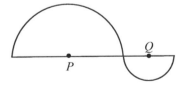

Since the ratio of the lengths of

$\dfrac{\text{semicircle P}}{\text{semicircle Q}} = \dfrac{1}{2}$,

The ratio of the areas $= \dfrac{1}{4}$. Therefore,

$\dfrac{x}{12\pi} = \dfrac{1}{4} \Rightarrow x = 3\pi$.

---

15. (B)

There are $100k$ paper cups. Therefore, the price for the one paper

cup$= \dfrac{100x}{100k}$ cents$= \dfrac{x}{k}$ cents.

---

16. (C)

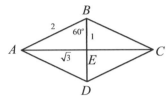

Let $AB = 2$, then $BD = 2$ and $AE = \sqrt{3}$.

$AC = 2\sqrt{3}$.

In the figure above, $\dfrac{BD}{AC} = \dfrac{2}{2\sqrt{3}} = \dfrac{1}{\sqrt{3}}$.

$\boxed{END}$

Dr. John Chung's SAT Math

TEST
7

SECTION 3

1 Ⓐ Ⓑ Ⓒ Ⓓ Ⓔ	11 Ⓐ Ⓑ Ⓒ Ⓓ Ⓔ	21 Ⓐ Ⓑ Ⓒ Ⓓ Ⓔ	31 Ⓐ Ⓑ Ⓒ Ⓓ Ⓔ
2 Ⓐ Ⓑ Ⓒ Ⓓ Ⓔ	12 Ⓐ Ⓑ Ⓒ Ⓓ Ⓔ	22 Ⓐ Ⓑ Ⓒ Ⓓ Ⓔ	32 Ⓐ Ⓑ Ⓒ Ⓓ Ⓔ
3 Ⓐ Ⓑ Ⓒ Ⓓ Ⓔ	13 Ⓐ Ⓑ Ⓒ Ⓓ Ⓔ	23 Ⓐ Ⓑ Ⓒ Ⓓ Ⓔ	33 Ⓐ Ⓑ Ⓒ Ⓓ Ⓔ
4 Ⓐ Ⓑ Ⓒ Ⓓ Ⓔ	14 Ⓐ Ⓑ Ⓒ Ⓓ Ⓔ	24 Ⓐ Ⓑ Ⓒ Ⓓ Ⓔ	34 Ⓐ Ⓑ Ⓒ Ⓓ Ⓔ
5 Ⓐ Ⓑ Ⓒ Ⓓ Ⓔ	15 Ⓐ Ⓑ Ⓒ Ⓓ Ⓔ	25 Ⓐ Ⓑ Ⓒ Ⓓ Ⓔ	35 Ⓐ Ⓑ Ⓒ Ⓓ Ⓔ
6 Ⓐ Ⓑ Ⓒ Ⓓ Ⓔ	16 Ⓐ Ⓑ Ⓒ Ⓓ Ⓔ	26 Ⓐ Ⓑ Ⓒ Ⓓ Ⓔ	36 Ⓐ Ⓑ Ⓒ Ⓓ Ⓔ
7 Ⓐ Ⓑ Ⓒ Ⓓ Ⓔ	17 Ⓐ Ⓑ Ⓒ Ⓓ Ⓔ	27 Ⓐ Ⓑ Ⓒ Ⓓ Ⓔ	37 Ⓐ Ⓑ Ⓒ Ⓓ Ⓔ
8 Ⓐ Ⓑ Ⓒ Ⓓ Ⓔ	18 Ⓐ Ⓑ Ⓒ Ⓓ Ⓔ	28 Ⓐ Ⓑ Ⓒ Ⓓ Ⓔ	38 Ⓐ Ⓑ Ⓒ Ⓓ Ⓔ
9 Ⓐ Ⓑ Ⓒ Ⓓ Ⓔ	19 Ⓐ Ⓑ Ⓒ Ⓓ Ⓔ	29 Ⓐ Ⓑ Ⓒ Ⓓ Ⓔ	39 Ⓐ Ⓑ Ⓒ Ⓓ Ⓔ
10 Ⓐ Ⓑ Ⓒ Ⓓ Ⓔ	20 Ⓐ Ⓑ Ⓒ Ⓓ Ⓔ	30 Ⓐ Ⓑ Ⓒ Ⓓ Ⓔ	40 Ⓐ Ⓑ Ⓒ Ⓓ Ⓔ

SECTION 5

1 Ⓐ Ⓑ Ⓒ Ⓓ Ⓔ	11 Ⓐ Ⓑ Ⓒ Ⓓ Ⓔ	21 Ⓐ Ⓑ Ⓒ Ⓓ Ⓔ	31 Ⓐ Ⓑ Ⓒ Ⓓ Ⓔ
2 Ⓐ Ⓑ Ⓒ Ⓓ Ⓔ	12 Ⓐ Ⓑ Ⓒ Ⓓ Ⓔ	22 Ⓐ Ⓑ Ⓒ Ⓓ Ⓔ	32 Ⓐ Ⓑ Ⓒ Ⓓ Ⓔ
3 Ⓐ Ⓑ Ⓒ Ⓓ Ⓔ	13 Ⓐ Ⓑ Ⓒ Ⓓ Ⓔ	23 Ⓐ Ⓑ Ⓒ Ⓓ Ⓔ	33 Ⓐ Ⓑ Ⓒ Ⓓ Ⓔ
4 Ⓐ Ⓑ Ⓒ Ⓓ Ⓔ	14 Ⓐ Ⓑ Ⓒ Ⓓ Ⓔ	24 Ⓐ Ⓑ Ⓒ Ⓓ Ⓔ	34 Ⓐ Ⓑ Ⓒ Ⓓ Ⓔ
5 Ⓐ Ⓑ Ⓒ Ⓓ Ⓔ	15 Ⓐ Ⓑ Ⓒ Ⓓ Ⓔ	25 Ⓐ Ⓑ Ⓒ Ⓓ Ⓔ	35 Ⓐ Ⓑ Ⓒ Ⓓ Ⓔ
6 Ⓐ Ⓑ Ⓒ Ⓓ Ⓔ	16 Ⓐ Ⓑ Ⓒ Ⓓ Ⓔ	26 Ⓐ Ⓑ Ⓒ Ⓓ Ⓔ	36 Ⓐ Ⓑ Ⓒ Ⓓ Ⓔ
7 Ⓐ Ⓑ Ⓒ Ⓓ Ⓔ	17 Ⓐ Ⓑ Ⓒ Ⓓ Ⓔ	27 Ⓐ Ⓑ Ⓒ Ⓓ Ⓔ	37 Ⓐ Ⓑ Ⓒ Ⓓ Ⓔ
8 Ⓐ Ⓑ Ⓒ Ⓓ Ⓔ	18 Ⓐ Ⓑ Ⓒ Ⓓ Ⓔ	28 Ⓐ Ⓑ Ⓒ Ⓓ Ⓔ	38 Ⓐ Ⓑ Ⓒ Ⓓ Ⓔ
9 Ⓐ Ⓑ Ⓒ Ⓓ Ⓔ	19 Ⓐ Ⓑ Ⓒ Ⓓ Ⓔ	29 Ⓐ Ⓑ Ⓒ Ⓓ Ⓔ	39 Ⓐ Ⓑ Ⓒ Ⓓ Ⓔ
10 Ⓐ Ⓑ Ⓒ Ⓓ Ⓔ	20 Ⓐ Ⓑ Ⓒ Ⓓ Ⓔ	30 Ⓐ Ⓑ Ⓒ Ⓓ Ⓔ	40 Ⓐ Ⓑ Ⓒ Ⓓ Ⓔ

Grid-in response boxes numbered 9, 10, 11, 12, 13, 14, 15, 16, 17, 18. Each contains a four-column grid with fraction bar (/) and decimal point (.) options in the top rows, followed by digits 0–9 in each column.

1	ⒶⒷⒸⒹⒺ	11	ⒶⒷⒸⒹⒺ	21	ⒶⒷⒸⒹⒺ	31	ⒶⒷⒸⒹⒺ
2	ⒶⒷⒸⒹⒺ	12	ⒶⒷⒸⒹⒺ	22	ⒶⒷⒸⒹⒺ	32	ⒶⒷⒸⒹⒺ
3	ⒶⒷⒸⒹⒺ	13	ⒶⒷⒸⒹⒺ	23	ⒶⒷⒸⒹⒺ	33	ⒶⒷⒸⒹⒺ
4	ⒶⒷⒸⒹⒺ	14	ⒶⒷⒸⒹⒺ	24	ⒶⒷⒸⒹⒺ	34	ⒶⒷⒸⒹⒺ
5	ⒶⒷⒸⒹⒺ	15	ⒶⒷⒸⒹⒺ	25	ⒶⒷⒸⒹⒺ	35	ⒶⒷⒸⒹⒺ
6	ⒶⒷⒸⒹⒺ	16	ⒶⒷⒸⒹⒺ	26	ⒶⒷⒸⒹⒺ	36	ⒶⒷⒸⒹⒺ
7	ⒶⒷⒸⒹⒺ	17	ⒶⒷⒸⒹⒺ	27	ⒶⒷⒸⒹⒺ	37	ⒶⒷⒸⒹⒺ
8	ⒶⒷⒸⒹⒺ	18	ⒶⒷⒸⒹⒺ	28	ⒶⒷⒸⒹⒺ	38	ⒶⒷⒸⒹⒺ
9	ⒶⒷⒸⒹⒺ	19	ⒶⒷⒸⒹⒺ	29	ⒶⒷⒸⒹⒺ	39	ⒶⒷⒸⒹⒺ
10	ⒶⒷⒸⒹⒺ	20	ⒶⒷⒸⒹⒺ	30	ⒶⒷⒸⒹⒺ	40	ⒶⒷⒸⒹⒺ

Math Scoring Worksheet

A. Section 3
_____   _____
numer of correct   number of incorrect

| |

B. Section 5 (1-8)
_____   _____
numer of correct   number of incorrect

+

C. Section 5 (9-18)
_____
numer of correct

+   +

D. Section 7
_____   _____
numer of correct   number of incorrect

=   =

E. Total Unrounded Raw Score
_____ − _____ ÷4 = _____
numer of correct   number of incorrect

F. Total Rounded Raw Score
_____ (See table)

Math Score Range = | — |

Math Conversion Table

Raw Score	Scaled Score	Raw Score	Scaled Score
54	800	23	490-550
53	780-800	22	480-540
52	760-800	21	470-530
51	740-800	20	460-520
50	720-780	19	450-510
49	700-760	18	450-510
48	690-750	17	440-500
47	680-740	16	430-490
46	670-730	15	420-480
45	660-720	14	420-480
44	650-710	13	410-470
43	650-710	12	400-460
42	640-700	11	390-450
41	630-690	10	380-440
40	620-680	9	390-430
39	610-670	8	380-420
38	610-670	7	370-410
37	600-660	6	360-400
36	590-650	5	340-380
35	580-640	4	320-370
34	570-630	3	310-360
33	560-620	2	300-350
32	560-620	1	270-320
31	550-610	0	240-300
30	540-600	-1	200-290
29	530-590	-2	200-270
28	530-590	-3	200-260
27	520-580	-4	200-240
26	510-570	-5	200-220
25	500-560	-6 and below	200
24	500-560		

SECTION 3
Time- 25 minutes
20 Questions

Turn to Section 3 (Page 1) of your answer sheet to answer the questions in this section.

Directions: For this section, solve each problem and decide which is the best of the choices given. Fill in the corresponding circle on the answer sheet. You may use any available space for scratchwork.

Notes

1. The use of a calculator is permitted.
2. All numbers used are real numbers.
3. Figures that accompany problems in this test are intended to provide information useful in solving the problems. They are drawn as accurately as possible EXCEPT when it is stated in a specific problem that the figure is not drawn to scale. All figure lie in a plane unless other indicated.
4. Unless otherwise specified, the domain of any function f is assumed to be set of all real numbers x for which $f(x)$ is a real number.

Reference Informatiom

$A = \pi r^2$ $A = \ell w$ $A = \dfrac{1}{2}bh$ $V = \ell wh$ $V = \pi r^2 h$ $c^2 = a^2 + b^2$ Special Right Triangles
$C = 2\pi r$

The numbers of degrees of arc in a circle is $360°$.

The sum of the measures in degrees of the angles is $180°$.

---

1. If $(x-2)^2 - 36 = 0$, which of the following could be the value of x?

   (A) -8
   (B) -6
   (C) -4
   (D) 4
   (E) 6

2. The length of a rectangle is 2 more than its width. If the perimeter of the rectangle is 24, what is the area of the rectangle?

   (A) 25
   (B) 30
   (C) 35
   (D) 40
   (E) 50

GO ON TO THE NEXT PAGE

3. If $y^{-2} = 3^{2x}$ and $y = 9$, what is the value of x?

(A) -2
(B) -1
(C) 0
(D) 1
(E) 2

Some of the students in an art class are juniors.

4. If the statement above is true, which of the following must also be true?

(A) All the students in the art class are juniors.
(B) All the students in the art class are seniors.
(C) Some students in the art class are seniors.
(D) The number of juniors in the art class is less than the number of seniors.
(E) Not all the students in the art class are seniors.

5. If the length of the side of a triangle is 24 and the lengths of the other two sides are both 15, what is the area of the triangle?

(A) 54
(B) 72
(C) 90
(D) 108
(E) 216

6. The figure above shows two views of the same cube. If each face of the cube has a different symbol on it, how many faces of the cube have NOT been shown in either view?

(A) One
(B) Two
(C) Three
(D) Four
(E) Five

GO ON TO THE NEXT PAGE

7. Mr. Lopez can clean the house in 5 hours. His son, Carl, can clean the house in 10 hours. How long will it take them to clean $\frac{2}{3}$ of the house, in hours, if they work together?

(A) 2

(B) $2\frac{2}{9}$

(C) 3

(D) $3\frac{1}{3}$

(E) 4

8. If $\dfrac{m+m}{m \times m} = 8$ and $m \neq 0$, what is the value of m?

(A) $\dfrac{1}{4}$

(B) $\dfrac{1}{2}$

(C) $\dfrac{2}{3}$

(D) $\dfrac{3}{4}$

(E) $\dfrac{4}{5}$

SALARY BY STUDENTS

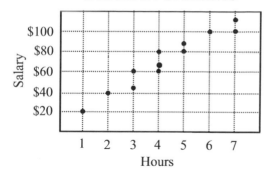

9. The scatterplot above shows the salary and the hours worked by 12 students in the library after school. Which of the following is true, according to this scatterplot?

(A) Five students worked more than the 5 hours.

(B) Only one student was paid $80.

(C) One student was paid more than $30 per hour.

(D) Most of the students worked less than 4 hours.

(E) The median of the worked hours by 12 students is 4 hours.

10. If imports at Computer Company X decreased by 15 percent and exports increased by 40 percent during last year, the ratio of imports to exports at the end of the year was how many times the ratio at the beginning of the year?

(A) $\dfrac{3}{8}$

(B) $\dfrac{2}{7}$

(C) $\dfrac{17}{28}$

(D) $\dfrac{3}{4}$

(E) $\dfrac{8}{3}$

GO ON TO THE NEXT PAGE

11. A piece of wire 4 feet long is cut into 3 pieces which are in the ratio 3:4:5. Which of the following is the length of the longest piece, in inches? (12 inches = 1 foot)

(A) 12
(B) 16
(C) 20
(D) 24
(E) 28

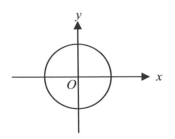

13. The figure above shows a circle with a radius of 5 in an *xy*-plane. Which of the following coordinates does not belong to the inside of the circle?

(A) $(-3, 3)$
(B) $(1, -4)$
(C) $(2, 4)$
(D) $(-3, -3.5)$
(E) $(-4, -3.1)$

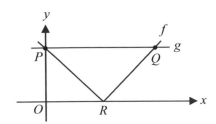

12. Let the function f be defined by $f(x) = |x - 5|$ and the function g be defined by $g(x) = k$, where k is a constant. What is the area of $\triangle PQR$?

(A) 10
(B) 15
(C) 20
(D) 25
(E) 50

14. Which of the following sets of numbers has the property that the sum of any two numbers in the set is also in the set?

I. $\{1, 2, 3, \ldots\}$
II. $\{1, 3, 5, \ldots\}$
III. $\{10, 20, 30, \ldots\}$

(A) I only
(B) II only
(C) III only
(D) I and III only
(E) II and III only

GO ON TO THE NEXT PAGE

15. If y is inversely proportional to x^2 and if $y = 6$ when $x = 5$, what is the value of y when $x = 10$?

(A) 12
(B) 5
(C) 3
(D) 2.5
(E) 1.5

16. Let the nth term of a sequence a_n be defined by $a_n = 4n + 2$. What is the value of $(a_{50} - a_{40})$?

(A) 40
(B) 36
(C) 32
(D) 28
(E) 24

17. If $\left| x - \dfrac{1}{2} \right| < \dfrac{3}{2}$, which of the following must be true?

(A) $x < 0$
(B) $x > 0$
(C) $-3 < x < 0$
(D) $-2 < x < 1$
(E) $-1 < x < 2$

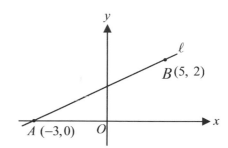

Note: Figure not drawn to scale.

18. In the figure above, what is the y-intercept of ℓ?

(A) $\dfrac{3}{4}$

(B) $\dfrac{4}{5}$

(C) $\dfrac{3}{2}$

(D) $\dfrac{4}{3}$

(E) $\dfrac{5}{4}$

GO ON TO THE NEXT PAGE

19. Let the function f be defined by

$f(x) = \left(g(x)\right)^2 - 2g(x) - 2$. If $f(2) = -3$,

what is the value of $g(2)$?

(A) -3
(B) -2
(C) -1
(D) 0
(E) 1

20. An urn above contains 4 white marbles and 5 black marbles, all of equal size. If two marbles are drawn at random with no replacement, what is the probability that two marbles are different colors?

(A) $\dfrac{4}{9}$

(B) $\dfrac{5}{9}$

(C) $\dfrac{5}{18}$

(D) $\dfrac{20}{81}$

(E) $\dfrac{40}{81}$

STOP

If you finish before time is called, you may check your work on this section only.
Do not turn to any other section in the test.

SECTION 5
Time- 25 minutes
18 Questions

Turn to Section 5 (Page 1) of your answer sheet to answer the questions in this section.

Directions: For this section, solve each problem and decide which is the best of the choices given. Fill in the corresponding circle on the answer sheet. You may use any available space for scratchwork.

1. If $|k-1| < 3$, how many integers k satisfy the inequality?

   (A) Two
   (B) Three
   (C) Four
   (D) Five
   (E) Six

$A = \{n \mid n = \text{ multiple of } 3\}$

$B = \{n \mid n = \text{ multiple of } 8\}$

$C - \{n \mid n = \text{ not divisible by } 16\}$

2. If n is a member of all of the sets A, B, and C above, which of the following could be the value of n?

   (A) 28
   (B) 36
   (C) 48
   (D) 72
   (E) 96

GO ON TO THE NEXT PAGE

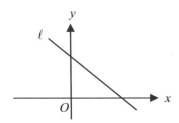

Note: Figure not drawn to scale.

3. The figure shows the graph of a linear function. Which of the following could be the equation of the line ℓ ?

(A) $x - y + 2 = 0$
(B) $x + y - 2 = 0$
(C) $x + y + 2 = 0$
(D) $2x - y + 1 = 0$
(E) $2x + y + 1 = 0$

Mr. Kay's Class After School
Distribution of registered program

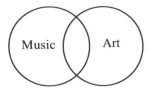

Note: Figure not drawn to scale

5. In Mr. Kay's class, 55 percent of the students have chosen Music and 70 percent of the students have chosen Art . If all the students chose either or both programs, and if there are 40 students, how many students have chosen both programs?

(A) 5
(B) 10
(C) 12
(D) 14
(E) 15

4. In the figure above, the length of $\overline{AB}$ is 10. If the length of $\overline{BC}$ is 5, and the length of $\overline{CD}$ is 2, where the points C and D are on the number line, which of the following cannot be the length of $\overline{AD}$? (Point C and D not shown)

(A) 3
(B) 7
(C) 13
(D) 14
(E) 17

6. A fisher man takes his boat out to sea at an average speed of 18 miles per hour and then back to the harbor along the same route at an average speed of 14 miles per hour. If his entire trip lasts 8 hours, what is the total number of miles in the round trip?

(A) 124
(B) 126
(C) 128
(D) 130
(E) 132

GO ON TO THE NEXT PAGE

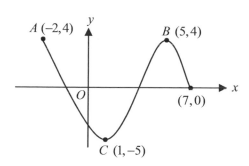

A (−2,4)   B (5,4)

(7,0)

C (1,−5)

7. The function f is graphed in the xy-plane above, where $-2 \le x \le 7$. If the function g is defined by $g(x) = 2f(x) + 2$, for how many values of k does $g(k)$ equal 10 ?

(A) None
(B) One
(C) Two
(D) Three
(E) Four

8. If $(p-1)^2 = 64$ and $(q-1)^2 = 36$, what is the greatest possible value of $(p-q)$?

(A) 16
(B) 14
(C) 12
(D) 10
(E) 8

GO ON TO THE NEXT PAGE

Directions: For Students-Produced Response questions 9-18, use the grid at the bottom of the answer sheet page on which you have answered questions 1-8.

Each of the remaining 10 questions requires you to solve the problem and enter your answer by making the circles in the special grid, as shown in the examples below. You may use any available space for scratchwork.

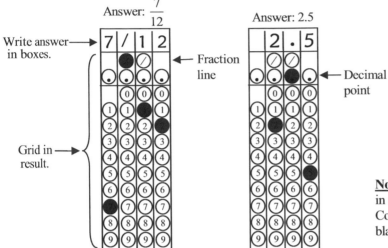

Answer: $\frac{7}{12}$

Write answer in boxes.

← Fraction line

Grid in result.

Answer: 2.5

← Decimal point

Answer: 201
Either position is correct.

Note: You may start your answers in any column, space permitting. Columns not needed should be left blank.

- Mark no more than one circle in any column.

- Because the answer sheet will be machine-scored, **you will receive credit only if the circles are filled in correctly.**

- Although not required, it is suggested that you write your answer in the boxes at the top of the columns to help you fill in the circles accurately.

- Some problems may have more than one correct answer. In such cases, grid only one answer.

- No question has a negative answer.

- **Mixed numbers** such as $3\frac{1}{2}$ must be gridded as 3.5 or 7/2. (If $\boxed{3\ 1\ /\ 2}$ is gridded, it will be interpreted as $\frac{31}{2}$, not $3\frac{1}{2}$.)

- **Decimal Answers:** If you obtain a decimal answer with more digits than the grid can accommodate, it may be either rounded or truncated, but it must fill the entire grid. For example, if you obtain an answer such as 0.6666..., you should record your result as .666 or .667. **A less accurate value such as .66 or .67 will be scored as incorrect.**

Acceptable ways to grid $\frac{2}{3}$ are:

9. If the value of four times a certain number equals the value of the number increased by 78, what is the number?

10. If $2x + y = 10$ and $y < 6$, what is the smallest integer value of x?

GO ON TO THE NEXT PAGE

$$h = vt - 16t^2$$

11. An object is thrown upward along a path which can be described by the equation above, with an initial velocity of v feet per second and time t, in seconds. If the object is thrown upward with an initial velocity of 64 feet per second, what will be its height, in feet, at $t = 2$?

12. If the lengths of the sides of a triangle are k, $2k+1$, and 7, where k is a positive integer, what is one possible value of k?

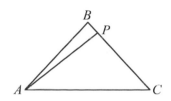

Note: Figure not drawn to scale.

13. In the triangle above, $AB = BC = 5$ and $AC = 6$. If $\overline{AP}$ is perpendicular to $\overline{BC}$, what is the length of $\overline{AP}$?

PRESIDENT ELECTION

Candidate	Votes
Alan	44
Bernard	31
Charles	n
David	48
Edward	52

14. The table above shows a student class election and the student receiving the greatest number of votes is elected. If every student in the class votes for 2 of 5 candidates, and Charles was elected president in the election, what is the minimum number of students in the class?

GO ON TO THE NEXT PAGE

15. What is the sum of 20 consecutive integers if the median of the list of the numbers is 40.5?

-30, -29, -28, -27,...

17. In the sequence above, each term after the first is one more than the previous term. If the sum of the terms in the sequence is 96, how many terms in the sequence?

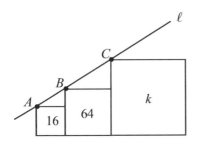

16. The figure above shows three squares with areas of 16, 64, and k, respectively. If A, B, and C are on line ℓ, what is the value of k?

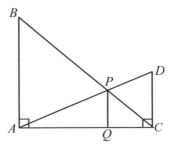

Note: Figure not drawn to scale.

18. In the figure above, $AB = 9$, $CD = 3$, and $AC = 12$, what is the length of $\overline{PQ}$?

STOP

If you finish before time is called, you may check your work on this section only.
Do not turn to any other section in the test.

SECTION 7
Time- 20 minutes
16 Questions

Turn to Section 7 (Page 2) of your answer sheet to answer the questions in this section.

Directions: For this section, solve each problem and decide which is the best of the choices given. Fill in the corresponding circle on the answer sheet. You may use any available space for scratchwork.

Notes

1. The use of a calculator is permitted.
2. All numbers used are real numbers.
3. Figures that accompany problems in this test are intended to provide information useful in solving the problems. They are drawn as accurately as possible EXCEPT when it is stated in a specific problem that the figure is not drawn to scale. All figure lie in a plane unless other indicated.
4. Unless otherwise specified, the domain of any function f is assumed to be set of all real numbers x for which $f(x)$ is a real number.

Reference Informatiom

$A = \pi r^2$
$C = 2\pi r$
$A = \ell w$
$A = \dfrac{1}{2}bh$
$V = \ell wh$
$V = \pi r^2 h$
$c^2 = a^2 + b^2$
Special Right Triangles

The numbers of degrees of arc in a circle is $360°$.

The sum of the measures in degrees of the angles is $180°$.

---

1. If $x = \dfrac{1}{3}$ and $y = 6$, what is the value of y in terms of x?

(A) $\dfrac{2}{x}$

(B) $12x$

(C) $6x + 3$

(D) $\dfrac{x+3}{2}$

(E) $\dfrac{3x+5}{3}$

2. If the price of a jacket was increased by 10% last week and then decreased 10% this week, what is the overall percent change from the original price?

(A) 0%
(B) 1%
(C) 2%
(D) 20%
(E) No change

GO ON TO THE NEXT PAGE

3. If $\sqrt{a^2 + b^2} = a + b$, which of the following must be true?

    (A) $a = 0$ and $b = 0$
    (B) $a + b = 0$
    (C) $ab = 0$
    (D) $ab \neq 0$
    (E) $a^2 - b^2 = 0$

4. Which of the following is equal to $\dfrac{1}{p}$ percent of k, for p and $k > 0$?

    (A) $\dfrac{k}{p}$

    (B) $100\,pk$

    (C) $\dfrac{100p}{k}$

    (D) $\dfrac{k}{100p}$

    (E) $\dfrac{pk}{100}$

5. Let $p \lozenge q$ be defined as $p \lozenge q = \dfrac{p+q}{p-q}$ for all numbers p and q. If $3 \lozenge a = 5$, what is the value of a?

    (A) 1
    (B) 2
    (C) 3
    (D) 4
    (E) 5

6. If $x^{-1}h = x^3 h^2$, what does h equal in terms of x?

    (A) $\dfrac{1}{x}$

    (B) $\dfrac{1}{x^2}$

    (C) $\dfrac{1}{x^3}$

    (D) $\dfrac{1}{x^4}$

    (E) x^4

GO ON TO THE NEXT PAGE

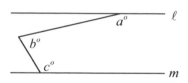

7. In the figure above, line ℓ is parallel to line m. What is the value of $a+b+c$?

(A) 180
(B) 270
(C) 300
(D) 360
(E) 420

8. If $k^2 + k = 42$, which of the following could be a value of $k^2 - k$?

(A) −30
(B) 10
(C) 20
(D) 56
(E) 84

9. Max walked one mile from his home to the mall to buy a scarf. He rested for 30 minutes and then walked home more slowly than before. Which of the following graphs could best represent his trip?

(A)

(B)

(C)

(D)

(E)

GO ON TO THE NEXT PAGE

10. There are 10 red, 10 black, 10 white, and 10 gray marbles in a jar. If all the marbles are identical, what is the least number of marbles that must be drawn out in order to ensure that among the marbles drawn out, 5 marbles will be the same color?

(A) 5
(B) 9
(C) 17
(D) 40
(E) 41

$$y = x^2 - 3$$
$$y = k$$

11. For which of the following values of k will the system of equations above have no solution?

(A) 0
(B) -2
(C) 2
(D) -4
(E) 4

12. Let the function f be defined by $f(x) = 2^{x+2}$ for all numbers x. Which of the following is equivalent to $f(a+b)$?

(A) $2(2^a + 2^b)$
(B) $4(2^a + 2^b)$
(C) $4 + 2^a + 2^b$
(D) $4 + (2^a \times 2^b)$
(E) $4 \times (2^a \times 2^b)$

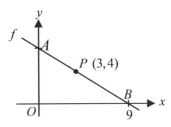

Note: Figure not drawn to scale.

13. The figure above shows the graph of function f. If point P is on the line, what is the area of $\triangle OAB$?

(A) 27
(B) 30
(C) 33
(D) 36
(E) 39

GO ON TO THE NEXT PAGE

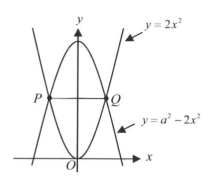

14. The figure above shows two graphs. If the length of $\overline{PQ} = 4$, what is the positive value of a ?

(A) 1
(B) 2
(C) 3
(D) 4
(E) 5

15. Which of the following is an equation of the line that is perpendicular to the line $y = \dfrac{1}{3}x + 5$ and has the same y-intercept as the line $y + 3 = -2x - 2$?

(A) $y = -3x + 6$
(B) $y = 3x - 3$
(C) $y = -4x + 5$
(D) $y = x + 3$
(E) $y = -3x - 5$

16. If the 5 cards shown above are placed in a row so that the card ▦ cannot be in the middle of the row, how many different arrangements are possible?

(A) 24
(B) 48
(C) 96
(D) 120
(E) 150

STOP

**If you finish before time is called, you may check your work on this section only.
Do not turn to any other section in the test.**

TEST 7	ANSWER KEY		
#	SECTION 3	SECTION 5	SECTION 7
1	C	D	A
2	C	D	B
3	A	B	C
4	E	D	D
5	D	B	B
6	B	B	D
7	B	C	D
8	A	B	D
9	E	26	E
10	C	3	C
11	C	64	D
12	D	3,4, or 5	E
13	E	$24/5$ or 4.8	A
14	D	114	D
15	E	810	E
16	A	256	C
17	E	64	
18	A	$9/4$ or 2.25	
19	E		
20	B		

TEST 7   SECTION 3

1. (C)

$(x-2)^2 = 36$

$\Rightarrow (x-2) = 6, \text{or} -6 \Rightarrow x = 8, \text{or} -4$.

---

2. (C)

Since width $= x$ and the length

$= x+2$, $2(x+x+2) = 24$.

$x = 5$. Therefore, the area $= 5 \times 7 = 35$.

---

3. (A)

Since $9^{-2} = \left(3^2\right)^{-2} = 3^{-4}$,

$3^{-4} = 3^{2x} \Rightarrow 2x = -4 \Rightarrow x = -2$.

---

4. (E)

Choices (A), (B), (C), and (D) are not always true. Only (E) is always true.

---

5. (D)

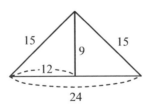

Since the triangle is an isosceles, the height is

9. Therefore, the area $= \frac{1}{2}(9 \times 24) = 108$.

---

6. (B)

From both figures, you can see symbols as follows.

■ △ □ #

The cube has six faces. Two faces are not seen in either view.

---

7. (B)

	Hours	Rate	Combined Rate
Mr.Lopez	5	$\dfrac{1}{5}$	$\dfrac{1}{5}+\dfrac{1}{10}=\dfrac{3}{10}$
His son	10	$\dfrac{1}{10}$	

Therefore, $\dfrac{2}{3}\div\dfrac{3}{10}=\dfrac{2}{3}\times\dfrac{10}{3}=2\dfrac{2}{9}$.

---

8. (A)

Since $\dfrac{m+m}{m^2}=\dfrac{2m}{m^2}=\dfrac{2}{m}$, $\dfrac{2}{m}=8$.

Therefore, $m=\dfrac{1}{4}$.

---

9. (E)

Only (E) is true.

---

10. (C)

If $K=\dfrac{\text{Import}}{\text{Export}}$, then at the end of the year,

$\dfrac{(1-0.15)\times\text{Import}}{(1+0.4)\times\text{Export}}=\dfrac{17}{28}\times K$.

Therefore, $\dfrac{17}{28}$ is the answer.

---

11. (C)

Since $3:4:5=3k,4k,5k$,

$12k=4(12)=48\Rightarrow k=4$.

Therefore, $5k=20$.

---

12. (D)

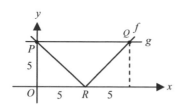

In the figure above, y-intercept is 5 and x-intercept is 5. $PQ=10$.

Therefore, the area of $\triangle PQR=\dfrac{5\times10}{2}=25$.

---

13. (E)

If $\sqrt{x^2+y^2}\le5$ or $x^2+y^2\le25$, then the point lies in the circle.

(E) $(-4)^2+(-3.1)^2=25.61>25$ → outside of the circle.

---

14. (D)

II $\Rightarrow 1+3=4$ is not in the same set.

---

15. (E)

Since $(y)(x^2)=K\,(\text{constant})$,

$(6)(5^2)=y(10^2)\Rightarrow y=1.5$

---

16. (A)

Since $a_n=4n+2$,

$a_{50}-a_{40}=202-162=40$.

---

17. (E)

$\left|x-\dfrac{1}{2}\right|<\dfrac{3}{2}\Rightarrow-\dfrac{3}{2}<x-\dfrac{1}{2}<\dfrac{3}{2}$

$\Rightarrow-1<x<2$.

---

18. (A)

Slope of the line $=\dfrac{2-0}{5-\,^-3}=\dfrac{1}{4}$. Thus, the equation is $y=\dfrac{1}{4}x+b$. Point $(5,2)$ lies on the line.

Therefore, $2=\dfrac{5}{4}+b\Rightarrow b=\dfrac{3}{4}$.

Or, let y-intercept be $(0,b)$. The slope between any two points is $\dfrac{1}{4}$. Therefore, the slope between $(0,b)$ and $(5,2)$ is $\dfrac{b-2}{0-5}=\dfrac{1}{4}$.

$b=\dfrac{3}{4}$.

---

19. (E)

$f(2)=\big(g(2)\big)^2-2g(2)-2=-3$. Let $g(2)=k$, then

$k^2-2k+1=0\Rightarrow(k-1)^2=0$. Therefore, $k=1$.

---

20. (B)

Two possible combinations are (W,B) or (B,W).

$$P(W,B)+P(B,W)=\left(\frac{4}{9}\times\frac{5}{8}\right)+\left(\frac{5}{9}\times\frac{4}{8}\right)=\frac{5}{9}$$

TEST 7   SECTION 5

1.   (D)
$|k-1|<3 \Rightarrow -3<k-1<3 \Rightarrow -2<x<4$.
Therefore, $x=-1,0,1,2,3$ (5 integers)

2.   (D)
n is a multiple of 24, but not divisible by 16.
72 is the correct number.

3.   (B)
The slope is negative and the y-intercept is positive.
(B) $y=-x+2$ is correct answer.

4.   (D)

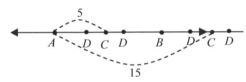

In the figure, the possible coordinates of point D are 3,7,13, and 17.

5.   (B)
$(55+70)-100=25\%$. Thus 25% of 40 is 10.

6.   (B)
Denote D is distance. Then, $\dfrac{D}{18}+\dfrac{D}{14}=8$.
$D=63$.
Therefore, $2D=126$ miles.

7.   (C)
Since $g(k)=2f(k)+2=10$, then $f(k)=4$.
Therefore, $k=-2$ and 5, where $-2\le k\le 7$.

8.   (B)
Since $(p-1)^2=64$, then $p=9$ or $p=-7$.
Since $(q-1)^2=36$, then $q=7$ or -5.
Therefore, the greatest possible value of $(p-q)$ is $9-(-5)=14$.

9.   26
$4n=n+78 \Rightarrow n=26$

10.   3
Since $y=10-2x<6$, $2x>4 \Rightarrow x>2$.
Therefore, $x=3$.

11.   64
$h=64(2)-16(2)^2=64$ feet.

12.   3, 4, or 5
From triangle inequality,
$(2k+1)-k<7<(2k+1)+k$. It follows that, $k<6$ and $k>2$. Therefore, $k=3,4,$ or 5.

13. $24/5$ or 4.8

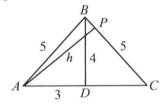

In the figure above, the area of
$\triangle ABC=\dfrac{1}{2}(6\times 4)=\dfrac{1}{2}(5\times h)$. Therefore, $h=4.8$.

14.   114
The sum of votes must be an even number.
Thus, the minimum number of votes is 228.
Therefore, the number of students is
$\dfrac{228}{2}=114$.

15.   810
For arithmetic sequence, the median is the mean.
Therefore, the sum is, $40.5\times 20=810$.

16.   256

The slope of the line is constant. That is,

$$\frac{4}{4} = \frac{\sqrt{k} - 8}{8} \Rightarrow k = 256.$$

17. 64

The sequence will be as follows.
$$-30, -29, -28, \ldots 0, 1, \ldots 29, 30, 31, 32, 33$$

61 numbers   +3=64 numbers.

18. $\frac{9}{4}$ or 2.25

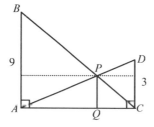

In the figure above, $\triangle ABP \sim \triangle CDP$. Thus
$\frac{AQ}{QC} = \frac{9}{3} = \frac{3}{1}$. That is, $AQ = 3k$ and $QC = k$.

$3k + k = 4k = 12 \Rightarrow k = 3$.

Since $\triangle ABC \sim QPC$, $\frac{AB}{QP} = \frac{AC}{QC}$.

Therefore $\frac{9}{QP} = \frac{12}{3} = 4$, it follows that

$QP = \frac{9}{4}$.

TEST 7   SECTION 7

1. (A)
$$y = \frac{2}{x} = \frac{2}{\frac{1}{3}} = 6$$

2. (B)
Final price
$= (1 + 0.1)(1 - 0.1)x = 0.99x \Rightarrow (1 - 0.01)x$.
Therefore, 1% changes.

3. (C)
$\sqrt{a^2 + b^2} = a + b \Rightarrow a^2 + b^2 = a^2 + 2ab + b^2$.
Therefore, $ab = 0$.

4. (D)
$$\frac{1}{p}\% \text{ of } k = \frac{1}{100p} \times k = \frac{k}{100p}.$$

5. (B)
$3 \lozenge a = \frac{3 + a}{3 - a} = 5$. Therefore, $a = 2$.

6. (D)
$$x^{-1} h = x^3 h^2 \Rightarrow h = \frac{x^{-1}}{x^3} = x^{-4} = \frac{1}{x^4}.$$

7. (D)

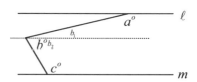

Since the sum of interior angles on the same side is 180°, $180^o + 180^o = 360^o$.

8. (D)
Since $k^2 + k - 42 = (k + 7)(k - 6)$, $k = -7$ or 6. Thus $k^2 - k = (-7)^2 - (-7) = 56$ or $6^2 - 6 = 30$.

9. (E)
The slope of the line represents the speed.

10. (C)
RRRR BBBB WWWW GGGG = 16 marbles. The next 17th marble ensures that 5 marbles are of the same color.

11. (D)
No solution means " no intersection". Thus,

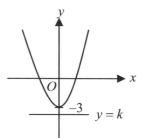

k must be less than -3. Therefore, $k = -4$ is the answer.

12. (E)
$$f(a+b) = 2^{a+b+2} = 4 \times 2^a \times 2^b = 4(2^a \times 2^b)$$

13. (A)

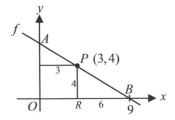

Since $\triangle PBR \sim \triangle ABO$, $\dfrac{4}{6} = \dfrac{OA}{9}$.

$OA = 6$. Therefore, the area of

$\triangle OAB = \dfrac{1}{2}(6 \times 9) = 27$.

14. (D)
The x-coordinate of point Q is 2. Therefore,
$2(2)^2 = a^2 - 2(2)^2$. $a^2 = 16 \Rightarrow a = 4$.

15. (E)
Since the slope must be -3 and y-intercept is
-5, the equation is, $y = -3x - 5$.

16. (C)

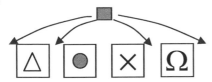

The number of arrangements of four
cards $= 4 \times 3 \times 2 \times 1 = 24$. The black card can
be assigned into 4 different places. Therefore,
$24 \times 4 = 96$.

$\boxed{END}$

Dr. John Chung's SAT Math

TEST

8

ANSWER SHEET　　TEST #:

SECTION 3

1	Ⓐ Ⓑ Ⓒ Ⓓ Ⓔ	11	Ⓐ Ⓑ Ⓒ Ⓓ Ⓔ	21	Ⓐ Ⓑ Ⓒ Ⓓ Ⓔ	31	Ⓐ Ⓑ Ⓒ Ⓓ Ⓔ
2	Ⓐ Ⓑ Ⓒ Ⓓ Ⓔ	12	Ⓐ Ⓑ Ⓒ Ⓓ Ⓔ	22	Ⓐ Ⓑ Ⓒ Ⓓ Ⓔ	32	Ⓐ Ⓑ Ⓒ Ⓓ Ⓔ
3	Ⓐ Ⓑ Ⓒ Ⓓ Ⓔ	13	Ⓐ Ⓑ Ⓒ Ⓓ Ⓔ	23	Ⓐ Ⓑ Ⓒ Ⓓ Ⓔ	33	Ⓐ Ⓑ Ⓒ Ⓓ Ⓔ
4	Ⓐ Ⓑ Ⓒ Ⓓ Ⓔ	14	Ⓐ Ⓑ Ⓒ Ⓓ Ⓔ	24	Ⓐ Ⓑ Ⓒ Ⓓ Ⓔ	34	Ⓐ Ⓑ Ⓒ Ⓓ Ⓔ
5	Ⓐ Ⓑ Ⓒ Ⓓ Ⓔ	15	Ⓐ Ⓑ Ⓒ Ⓓ Ⓔ	25	Ⓐ Ⓑ Ⓒ Ⓓ Ⓔ	35	Ⓐ Ⓑ Ⓒ Ⓓ Ⓔ
6	Ⓐ Ⓑ Ⓒ Ⓓ Ⓔ	16	Ⓐ Ⓑ Ⓒ Ⓓ Ⓔ	26	Ⓐ Ⓑ Ⓒ Ⓓ Ⓔ	36	Ⓐ Ⓑ Ⓒ Ⓓ Ⓔ
7	Ⓐ Ⓑ Ⓒ Ⓓ Ⓔ	17	Ⓐ Ⓑ Ⓒ Ⓓ Ⓔ	27	Ⓐ Ⓑ Ⓒ Ⓓ Ⓔ	37	Ⓐ Ⓑ Ⓒ Ⓓ Ⓔ
8	Ⓐ Ⓑ Ⓒ Ⓓ Ⓔ	18	Ⓐ Ⓑ Ⓒ Ⓓ Ⓔ	28	Ⓐ Ⓑ Ⓒ Ⓓ Ⓔ	38	Ⓐ Ⓑ Ⓒ Ⓓ Ⓔ
9	Ⓐ Ⓑ Ⓒ Ⓓ Ⓔ	19	Ⓐ Ⓑ Ⓒ Ⓓ Ⓔ	29	Ⓐ Ⓑ Ⓒ Ⓓ Ⓔ	39	Ⓐ Ⓑ Ⓒ Ⓓ Ⓔ
10	Ⓐ Ⓑ Ⓒ Ⓓ Ⓔ	20	Ⓐ Ⓑ Ⓒ Ⓓ Ⓔ	30	Ⓐ Ⓑ Ⓒ Ⓓ Ⓔ	40	Ⓐ Ⓑ Ⓒ Ⓓ Ⓔ

SECTION 5

1	Ⓐ Ⓑ Ⓒ Ⓓ Ⓔ	11	Ⓐ Ⓑ Ⓒ Ⓓ Ⓔ	21	Ⓐ Ⓑ Ⓒ Ⓓ Ⓔ	31	Ⓐ Ⓑ Ⓒ Ⓓ Ⓔ
2	Ⓐ Ⓑ Ⓒ Ⓓ Ⓔ	12	Ⓐ Ⓑ Ⓒ Ⓓ Ⓔ	22	Ⓐ Ⓑ Ⓒ Ⓓ Ⓔ	32	Ⓐ Ⓑ Ⓒ Ⓓ Ⓔ
3	Ⓐ Ⓑ Ⓒ Ⓓ Ⓔ	13	Ⓐ Ⓑ Ⓒ Ⓓ Ⓔ	23	Ⓐ Ⓑ Ⓒ Ⓓ Ⓔ	33	Ⓐ Ⓑ Ⓒ Ⓓ Ⓔ
4	Ⓐ Ⓑ Ⓒ Ⓓ Ⓔ	14	Ⓐ Ⓑ Ⓒ Ⓓ Ⓔ	24	Ⓐ Ⓑ Ⓒ Ⓓ Ⓔ	34	Ⓐ Ⓑ Ⓒ Ⓓ Ⓔ
5	Ⓐ Ⓑ Ⓒ Ⓓ Ⓔ	15	Ⓐ Ⓑ Ⓒ Ⓓ Ⓔ	25	Ⓐ Ⓑ Ⓒ Ⓓ Ⓔ	35	Ⓐ Ⓑ Ⓒ Ⓓ Ⓔ
6	Ⓐ Ⓑ Ⓒ Ⓓ Ⓔ	16	Ⓐ Ⓑ Ⓒ Ⓓ Ⓔ	26	Ⓐ Ⓑ Ⓒ Ⓓ Ⓔ	36	Ⓐ Ⓑ Ⓒ Ⓓ Ⓔ
7	Ⓐ Ⓑ Ⓒ Ⓓ Ⓔ	17	Ⓐ Ⓑ Ⓒ Ⓓ Ⓔ	27	Ⓐ Ⓑ Ⓒ Ⓓ Ⓔ	37	Ⓐ Ⓑ Ⓒ Ⓓ Ⓔ
8	Ⓐ Ⓑ Ⓒ Ⓓ Ⓔ	18	Ⓐ Ⓑ Ⓒ Ⓓ Ⓔ	28	Ⓐ Ⓑ Ⓒ Ⓓ Ⓔ	38	Ⓐ Ⓑ Ⓒ Ⓓ Ⓔ
9	Ⓐ Ⓑ Ⓒ Ⓓ Ⓔ	19	Ⓐ Ⓑ Ⓒ Ⓓ Ⓔ	29	Ⓐ Ⓑ Ⓒ Ⓓ Ⓔ	39	Ⓐ Ⓑ Ⓒ Ⓓ Ⓔ
10	Ⓐ Ⓑ Ⓒ Ⓓ Ⓔ	20	Ⓐ Ⓑ Ⓒ Ⓓ Ⓔ	30	Ⓐ Ⓑ Ⓒ Ⓓ Ⓔ	40	Ⓐ Ⓑ Ⓒ Ⓓ Ⓔ

9　10　11　12　13

14　15　16　17　18

SECTION

7

1	ⒶⒷⒸⒹⒺ	11	ⒶⒷⒸⒹⒺ	21	ⒶⒷⒸⒹⒺ	31	ⒶⒷⒸⒹⒺ
2	ⒶⒷⒸⒹⒺ	12	ⒶⒷⒸⒹⒺ	22	ⒶⒷⒸⒹⒺ	32	ⒶⒷⒸⒹⒺ
3	ⒶⒷⒸⒹⒺ	13	ⒶⒷⒸⒹⒺ	23	ⒶⒷⒸⒹⒺ	33	ⒶⒷⒸⒹⒺ
4	ⒶⒷⒸⒹⒺ	14	ⒶⒷⒸⒹⒺ	24	ⒶⒷⒸⒹⒺ	34	ⒶⒷⒸⒹⒺ
5	ⒶⒷⒸⒹⒺ	15	ⒶⒷⒸⒹⒺ	25	ⒶⒷⒸⒹⒺ	35	ⒶⒷⒸⒹⒺ
6	ⒶⒷⒸⒹⒺ	16	ⒶⒷⒸⒹⒺ	26	ⒶⒷⒸⒹⒺ	36	ⒶⒷⒸⒹⒺ
7	ⒶⒷⒸⒹⒺ	17	ⒶⒷⒸⒹⒺ	27	ⒶⒷⒸⒹⒺ	37	ⒶⒷⒸⒹⒺ
8	ⒶⒷⒸⒹⒺ	18	ⒶⒷⒸⒹⒺ	28	ⒶⒷⒸⒹⒺ	38	ⒶⒷⒸⒹⒺ
9	ⒶⒷⒸⒹⒺ	19	ⒶⒷⒸⒹⒺ	29	ⒶⒷⒸⒹⒺ	39	ⒶⒷⒸⒹⒺ
10	ⒶⒷⒸⒹⒺ	20	ⒶⒷⒸⒹⒺ	30	ⒶⒷⒸⒹⒺ	40	ⒶⒷⒸⒹⒺ

Math Scoring Worksheet

A. Section 3

_____   _____
numer of correct   number of incorrect

+   +

B. Section 5 (1-8)

_____   _____
numer of correct   number of incorrect

+

C. Section 5 (9-18)

_____
numer of correct

+   +

D. Section 7

_____   _____
numer of correct   number of incorrect

=   =

E. Total Unrounded Raw Score

_____   –   _____   ÷4 = _____
numer of correct   number of incorrect

F. Total Rounded Raw Score   _____   (See table)

Math Score Range =   [_____ — _____]

Math Conversion Table

Raw Score	Scaled Score	Raw Score	Scaled Score
54	800	23	490-550
53	780-800	22	480-540
52	760-800	21	470-530
51	740-800	20	460-520
50	720-780	19	450-510
49	700-760	18	450-510
48	690-750	17	440-500
47	680-740	16	430-490
46	670-730	15	420-480
45	660-720	14	420-480
44	650-710	13	410-470
43	650-710	12	400-460
42	640-700	11	390-450
41	630-690	10	380-440
40	620-680	9	390-430
39	610-670	8	380-420
38	610-670	7	370-410
37	600-660	6	360-400
36	590-650	5	340-380
35	580-640	4	320-370
34	570-630	3	310-360
33	560-620	2	300-350
32	560-620	1	270-320
31	550-610	0	240-300
30	540-600	-1	200-290
29	530-590	-2	200-270
28	530-590	-3	200-260
27	520-580	-4	200-240
26	510-570	-5	200-220
25	500-560	-6 and below	200
24	500-560		

SECTION 3
Time- 25 minutes
20 Questions

> Turn to Section 3 (Page 1) of your answer sheet to answer the questions in this section.

Directions: For this section, solve each problem and decide which is the best of the choices given. Fill in the corresponding circle on the answer sheet. You may use any available space for scratchwork.

Notes

1. The use of a calculator is permitted.
2. All numbers used are real numbers.
3. Figures that accompany problems in this test are intended to provide information useful in solving the problems. They are drawn as accurately as possible EXCEPT when it is stated in a specific problem that the figure is not drawn to scale. All figure lie in a plane unless other indicated.
4. Unless otherwise specified, the domain of any function f is assumed to be set of all real numbers x for which $f(x)$ is a real number.

Reference Informatiom

$A - \pi r^2$
$C = 2\pi r$
$A = \ell w$
$A = \frac{1}{2}bh$
$V = \ell wh$
$V = \pi r^2 h$
$c^2 = a^2 + b^2$
Special Right Triangles

The numbers of degrees of arc in a circle is 360°.

The sum of the measures in degrees of the angles is 180°.

---

1. If p pencils cost d dollars, x pencils cost how many dollars?

(A) $\dfrac{dp}{x}$

(B) $\dfrac{x+d}{p}$

(C) $\dfrac{x+p}{d}$

(D) $\dfrac{xp}{d}$

(E) $\dfrac{dx}{p}$

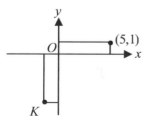

2. If the two rectangles in the figure are identical, which of the following are the coordinates of point K ?

(A) $(-5,1)$

(B) $(-5,5)$

(C) $(-1,1)$

(D) $(-1,-5)$

(E) $(-5,-1)$

> **GO ON TO THE NEXT PAGE**

3. If p percent of 40 percent of 1000 is k, what is p in terms of k?

(A) $400k$

(B) $\dfrac{k}{4}$

(C) $\dfrac{40}{k}$

(D) $40k$

(E) $\dfrac{400}{k}$

$$A = \{2, 4, 6\}$$
$$B = \{1, 2, 3, 4\}$$

4. Set A and B are shown above. If a is a member of set A and b is a member of set B, how many different values are possible for ab?

(A) 6
(B) 8
(C) 10
(D) 12
(E) 14

5. If $-1 \le p \le 1$, and $-3 \le q \le -2$, and $r = (p-q)^2$, what is the smallest possible value of r?

(A) 0
(B) 1
(C) 4
(D) 9
(E) 16

6. If a and b are integers and $a^2 - b^2 = 16$, which of the following cannot be a value of a?

(A) -5
(B) -4
(C) 0
(D) 4
(E) 5

GO ON TO THE NEXT PAGE

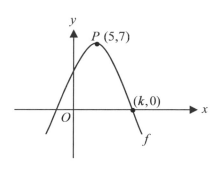

Note: Figure not drawn to scale.

7. The figure above shows the graph of
$f(x) = a(x-5)^2 + 7$. Which of the following
could be a possible value of k?

(A) 7
(B) 8
(C) 9
(D) 10
(E) 11

8. If $\dfrac{m+m}{m \times m} = \dfrac{1}{5}$, which of the following must be
true?

     I. $m = 0$
     II. $m = 5$
     III. $m = 10$

(A) I only
(B) II only
(C) III only
(D) I and II only
(E) I and III only

Questions 9-10 refer to the following definition.

Let $\otimes$ be defined by $a \otimes b = a + 2b + 4ab$ for
all numbers a and b.

9. If $5 \otimes k = 93$, what is the value of k?

(A) 3
(B) 4
(C) 5
(D) 6
(E) 7

———————————————

10. For what value of x is $x \otimes y = x$ could be true?

(A) -1
(B) $-\dfrac{1}{2}$
(C) 0
(D) $\dfrac{1}{2}$
(E) 1

11. The weight of a box of apples ranges from 1.75
pounds to 2.25 pounds. If p is the weight, in
pounds, of the box, which of the following must
be true?

(A) $|p| \le 2$
(B) $|p-2| \le 2$
(C) $|p+2| \le 0.25$
(D) $|p-2| \le 0.25$
(E) $|p-17.5| \le 0.25$

GO ON TO THE NEXT PAGE

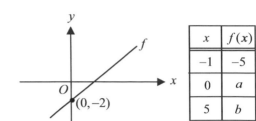

12. The table above shows the coordinates of some points on the function f graphed in the xy-coordinate plane. Which of the following is the value of b?

(A) 5
(B) 10
(C) 13
(D) 15
(E) 20

13. There are 3 Republicans and 2 Democrats on a Senate committee. If a 3-person subcommittee is to be formed from this committee, what is the probability of selecting two Republicans and one Democrat?

(A) $\dfrac{1}{20}$

(B) $\dfrac{3}{20}$

(C) $\dfrac{3}{10}$

(D) $\dfrac{3}{5}$

(E) $\dfrac{2}{3}$

14. Admission to the local movie theater for a group of 12 people is $3 for each child and $7 for each adult. If the group pays $64, which of the following could be the number of children in the group?

(A) 4
(B) 5
(C) 6
(D) 7
(E) 8

15. When two people shake hands with one another, that counts as one handshake. Every person in a room shakes hands with each other person in the room exactly once. If there are total of 15 handshakes, which of following could be the number of people in the room?

(A) 5
(B) 6
(C) 7
(D) 8
(E) 9

GO ON TO THE NEXT PAGE

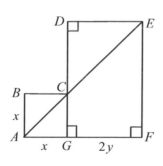

Note: Figure not drawn to scale.

16. In the figure above, the length of $\overline{AB}$, a side of the square, is x and the length of $\overline{FG}$ is $2y$. Which of the following represents the area of the figure $ABCDEFG$?

(A) $x^2 + xy$

(B) $x^2 + 4y^2$

(C) $x^2 + xy + 4y^2$

(D) $(x + 2y)^2$

(E) $x^2 + 2xy + 4y^2$

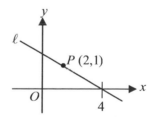

Note: Figure not drawn to scale.

17. The figure above shows line ℓ in the xy-coordinate plane. Line m (not shown) has the equation $y = ax + b$, where a and b are constants. If line m is perpendicular to line ℓ, and passes through the point P, which of the following must be true?

(A) $ab < 0$

(B) $a < 0$

(C) $b > 0$

(D) $b = 0$

(E) $ab > 0$

18. If $(x + y)(x - y) = 0$, which of the following must be true?

(A) $x = 0$ and $y = 0$

(B) $x = y$

(C) $x = -y$

(D) $x^3 = y^3$

(E) $x^2 = y^2$

GO ON TO THE NEXT PAGE

MATH TEST RESULTS

Scores	Number of Students
100	2
90	4
80	6
70	n
60	5
50	1

19. The table above shows the number of students and the scores on a math test. If the median of the test is 75, which of the following could be the value of n?

(A) 8
(B) 7
(C) 6
(D) 5
(E) 4

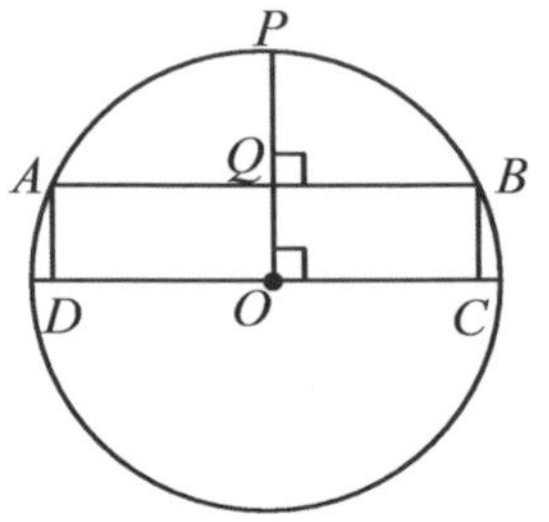

Note: Figure not drawn to scale.

20. In the figure above, the area of $ABCD$ is 64, and $AQ = 2AD$. What is the area of the circle?

(A) 52π
(B) 60π
(C) 80π
(D) 84π
(E) 90π

STOP

If you finish before time is called, you may check your work on this section only.
Do not turn to any other section in the test.

SECTION 5
Time- 25 minutes
18 Questions

Turn to Section 5 (Page 1) of your answer sheet to answer the questions in this section.

Directions: For this section, solve each problem and decide which is the best of the choices given. Fill in the corresponding circle on the answer sheet. You may use any available space for scratchwork.

Notes

1. The use of a calculator is permitted.
2. All numbers used are real numbers.
3. Figures that accompany problems in this test are intended to provide information useful in solving the problems. They are drawn as accurately as possible EXCEPT when it is stated in a specific problem that the figure is not drawn to scale. All figure lie in a plane unless other indicated.
4. Unless otherwise specified, the domain of any function f is assumed to be set of all real numbers x for which $f(x)$ is a real number.

Reference Informatiom

$A = \pi r^2$
$C = 2\pi r$
$A = \ell w$
$A = \frac{1}{2}bh$
$V = \ell wh$
$V = \pi r^2 h$
$c^2 = a^2 + b^2$
Special Right Triangles

The numbers of degrees of arc in a circle is $360°$.

The sum of the measures in degrees of the angles is $180°$.

---

1. Which of the following triples (a, b, c) does not satisfy the equation $a^{b+c} = 16$?

(A) (2, 1, 3)
(B) (4, 1, 1)
(C) (2, 2, 2)
(D) (16, 0, 1)
(E) (8, 2, 0)

2. A large cube, $5\,cm$ by $5\,cm$ by $5\,cm$ is painted orange on all six faces, and then it is cut into 125 small cubes, each $1\,cm$ by $1\,cm$ by $1\,cm$. How many of the small cubes are not painted orange on any face?

(A) 125
(B) 64
(C) 27
(D) 24
(E) 9

GO ON TO THE NEXT PAGE

3. The figure above shows five office rooms and each room is occupied by only one person. Room 1 is occupied by Adam, Bernard will be next to Charles, David will be next to Bernard and Edward, and Edward will be next to Adam. In which room could Bernard be occupied?

   (A) 2 only
   (B) 3 only
   (C) 4 only
   (D) 4 and 5
   (E) 3 and 4

4. A painter takes 8 hours to paint the entire house. How many hours will it take 3 painters working at the same rate to complete painting the house?

   (A) 2 hours
   (B) 2 hours 40 minutes
   (C) 3 hours
   (D) 3 hours 30 minutes
   (E) 4 hours

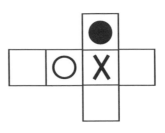

5. If the pattern above can be folded into a cube, which of the following cannot be the cube?

   (A)

   (B)

   (C)

   (D)

   (E)

GO ON TO THE NEXT PAGE

6. Josephine invests \$1,000 in a savings account paying 12 percent annual interest. If she invests n years, which of the following functions represents the amount in her account , y , in dollars, after n years?

(A) $y(n) = 1,000 + 0.12n$

(B) $y(n) = 1,000 + 1,000(1 + 0.12n)$

(C) $y(n) = 1,000(0.88)^n$

(D) $y(n) = 1,000(1.12)^n$

(E) $y(n) = 1,000(1.12n)$

7. If all four sides of quadrilateral P have the same lengths, which of the following statements must be true?

   I. All angles of P are equal.
   II. The diagonals of P are perpendicular.
   III. The lengths of diagonals of P are equal.

(A) None

(B) I only

(C) II only

(D) I and II only

(E) II and III only

8. If k, n, x, and y are positive numbers satisfying $x^{-\frac{1}{3}} = k^2$ and $y^{\frac{1}{3}} = n^2$, what is $(xy)^{-\frac{2}{3}}$ in terms of n and k ?

(A) $\dfrac{1}{nk}$

(B) $\dfrac{n}{k}$

(C) $n^4 k^4$

(D) $\dfrac{k^4}{n^4}$

(E) kn^4

GO ON TO THE NEXT PAGE

5 ☐☐☐ 5 ☐☐☐ 5 ☐☐☐ 5

Directions: For Students-Produced Response questions 9-18, use the grid at the bottom of the answer sheet page on which you have answered questions 1-8.

Each of the remaining 10 questions requires you to solve the problem and enter your answer by making the circles in the special grid, as shown in the examples below. You may use any available space for scratchwork.

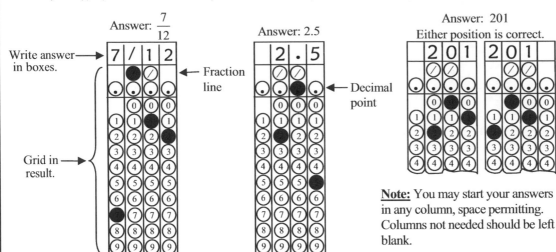

Answer: $\frac{7}{12}$

Write answer in boxes.

Fraction line

Grid in result.

Answer: 2.5

Decimal point

Answer: 201
Either position is correct.

Note: You may start your answers in any column, space permitting. Columns not needed should be left blank.

- Mark no more than one circle in any column.

- Because the answer sheet will be machine-scored, **you will receive credit only if the circles are filled in correctly.**

- Although not required, it is suggested that you write your answer in the boxes at the top of the columns to help you fill in the circles accurately.

- Some problems may have more than one correct answer. In such cases, grid only one answer.

- No question has a negative answer.

- **Mixed numbers** such as $3\frac{1}{2}$ must be gridded as 3.5 or 7/2. (If `3 1 / 2` is gridded, it will be interpreted as $\frac{31}{2}$, not $3\frac{1}{2}$.)

- **Decimal Answers:** If you obtain a decimal answer with more digits than the grid can accommodate, it may be either rounded or truncated, but it must fill the entire grid. For example, if you obtain an answer such as 0.6666..., you should record your result as .666 or .667. **A less accurate value such as .66 or .67 will be scored as incorrect.**

Acceptable ways to grid $\frac{2}{3}$ are:

9. If $\dfrac{x-y}{2} = 1 - \dfrac{y}{2}$, what is the value of x?

10. The population of students in a certain high school increased by 50% every 3 years since 1990. If the population of students in the school in 1999 is 675, what was the population in 1990?

GO ON TO THE NEXT PAGE

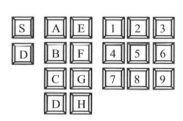

11. A combined snack and drink machine has buttons arranged as shown above. If a selection is made by choosing snack (S) or drink (D), followed by a letter, and followed by a one-digit number, what is the greatest number of different selections that can be made?

12. If the diagonal of a square-shaped field is 40 feet, what is the area of the field in square feet?

13. If $\left(7.6\times10^{k}\right)+\left(3\times10^{9}\right)=3.076\times10^{9}$, what is the value of k?

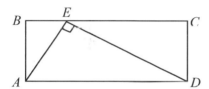

Note: Figure not drawn to scale.

14. The figure above shows a rectangle $ABCD$ and a right triangle AED. If $AE = 6$ and $ED = 8$, what is the area of rectangle $ABCD$?

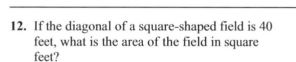

GO ON TO THE NEXT PAGE ⟩

15. If $15x - 7 + 6y = 20$, what is the value of $10x + 4y - 5$?

16. If x is divided by x^{-2} is $\dfrac{1}{27}$, what is the value of x?

17. A grocery customer spent a total of $17.50 for butter and peanut. The peanut cost 3 times as much per pound as the butter, and the customer bought 4 times as many pounds of butter as pounds of peanut. How much, in dollars, did the customer spend on butter? (Discard the $ sign when gridding your answer.)

18. If $(m^2)^3 = x^2$, where $x > 1$, for what value of y does $m^{3(y+2)} = x^8$?

STOP

If you finish before time is called, you may check your work on this section only.
Do not turn to any other section in the test.

SECTION 7
Time- 20 minutes
16 Questions

Turn to Section 7 (Page 2) of your answer sheet to answer the questions in this section.

Directions: For this section, solve each problem and decide which is the best of the choices given. Fill in the corresponding circle on the answer sheet. You may use any available space for scratchwork.

Notes

1. The use of a calculator is permitted.
2. All numbers used are real numbers.
3. Figures that accompany problems in this test are intended to provide information useful in solving the problems. They are drawn as accurately as possible EXCEPT when it is stated in a specific problem that the figure is not drawn to scale. All figure lie in a plane unless other indicated.
4. Unless otherwise specified, the domain of any function f is assumed to be set of all real numbers x for which $f(x)$ is a real number.

Reference Informatiom

$A = \pi r^2$
$C = 2\pi r$
$A = \ell w$
$A = \dfrac{1}{2}bh$
$V = \ell wh$
$V = \pi r^2 h$
$c^2 = a^2 + b^2$
Special Right Triangles

The numbers of degrees of arc in a circle is $360°$.

The sum of the measures in degrees of the angles is $180°$.

1. A function f is defined by $f(x) = 5g(x) + 3$. If $f(2) = 13$, what is the value of $g(2)$?

   (A) 0
   (B) 1
   (C) 2
   (D) 3
   (E) 4

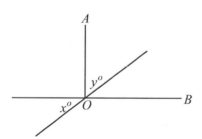

Note: Figure not drawn to scale.

2. In the figure above, $\overline{AO} \perp \overline{BO}$ and $y = 3x$. What is the value of y?

   (A) 22.5
   (B) 45
   (C) 60
   (D) 67.5
   (E) 72

GO ON TO THE NEXT PAGE

SHOE SIZES AT COMPANY X

Size	Length in inches
7	10
8	a
9	b
10	c
11	d
12	14

3. The table above shows shoe sizes and their corresponding lengths at Company X. If the lengths of shoe sizes increase by a constant length for each size, what is the value of c ?

(A) 11.4
(B) 12.4
(C) 12.6
(D) 13.0
(E) 13.2

4, 16, 52, 160,…

4. In the sequence above, the first is 4 and each number after the first is m more than p times the preceding number. What is the value of $m+p$?

(A) 5
(B) 6
(C) 7
(D) 8
(E) 9

5. If $x < -16$ or $x > 16$, which of the following must be true?

I. $x^2 > 16$
II. $|x| > 16$
III. $x^3 > 16$

(A) I only
(B) II only
(C) I and III only
(D) II and III only
(E) I and II only

6. If the average (arithmetic mean) of a and b is 6, the average of b and c is 8, and the average of a and c is 10, What is the average of a, b, and c ?

(A) 12
(B) 10
(C) 9
(D) 8
(E) 7

GO ON TO THE NEXT PAGE

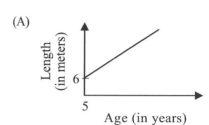

Note: Figure not drawn to scale.

7. In the figure above, the pyramid has a square base and all four triangular faces are congruent. If $AB = 4$ and the total surface area of the pyramid is 56, what is the length of $\overline{DE}$?

(A) 4
(B) 5
(C) $5\sqrt{2}$
(D) 6
(E) $6\sqrt{3}$

8. If $a + b = 4$ and $ab = 3$, what is the value of $a^2 + b^2$?

(A)   7
(B)   8
(C) 10
(D) 12
(E) 14

HEIGHT OF A TREE

Age (in years)	5	6	7	8	9
Height (in meters)	6	7	7.6	8.1	8.2

9. Which of the following graphs best represents the information in the table above?

(A)

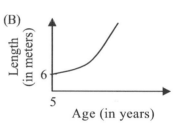

(B)

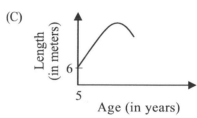

(C)

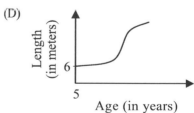

(D)

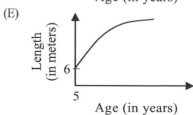

(E)

GO ON TO THE NEXT PAGE

10. If k is a positive integer, which of the following is equivalent to $3^k + 3^k + 3^k$?

(A) 3^{3k}

(B) 3^{k+2}

(C) 3^{k+1}

(D) 9^{2k}

(E) 9^{3k}

11. The ratio of a to b is 3 to 8. If $a = 3c$, what is the ratio of b to c ?

(A) 1 to 8

(B) 2 to 5

(C) 5 to 2

(D) 8 to 1

(E) 8 to 3

12. If $4\sqrt{8} = m\sqrt{n}$, and if n and m are positive integers, which of the following cannot be the value of $m + n$?

(A) 129

(B) 34

(C) 24

(D) 12

(E) 10

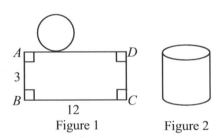

Figure 1   Figure 2

13. The pattern in figure 1 above is folded and forms a cylinder in figure 2 above. If $AB = 3$ and $BC = 12$, what is the volume of the cylinder?

(A) 36π

(B) 24π

(C) $\dfrac{54}{\pi}$

(D) $\dfrac{72}{\pi}$

(E) $\dfrac{108}{\pi}$

GO ON TO THE NEXT PAGE

14. In $\triangle ABC$, the length of $\overline{AB}$ is x and the length of $\overline{BC}$ is $x+5$. If the length of $\overline{AC}$ is 16, what is the smallest possible integer value of x?

(A) 5
(B) 6
(C) 7
(D) 8
(E) 9

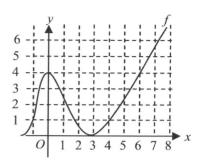

16. The figure above shows the graph of function f in the xy-coordinate plane, and a function g is defined by $g(x) = 2f(x) - 3$. If $-1 \le x \le 7$, for how many values of k does $g(k) = 3$?

(A) None
(B) One
(C) Two
(D) Three
(E) Four

15. In the xy-coordinate plane, the graph of $y = 2x^2 - 10$ intersects line ℓ at $(m,\ 8)$ and $(n,\ -8)$. What is the greatest possible value of the slope of line ℓ?

(A) 4
(B) 5
(C) 6
(D) 7
(E) 8

STOP

**If you finish before time is called, you may check your work on this section only.
Do not turn to any other section in the test.**

#	SECTION 3	SECTION 5	SECTION 7
1	E	E	C
2	D	C	D
3	B	C	B
4	B	B	C
5	B	D	E
6	C	D	D
7	E	C	B
8	C	D	C
9	B	2	E
10	B	200	C
11	D	144	D
12	C	800	C
13	D	7	E
14	B	48	B
15	B	13	E
16	E	$\frac{1}{3}$ or .333, .334	D
17	A	10	
18	E	6	
19	C		
20	C		

TEST 8   SECTION 3

1. (E)

Proportion: $\dfrac{p}{d} = \dfrac{x}{y} \Rightarrow y = \dfrac{dx}{p}$.

2. (D)

Rotation: $(5,1) \rightarrow (-1,-5)$.

3. (B)

Since $\dfrac{p}{100} \times \dfrac{40}{100} \times 1000 = k$, $p = \dfrac{k}{4}$.

4. (B)

$A \times B = 2, 4, 6, 8$

$\quad\quad\quad \cancel{8}, \cancel{12}, 12, 16$

$\quad\quad\quad \cancel{6}, \cancel{12}, 18, 24$

Therefore, 8 different values.

5. (B)

$-1 - {}^{-}2 \le p - q \le 1 - {}^{-}3 \Rightarrow 1 \le p - q \le 4$.

therefore, the least value of $(p-q)^2 = 1^2 = 1$.

6. (C)

$a^2 - b^2 = 16$ *or* $b^2 = a^2 - 16$

If $a = -5$, $b^2 = 25 - 16 = 9$, $b = \pm 3$ (OK)

If $a = 0$, $b^2 = 0 - 16 = -16$, b^2 CANNOT

be negative. The answer is C.

7. (E)

The length of $\overline{QR}$ is greater than 5.

Quadratic functions are exactly symmetrical with respect to the axis of symmetry.

That is, $k - Q > 5$, $k > Q + 5 = 10$.

The answer is (E).

8. (C)

Since $\dfrac{2m}{m^2} = \dfrac{1}{5}$, $m = 10$.

9. (B)

$5 \otimes k = 5 + 2k + 4(5)(k) = 93 \Rightarrow 22k = 88$

Therefore, $k = 4$.

10. (B)

$x + 2y + 4xy = x \Rightarrow 2y + 4xy = 0$. Thus,

$2y(1 + 2x) = 0 \Rightarrow y = 0$ or $x = -\dfrac{1}{2}$.

11. (D)

Midpoint=$\dfrac{1.75 + 2.25}{2} = 2$. Distance from the

midpoint is 0.25. Therefore, $|p - 2| \le 0.25$.

12. (C)

Since $a = -2$,

the slope=$\dfrac{-1 - (-5)}{0 - (-1)} = \dfrac{b - (-2)}{5 - 0} \Rightarrow b = 13$.

13. (D)

Select two out of three republicans (A, B, C).

There are six different combinations as follows.

AB, AC, BC --- 3 combinations

Select one out of two democrats (P, Q).

P, Q ---- 2 combinations

Therefore $3 \times 2 = 6$ combinations

The number of all possible combinations is 10.

Or,

$_5C_2 = \dfrac{5 \times 4}{2} = 10$

The probability is $\dfrac{6}{10} = \dfrac{3}{5}$

The answer is (D).

14. (B)

$3n + 7(12 - n) = 64 \Rightarrow n = 5$.

15. (B)

$1 + 2 + 3 + 4 + 5 = 15$.

Therefore, 6 people is correct.

Or, let n be the number of people. Then

$_nC_2 = \dfrac{n(n-1)}{2} = 15$.

$n^2 - n - 30 = 0 \rightarrow (n - 6)(n + 5) = 0$

Therefore $n = 6$.

16. (E)

The area$= x^2 + 2y(2y + x) = x^2 + 2xy + 4y^2$.

17. (A)

The slope of line $\ell = \dfrac{1 - 0}{2 - 4} = -\dfrac{1}{2}$.

Thus the slope of line $m = 2$.

Therefore, $a = 2 \Rightarrow y = 2x + b$. Then

substitute $(2, 1)$ into the equation,

$1 = 2(2) + b \Rightarrow b = -3 \Rightarrow y = 2x - 3$.

$a = 2$ and $b = -2 \Rightarrow ab < 0$.

18. (E)

From $(x + y)(x - y) = 0$

$x = -y$ or $x = y$

"MUST be TRUE" means that the answer must satisfy both of the two solutions.

For example, (B) is not always true, because when $x = -y$, the equation is 0.

Therefore, $x^2 - y^2 = 0$ implies $x = -y$ and $x = y$.

The answer is (E).

19. (C)

To be in the middle, $2 + 4 + 6 = 1 + 5 + n$.

$n = 6$.

20. (C)

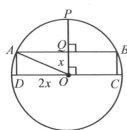

In the figure above, the area of $ABCD = 4x^2 = 64$, then $x = 4$. Therefore, $AO = \sqrt{4^2 + 8^2} = \sqrt{80}$. The area of the circle $= \pi\left(\sqrt{80}\right)^2 = 80\pi$.

TEST 8   SECTION 5

1. (E)

$8^{2+0} \neq 16$

2. (C)

Except outer layers, there are $3 \times 3 \times 3 = 27$ unpainted cubes.

3. (C)

Adam, Edward, David, Bernard, Charles

4. (B)

Inverse proportion:

$8 \times 1 = 3 \times x$

$\Rightarrow x = \dfrac{8}{3} \text{ hour} = 2 \text{ hours } 40 \text{ minutes.}$

5. (D)

6. (D)

$y = 1000(1 + 0.12)^n = 1000(1.12)^n$

7. (C)

The quadrilateral is a rhombus. Rhombus has (1) 4 sides that are congruent. (2) The diagonals are perpendicular and bisect each other.

8. (D)

$\left(x^{-\frac{1}{3}}\right)^{-3} = \left(k^2\right)^{-3} \Rightarrow x = k^{-6}$ and

$\left(y^{\frac{1}{3}}\right)^{3} = \left(n^2\right)^{3} \Rightarrow y = n^6$. Therefore,

$(xy)^{-\frac{2}{3}} = \left(k^{-6}n^6\right)^{-\frac{2}{3}} = k^4 n^{-4} = \dfrac{k^4}{n^4}$

9. 2

$\dfrac{x-y}{2} = 1 - \dfrac{y}{2} \Rightarrow \dfrac{x-y}{2} = \dfrac{2-y}{2}$. Therefore,

$x - y = 2 - y \Rightarrow x = 2$.

10. 200

Every 3 years increased by 50% since 1990. In 1999 the population is 675, then 1996 the population will be $\dfrac{675}{1.5} = 450$, and in 1993, $\dfrac{450}{1.5} = 300$. Therefore, in 1990, the population is $\dfrac{300}{1.5} = 200$.

11. 144

$2 \times 8 \times 9 = 144$.

12. 800

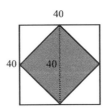

The area of the square $= \dfrac{40 \times 40}{2} = 800$

Or, since the length of a side is $20\sqrt{2}$, the area is $20\sqrt{2} \times 20\sqrt{2} = 800$.

Or,

There are four right triangles.

$\dfrac{20 \times 20}{2} \times 4 = 800$

13. 7

$3.076 \times 10^9 = 3 \times 10^9 + 0.076 \times 10^9$. Therefore,
$0.076 \times 10^9 = 7.6 \times 10^7$. $k = 7$.

---

14. 48

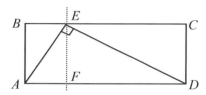

$AE = 6$, $ED = 8$, and $AE \perp ED$

The area of $\triangle AED = \dfrac{6 \times 8}{2} = 24$.

The area of $ABCD$ is twice the area of $\triangle AED$.

The area of $ABCD = 24 \times 2 = 48$.

Or, $\dfrac{EF \times AD}{2} = 24$, $EF = 4.8$ and $AD = 10$.

Therefore, the area $= 10 \times 4.8 = 48$.

---

15. 13

$15x + 6y = 27$ or $5x + 2y = 9$

$10x + 4y - 5 = 2(5x + 2y) - 5 = 2(9) - 5 = 13$

---

16. $\dfrac{1}{3}$

$\dfrac{x}{x^{-2}} = \dfrac{1}{27} \Rightarrow x = \dfrac{1}{3}$.

---

17. 10

	Price	Pound	Sub total
Butter	a	$4b$	$4ab$
Peanut	$3a$	b	$3ab$

Since $4ab + 3ab = 17.50$,

$7ab = 17.50 \Rightarrow ab = 2.50$.

$\therefore$ The amount spent on butter $= 4ab$

$4ab = 4 \times 2.5 = \$10$

---

18. 6

$(m^2)^3 = x^2$ or $m^6 = x^2$

$m^{3(y+2)} = m^{3y+6}$

$x^8 = \left(x^2\right)^4 = \left(m^6\right)^4 = m^{24}$

Therefore,

$m^{3y+6} = m^{24}$, or $3y + 6 = 24$

Then $y = 6$

---

1. (C)

Since $f(2) = 5g(2) + 3 = 13$, $g(2) = 2$.

---

2. (D)

$y = 3x$ and $3x + x = 90$. $x = 22.5$. Therefore, $y = 3x = 67.5$.

---

3. (B)

Constant increase is $\dfrac{14 - 10}{12 - 7} = 0.8$.

Therefore, $c = 10 + 3 \times 0.8 = 12.4$.

---

4. (C)

In the sequence, $p = 3$ and $m = 4$. Therefore, $m + p = 7$.

---

5. (E)

For $x < -16$ or $x > 16$, x^2 is always greater than 16.

When $x < -16$, $x^3 < 16$. III is not true.

---

6. (D)

Since $\dfrac{a+b}{2} = 6$, $\dfrac{b+c}{2} = 8$, and $\dfrac{a+c}{2} = 10$.

$\left.\begin{array}{l} a + b = 12 \\ b + c = 16 \\ c + a = 20 \end{array}\right\}$

$2(a+b+c) = 48 \Rightarrow a+b+c = 24$

Therefore, $\dfrac{a+b+c}{3} = 8$.

---

7. (B)

The area of four lateral triangles $= 56 - 16 = 40$. The area of one triangle $= \dfrac{40}{4} = 10$. Therefore,

$\dfrac{ED \times AB}{2} = \dfrac{4 \times ED}{2} = 10$. That is, $ED = 5$.

---

8. (C)

Since $a + b = 4$, $ab = 3$,

$(a+b)^2 = a^2 + b^2 + 2ab = 16$.

Therefore, $a^2 + b^2 = 10$.

---

9. (E)

The rate of increase in height is getting slower.

---

10. (C)

$3^k + 3^k + 3^k = 3 \times 3^k = 3^{k+1}$.

---

11. (D)

Since $\dfrac{a}{b} = \dfrac{3}{8} \Rightarrow \dfrac{3c}{b} = \dfrac{3}{8} \Rightarrow \dfrac{b}{c} = \dfrac{8}{1}$.

---

12. (C)

$4\sqrt{8} = m\sqrt{n}$

Four different possibilities as follows.

(1) $4\sqrt{8} = 4\sqrt{8} \Rightarrow m = 4, \ n = 8$

(2) $2 \times 2\sqrt{8} = 2\sqrt{4 \times 8} = 2\sqrt{32}$
$\qquad m = 2, \ n = 32$

(3) $\sqrt{16} \times \sqrt{8} = \sqrt{128} \qquad m = 1, \ n = 128$

(4) $4\sqrt{8} = 4 \times \sqrt{2 \times 2 \times 2} = 4 \times 2\sqrt{2} = 8\sqrt{2}$
$\qquad m = 8, \ n = 2$

Therefore, $m + n = 12, \ 34, \ 129$, or 10.

---

13. (E)

Since $12 = 2\pi r \Rightarrow r = \dfrac{6}{\pi}$,

$V = \pi r^2 h = \pi \left(\dfrac{6}{\pi}\right)^2 (3) = \dfrac{108}{\pi}$.

---

14. (B)

Triangle inequality: $x, \ x+5, \ 16$

$\left. \begin{array}{l} x < x + 21 \\ x + 5 < x + 16 \\ 16 < 2x + 5 \end{array} \right\} \Rightarrow x > 5.5$

Therefore, the smallest integer value of x is 6.

Or, $(x+5) - x < 16 < x + (x+5) \ \rightarrow$
$5 < 16 < 2x + 5 \ \rightarrow 2x > 11 \rightarrow x > 5.5$

---

15. (E)

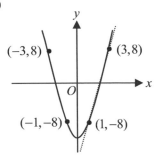

$2m^2 - 10 = 8 \Rightarrow m = \pm 3$. Therefore, the intersections are $(3,8)$ and $(-3,8)$

$2n^2 - 10 = -8 \Rightarrow n = \pm 1$. Therefore, the intersections are $(-1,-8)$ and $(1,-8)$. The greatest value of the slope $= \dfrac{8 - (-8)}{3 - 1} = 8$.

---

16. (D)

Since $g(k) = 2f(k) - 3 = 3$,

$f(k) = 3$. Therefore, there are three intersections with $y = 3$.

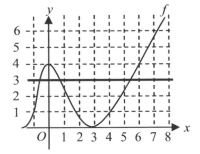

$\boxed{END}$

Dr. John Chung's SAT Math

TEST
9

ANSWER SHEET   TEST #:

SECTION 3

1	ⒶⒷⒸⒹⒺ	11	ⒶⒷⒸⒹⒺ	21	ⒶⒷⒸⒹⒺ	31	ⒶⒷⒸⒹⒺ
2	ⒶⒷⒸⒹⒺ	12	ⒶⒷⒸⒹⒺ	22	ⒶⒷⒸⒹⒺ	32	ⒶⒷⒸⒹⒺ
3	ⒶⒷⒸⒹⒺ	13	ⒶⒷⒸⒹⒺ	23	ⒶⒷⒸⒹⒺ	33	ⒶⒷⒸⒹⒺ
4	ⒶⒷⒸⒹⒺ	14	ⒶⒷⒸⒹⒺ	24	ⒶⒷⒸⒹⒺ	34	ⒶⒷⒸⒹⒺ
5	ⒶⒷⒸⒹⒺ	15	ⒶⒷⒸⒹⒺ	25	ⒶⒷⒸⒹⒺ	35	ⒶⒷⒸⒹⒺ
6	ⒶⒷⒸⒹⒺ	16	ⒶⒷⒸⒹⒺ	26	ⒶⒷⒸⒹⒺ	36	ⒶⒷⒸⒹⒺ
7	ⒶⒷⒸⒹⒺ	17	ⒶⒷⒸⒹⒺ	27	ⒶⒷⒸⒹⒺ	37	ⒶⒷⒸⒹⒺ
8	ⒶⒷⒸⒹⒺ	18	ⒶⒷⒸⒹⒺ	28	ⒶⒷⒸⒹⒺ	38	ⒶⒷⒸⒹⒺ
9	ⒶⒷⒸⒹⒺ	19	ⒶⒷⒸⒹⒺ	29	ⒶⒷⒸⒹⒺ	39	ⒶⒷⒸⒹⒺ
10	ⒶⒷⒸⒹⒺ	20	ⒶⒷⒸⒹⒺ	30	ⒶⒷⒸⒹⒺ	40	ⒶⒷⒸⒹⒺ

SECTION 5

1	ⒶⒷⒸⒹⒺ	11	ⒶⒷⒸⒹⒺ	21	ⒶⒷⒸⒹⒺ	31	ⒶⒷⒸⒹⒺ
2	ⒶⒷⒸⒹⒺ	12	ⒶⒷⒸⒹⒺ	22	ⒶⒷⒸⒹⒺ	32	ⒶⒷⒸⒹⒺ
3	ⒶⒷⒸⒹⒺ	13	ⒶⒷⒸⒹⒺ	23	ⒶⒷⒸⒹⒺ	33	ⒶⒷⒸⒹⒺ
4	ⒶⒷⒸⒹⒺ	14	ⒶⒷⒸⒹⒺ	24	ⒶⒷⒸⒹⒺ	34	ⒶⒷⒸⒹⒺ
5	ⒶⒷⒸⒹⒺ	15	ⒶⒷⒸⒹⒺ	25	ⒶⒷⒸⒹⒺ	35	ⒶⒷⒸⒹⒺ
6	ⒶⒷⒸⒹⒺ	16	ⒶⒷⒸⒹⒺ	26	ⒶⒷⒸⒹⒺ	36	ⒶⒷⒸⒹⒺ
7	ⒶⒷⒸⒹⒺ	17	ⒶⒷⒸⒹⒺ	27	ⒶⒷⒸⒹⒺ	37	ⒶⒷⒸⒹⒺ
8	ⒶⒷⒸⒹⒺ	18	ⒶⒷⒸⒹⒺ	28	ⒶⒷⒸⒹⒺ	38	ⒶⒷⒸⒹⒺ
9	ⒶⒷⒸⒹⒺ	19	ⒶⒷⒸⒹⒺ	29	ⒶⒷⒸⒹⒺ	39	ⒶⒷⒸⒹⒺ
10	ⒶⒷⒸⒹⒺ	20	ⒶⒷⒸⒹⒺ	30	ⒶⒷⒸⒹⒺ	40	ⒶⒷⒸⒹⒺ

9   10   11   12   13

14   15   16   17   18

Math Scoring Worksheet

A. Section 3

    _____   _____
    numer of correct   *number of incorrect*

    +   +

B. Section 5 (1-8)

    _____   _____
    numer of correct   *number of incorrect*

    +

C. Section 5 (9-18)

    _____
    numer of correct

    +   +

D. Section 7

    _____   _____
    numer of correct   *number of incorrect*

    =   =

E. Total Unrounded Raw Score

    _____ − _____ ÷4 = _____
    numer of correct   *number of incorrect*

F. Total Rounded Raw Score   _____ (See table)

         Math Score Range = | ⸺ |

Math Conversion Table

Raw Score	Scaled Score	Raw Score	Scaled Score
54	800	23	490-550
53	780-800	22	480-540
52	760-800	21	470-530
51	740-800	20	460-520
50	720-780	19	450-510
49	700-760	18	450-510
48	690-750	17	440-500
47	680-740	16	430-490
46	670-730	15	420-480
45	660-720	14	420-480
44	650-710	13	410-470
43	650-710	12	400-460
42	640-700	11	390-450
41	630-690	10	380-440
40	620-680	9	390-430
39	610-670	8	380-420
38	610-670	7	370-410
37	600-660	6	360-400
36	590-650	5	340-380
35	580-640	4	320-370
34	570-630	3	310-360
33	560-620	2	300-350
32	560-620	1	270-320
31	550-610	0	240-300
30	540-600	-1	200-290
29	530-590	-2	200-270
28	530-590	-3	200-260
27	520-580	-4	200-240
26	510-570	-5	200-220
25	500-560	-6 and below	200
24	500-560		

SECTION 3
Time- 25 minutes
20 Questions

Turn to Section 3 (Page 1) of your answer sheet to answer the questions in this section.

Directions: For this section, solve each problem and decide which is the best of the choices given. Fill in the corresponding circle on the answer sheet. You may use any available space for scratchwork.

Notes

1. The use of a calculator is permitted.
2. All numbers used are real numbers.
3. Figures that accompany problems in this test are intended to provide information useful in solving the problems. They are drawn as accurately as possible EXCEPT when it is stated in a specific problem that the figure is not drawn to scale. All figure lie in a plane unless other indicated.
4. Unless otherwise specified, the domain of any function f is assumed to be set of all real numbers x for which $f(x)$ is a real number.

Reference Informatiom

$A = \pi r^2$ $A = \ell w$ $A = \frac{1}{2}bh$ $V = \ell wh$ $V = \pi r^2 h$ $c^2 = a^2 + b^2$ Special Right Triangles
$C = 2\pi r$

The numbers of degrees of arc in a circle is $360°$.

The sum of the measures in degrees of the angles is $180°$.

---

1. If $3x - 5 = 12$, what is the value of $6x - 5$?

(A) 24
(B) 25
(C) 27
(D) 29
(E) 39

2. Ten to fifteen tires out of 300 are defective. If there are 1800 tires in a shipment, which of the following could be the number of defective tires?

(A) 30
(B) 35
(C) 40
(D) 50
(E) 80

GO ON TO THE NEXT PAGE

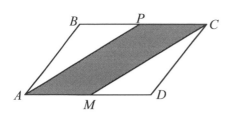

Note: Figure not drawn to scale.

3. In the figure above, $ABCD$ is a parallelogram, M is the midpoint of $\overline{AD}$, and P is the midpoint of $\overline{BC}$. If $AB = 8$, $AD = 12$, and $\angle ADC = 150^o$, what is the area of the shaded region?

(A) 24
(B) 30
(C) 42
(D) 48
(E) 60

4. The average (arithmetic mean) of $2a$ and b is 15 and the average of $2a$, b, and $3c$ is 20. What is the value of c?

(A) 5
(B) 10
(C) 15
(D) 20
(E) 25

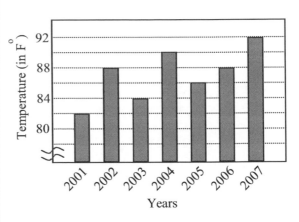

5. The graph above shows how the average temperature in July was changed for the years 2001 through 2007 in J.C town. According to the graph, which of the following two consecutive years had the greatest rate of change in temperature?

(A) From 2001 to 2002
(B) From 2002 to 2003
(C) From 2004 to 2005
(D) From 2005 to 2006
(E) From 2006 to 2007

GO ON TO THE NEXT PAGE

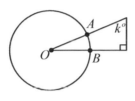

Note: Figure not drawn to scale.

6. The figure above shows a circle O with a radius of 10. If the length of arc AB is π, what is the value of k ?

(A) 50
(B) 65
(C) 70
(D) 72
(E) 80

7. In the figure above, a square is inscribed in a circle. If the length of $\overline{AB}$ is x, what is the area, in terms of x, of the shaded region?

(A) $\dfrac{(\pi-2)x^2}{2}$

(B) $\dfrac{(\pi-4)x^2}{2}$

(C) $\pi(x^2-2)$

(D) $\dfrac{\pi(x^2-2)}{2}$

(E) $\dfrac{\pi(x^2-4)}{2}$

8. If $r^2 s^3 t^4 u > 0$, then which of the following must be true?

(A) $r > 0$
(B) $s < 0$
(C) $t < 0$
(D) $su > 0$
(E) $st > 0$

9. When positive integers a and b are divided by 7, the remainders are 4 and 2 respectively. If ab is divided by 7, what is the remainder?

(A) 1
(B) 2
(C) 3
(D) 5
(E) 6

GO ON TO THE NEXT PAGE

10. If $2^x + 2^x + 2^x + 2^x = 8^{x-2}$, what is the value of x?

(A) 4
(B) 5
(C) 6
(D) 7
(E) 8

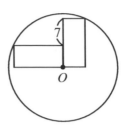

Note: Figure not drawn to scale.

12. In the figure above, two identical rectangles are inside a circle with center at point O. If the area of the rectangle is 60, what is the area of the circle?

(A) 49π
(B) 64π
(C) 81π
(D) 144π
(E) 169π

11. If the difference of the squares of two numbers is 24 and the difference of the two numbers is 4, what is the sum of these two numbers?

(A) 4
(B) 6
(C) 8
(D) 10
(E) 12

13. If x is $\dfrac{2}{5}$ of y and y is $\dfrac{5}{12}$ of z, what is the value of $\dfrac{x}{z}$?

(A) $\dfrac{1}{12}$

(B) $\dfrac{1}{6}$

(C) $\dfrac{1}{3}$

(D) $\dfrac{1}{2}$

(E) $\dfrac{2}{3}$

GO ON TO THE NEXT PAGE ⟩

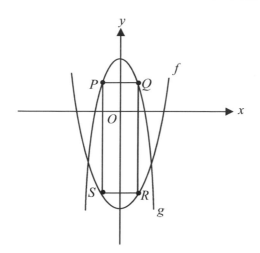

Note: Figure not drawn to scale.

14. In the figure above, the function f is defined by $f(x) = 2x^2 - 32$ and the function g is defined by $g(x) = -x^2 + 9$. If the length of $\overline{PQ}$ is 4, what is the perimeter of rectangle $PQRS$?

(A) 42
(B) 48
(C) 66
(D) 78
(E) 90

x	$f(x)$
a	10
3	16
b	22

15. The table above shows some values for the function f. If f is a linear function, what is the value of $a + b$?

(A) 4
(B) 5
(C) 6
(D) 7
(E) 8

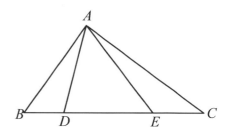

Note: Figure not drawn to scale.

16. The figure above consists of three triangles. If the ratio of the lengths of $BD : DE : EC$ is $1 : 3 : 1$ and the area of $\triangle ADE$ is 12, what is the sum of the areas of $\triangle ABD$ and $\triangle ACE$?

(A) 4
(B) 6
(C) 8
(D) 10
(E) 12

GO ON TO THE NEXT PAGE

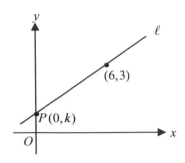

Note: Figure not drawn to scale.

17. In the *xy*-coordinate plane above, line ℓ is perpendicular to $y = -3x - 9$ (not shown). What is the value of k ?

(A) 1
(B) 2
(C) 2.5
(D) 3
(E) 3.5

18. If p is a multiple of 3 and q is a multiple of 5, which of the following must be true?

    I. $p + q$ is even.

    II. pq is odd.

    III. $5p + 3q$ is a multiple of 15.

(A) I only
(B) II only
(C) III only
(D) I and II only
(E) II and III only

19. A drawer has 10 red, 10 blue, and 10 green socks. If k is the smallest number of socks which must be chosen in order to ensure at least one pair of each color, which of the following could be the value of k ?

(A) 6
(B) 10
(C) 13
(D) 21
(E) 22

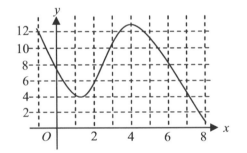

20. In the *xy*-coordinate plane above, the figure shows the graph of function f , where $-1 \le x \le 8$, and the function g is defined by $g(x) = 3f(x) - 10$. If $g(a) = 14$, which of the following could be the value of a ?

(A) 1
(B) 2
(C) 4
(D) 6
(E) 7

STOP

If you finish before time is called, you may check your work on this section only.
Do not turn to any other section in the test.

SECTION 5
Time- 25 minutes
18 Questions

Turn to Section 5 (Page 1) of your answer sheet to answer the questions in this section.

Directions: For this section, solve each problem and decide which is the best of the choices given. Fill in the corresponding circle on the answer sheet. You may use any available space for scratchwork.

Notes

1. The use of a calculator is permitted.
2. All numbers used are real numbers.
3. Figures that accompany problems in this test are intended to provide information useful in solving the problems. They are drawn as accurately as possible EXCEPT when it is stated in a specific problem that the figure is not drawn to scale. All figure lie in a plane unless other indicated.
4. Unless otherwise specified, the domain of any function f is assumed to be set of all real numbers x for which $f(x)$ is a real number.

Reference Information

$A = \pi r^2$
$C = 2\pi r$
$A = \ell w$
$A = \frac{1}{2}bh$
$V = \ell wh$
$V = \pi r^2 h$
$c^2 = a^2 + b^2$
Special Right Triangles

The numbers of degrees of arc in a circle is $360°$.

The sum of the measures in degrees of the angles is $180°$.

---

1. If $a = 5b + 3$, how much greater is a when $b = 9$ than when $b = 7$?

   (A) 10
   (B) 12
   (C) 14
   (D) 16
   (E) 18

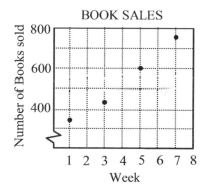

BOOK SALES

2. The graph above describes book sales at K.G book store during the first 8 weeks since the store has opened. According to the graph, which of the following is the best estimate of the number of books sold by the 6th week?

   (A) 610   (B) 650   (C) 680
   (D) 700   (E) 720

GO ON TO THE NEXT PAGE

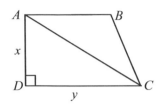

Note: Figure not drawn to scale.

3. In the figure above, if $AB = BC = 4$ and $\angle BAC = 30°$, what is the value of $x^2 + y^2$?

   (A) 28
   (B) 32
   (C) 36
   (D) 48
   (E) 72

4. If $2 < |x - 4| < 5$, which of the following could be the value of x?

   (A) -6
   (B) -7
   (C) 6
   (D) 8
   (E) 9

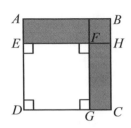

Note: Figure not drawn to scale.

5. In the figure above, $ABCD$ is a square, $EF = 10$, and $FH = 2$. If the area of the shaded region is 54, what is the area of $EFGD$?

   (A) 80
   (B) 84
   (C) 90
   (D) 96
   (E) 98

6. When a player rolls a die twice, the fraction $\dfrac{a}{b}$ is formed, where a is the first number after the first roll and b is the second number after the second roll. What is the probability that the fraction of $\dfrac{a}{b}$ is greater than 2?

   (A) $\dfrac{1}{6}$

   (B) $\dfrac{2}{9}$

   (C) $\dfrac{1}{4}$

   (D) $\dfrac{1}{3}$

   (E) $\dfrac{5}{12}$

GO ON TO THE NEXT PAGE

7. Which of the following tables shows a relationship in which y is directly proportional to x?

(A)

x	y
1	2
2	3
3	4

(B)

x	y
2	3
3	4
4	5

(C)

x	y
−3	6
−4	8
−5	10

(D)

x	y
4	6
5	8
6	10

(E)

x	y
5	10
10	15
15	20

8. The price of a certain item is $\frac{p}{3}$ dollars. A sales tax of s percent is added to the price to obtain the final cost. If p is a positive integer, what is the final cost, in dollars?

(A) $\dfrac{ps}{100}$

(B) $\dfrac{100s}{p}$

(C) $\dfrac{p(1+s)}{300}$

(D) $\dfrac{p(100+s)}{300}$

(E) $\dfrac{s(100+p)}{300}$

GO ON TO THE NEXT PAGE

Directions: For Students-Produced Response questions 9-18, use the grid at the bottom of the answer sheet page on which you have answered questions 1-8.

Each of the remaining 10 questions requires you to solve the problem and enter your answer by making the circles in the special grid, as shown in the examples below. You may use any available space for scratchwork.

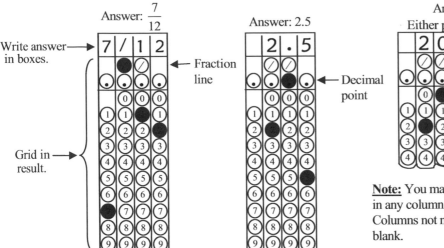

Answer: $\frac{7}{12}$

Write answer in boxes. → Fraction line

Grid in result.

Answer: 2.5

← Decimal point

Answer: 201
Either position is correct.

Note: You may start your answers in any column, space permitting. Columns not needed should be left blank.

- Mark no more than one circle in any column.

- Because the answer sheet will be machine-scored, **you will receive credit only if the circles are filled in correctly.**

- Although not required, it is suggested that you write your answer in the boxes at the top of the columns to help you fill in the circles accurately.

- Some problems may have more than one correct answer. In such cases, grid only one answer.

- No question has a negative answer.

- **Mixed numbers** such as $3\frac{1}{2}$ must be gridded as 3.5 or 7/2. (If [3 1 / 2] is gridded, it will be interpreted as $\frac{31}{2}$, not $3\frac{1}{2}$.)

- **Decimal Answers:** If you obtain a decimal answer with more digits than the grid can accommodate, it may be either rounded or truncated, but it must fill the entire grid. For example, if you obtain an answer such as 0.6666..., you should record your result as .666 or .667. **A less accurate value such as .66 or .67 will be scored as incorrect.**

Acceptable ways to grid $\frac{2}{3}$ are:

9. What is the smallest positive integer k such that k^2 is greater than or equal to 165?

10. If $p^2 + q^2 = 20$ and $p = 2q$, what is the positive value of q?

GO ON TO THE NEXT PAGE ⟹

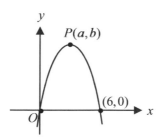

Note: Figure not drawn to scale.

11. The quadratic function f is graphed in the xy-coordinate plane above. If $b = 3a$, what is the value of b ?

12. The sum of 7 consecutive odd integers is 315. What is the value of the greatest of these integers?

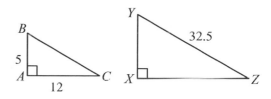

Note: Figures not drawn to scale.

13. Each angle of $\triangle ABC$ above has the same measures as an angle in $\triangle XYZ$. What is the perimeter of $\triangle XYZ$?

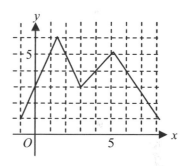

14. The figure above shows the graph of $y = f(x)$. If the function g is defined by
$$g(x) = 2f\left(\frac{x}{2}\right) + 5,$$
what is the value of $g(6)$?

GO ON TO THE NEXT PAGE

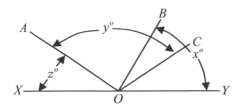

Note: Figure not drawn to scale.

15. In the figure above, $x = 60$ and $y = 100$.
If $z = 40$, what is the value of $\angle BOC$?

$$a \triangle b = a^2 - 5b$$
$$a \square b = a^2 + 5b$$

16. Let the operations $\triangle$ and $\square$ above be defined
for all real numbers a and b.
If $3(k \triangle 1) = 2(k \square 1)$, what is the positive value
of k?

17. In the xy-coordinate plane, the graph of
$y = x^2 - 9$ intersects line ℓ at point $P(a, 0)$
and point $Q(b, 16)$. What is the greatest possible
value of the slope of the line ℓ?

18. How many ways can seven books be arranged
on a shelf if two of them are math books and
must be kept together?

STOP

If you finish before time is called, you may check your work on this section only.
Do not turn to any other section in the test.

SECTION 7
Time- 20 minutes
16 Questions

Turn to Section 7 (Page 2) of your answer sheet to answer the questions in this section.

Directions: For this section, solve each problem and decide which is the best of the choices given. Fill in the corresponding circle on the answer sheet. You may use any available space for scratchwork.

Notes

1. The use of a calculator is permitted.
2. All numbers used are real numbers.
3. Figures that accompany problems in this test are intended to provide information useful in solving the problems. They are drawn as accurately as possible EXCEPT when it is stated in a specific problem that the figure is not drawn to scale. All figure lie in a plane unless other indicated.
4. Unless otherwise specified, the domain of any function f is assumed to be set of all real numbers x for which $f(x)$ is a real number.

Reference Informatiom

       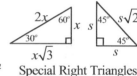

$A = \pi r^2$
$C = 2\pi r$
$A = \ell w$
$A = \dfrac{1}{2}bh$
$V = \ell wh$
$V = \pi r^2 h$
$c^2 = a^2 + b^2$
Special Right Triangles

The numbers of degrees of arc in a circle is $360°$.

The sum of the measures in degrees of the angles is $180°$.

---

1. If $pq = 1$ and $3q = 9$, what is the value of p?

(A) 2
(B) $\dfrac{1}{2}$
(C) 3
(D) $\dfrac{1}{3}$
(E) $\dfrac{1}{9}$

2. If k percent of a number n is 56, which of the following is the number n?

(A) $\dfrac{56}{k}$
(B) $\dfrac{k}{56}$
(C) $56 \times 0.k$
(D) $\dfrac{56k}{100}$
(E) $\dfrac{56 \times 100}{k}$

GO ON TO THE NEXT PAGE

3. If $3(a+2b) = 27$, what is the value of $5a+10b$?

(A) 18
(B) 35
(C) 45
(D) 60
(E) 72

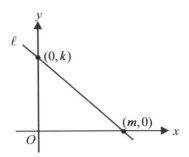

Note: Figure not drawn to scale.

5. In the figure above, the slope of line ℓ is $-\dfrac{2}{3}$. What is the value of k in terms of m?

(A) $\dfrac{3}{2}m$

(B) $-\dfrac{3}{2}m$

(C) $\dfrac{2}{3}m$

(D) $-\dfrac{2}{3}m$

(E) m

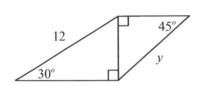

Note: Figure not drawn to scale.

4. In the figure above, what is the value of y?

(A) $16\sqrt{2}$
(B) $12\sqrt{2}$
(C) $8\sqrt{2}$
(D) $6\sqrt{2}$
(E) $4\sqrt{2}$

6. If $a+\dfrac{1}{a} = 2+\dfrac{1}{2}$, which of the following is the value of a?

  I. 2

  II. $\dfrac{1}{2}$

  III. -2

(A) I only
(B) II only
(C) III only
(D) I and II only
(E) I and III only

GO ON TO THE NEXT PAGE

Ages	Number of Students
18	2
17	6
16	12
15	2

7. The chart above shows the ages of 22 students enrolled in an art class. Which of the following is true?

(A) median > mode

(B) mode > median

(C) median = mean

(D) median = mode

(E) mode = mean

8. A fashion designer's total expenses E, in dollars, of the number of garments is given by $E(n) = 60n + 400$, where n is the number of garments. If the total expenses are t dollars, what is the value of n in terms of t ?

(A) $60t + 400$

(B) $\dfrac{t - 400}{60}$

(C) $\dfrac{60}{t - 400}$

(D) $\dfrac{60t}{400}$

(E) $\dfrac{400t}{60}$

9. If $(x + y)^2 = 30$ and $xy = 7$, what is the value of $(x - y)^2$?

(A) 1
(B) 2
(C) 10
(D) 15
(E) 20

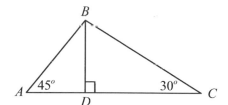

Note: Figure not drawn to scale.

10. In the figure above, if $AB = 2\sqrt{2}$, what is the area of $\triangle BCD$?

(A) 2
(B) $2\sqrt{2}$
(C) $3\sqrt{2}$
(D) $2\sqrt{3}$
(E) $3\sqrt{3}$

GO ON TO THE NEXT PAGE

11. If $\dfrac{4}{p}$ percent of a positive number is equal to p percent of the same number, what is the value of p?

(A) 1
(B) 2
(C) 3
(D) 4
(E) 5

12. In the figure above, the vertices of each square are on the midpoint of a side of the square which the vertex intersects. If the area of $ABCD$ is 48, what is the area of the shaded region?

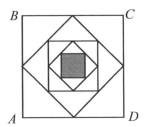

(A) 3
(B) 4
(C) 6
(D) 8
(E) 12

13. If p, q, and r are positive, which of the following represents p percent of q percent of r?

(A) $0.0001pqr$
(B) $0.001pqr$
(C) $0.1pqr$
(D) $100pqr$
(E) $10000pqr$

14. In the xy-plane above, a circle is at center O. Which of the following points is inside the circle?

(A) $(6,9)$
(B) $(7,8)$
(C) $(-8,-8)$
(D) $(-9,4)$
(E) $(5,-9)$

GO ON TO THE NEXT PAGE

15. If $p < q < 0$, which of the following must be

less than $\dfrac{p}{q}$?

(A) $-p$
(B) $-q$
(C) pq
(D) 2
(E) 1

16. Let function f be defined by
$f(x) = (x-h)^2 + k$, where h and k are constants and function g be defined by
$g(x) = f(x-3)$. If $g(3) = 5$ and $g(4) = 2$, what is the value of h ?

(A) 2
(B) 4
(C) 6
(D) 8
(E) 10

STOP

If you finish before time is called, you may check your work on this section only.
Do not turn to any other section in the test.

TEST 9 ANSWER KEY

#	SECTION 3	SECTION 5	SECTION 7
1	D	A	D
2	E	C	E
3	A	D	C
4	B	D	D
5	A	C	C
6	D	A	D
7	A	C	D
8	D	D	B
9	A	13	B
10	A	2	D
11	B	9	B
12	E	51	A
13	B	75	A
14	C	11	D
15	C	20	E
16	C	5	A
17	A	8	
18	C	1440	
19	E		
20	D		

TEST 9   SECTION 3

1. (D)
Since
$3x = 17$, $6x - 5 = 2(3x) - 5 = 2(17) - 5 = 29$.

2. (E)
Since $10 \leq$ Defective tires (out of 300) ≤ 15,
$10 \times 6 \leq$ Defective tires $(300 \times 6) \leq 15 \times 6$.
Therefore,
$60 \leq$ Defective tires ≤ 90.

3. (A)

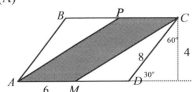

In the figure above, the parallelogram has a

base of 6 and a height of 4. Therefore, the area is $6 \times 4 = 24$.

4. (B)
Since $\dfrac{2a + b}{2} = 15$ and $\dfrac{2a + b + 3c}{3} = 20$,
$2a + b = 30$.
Thus, $\dfrac{30 + 3c}{3} = 20 \Rightarrow c = 10$.

5. (A)

(A) $\dfrac{88 - 82}{88} = 0.073$

(B) $\dfrac{84 - 88}{88} = |-0.045| = 0.045$

(C) $\dfrac{86 - 90}{90} = |-0.044| = 0.044$

(D) $\dfrac{88 - 86}{86} = 0.023$

(E) $\dfrac{92-88}{88}=0.045$

6. (D)

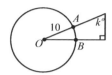

Since the circumference is $2\pi r = 20\pi$,

$\angle AOB = 360^{\circ} \times \dfrac{\pi}{20\pi} = 18^{\circ}$. Therefore,

$k = 72^{\circ}$.

7. (A)

Since $AB = x$, then $AC = x\sqrt{2}$ and the area

of the circle is $\pi\left(\dfrac{x\sqrt{2}}{2}\right)^{2} = \dfrac{\pi x^{2}}{2}$. Therefore,

the area of the shaded region =

$\dfrac{\pi x^{2}}{2} - x^{2} = \dfrac{(\pi-2)x^{2}}{2}$.

8. (D)

$\dfrac{r^{2}s^{3}t^{4}u}{r^{2}s^{2}t^{4}} > 0 \Rightarrow su > 0$

9. (A)

Select $a = 11$ and $b = 9$, then $ab = 99$.
When 99 is divided by 7, the remainder is 1.
Or, algebraically, let $a = 7q_{1} + 4$ and
$b = 7q_{2} + 2$.
Thus, $ab = 49q_{1}q_{2} + 14q_{1} + 28q_{2} + 8 =$
$7(7q_{1}q_{2} + 2q_{1} + 4q_{2} + 1) + 1$.
Therefore, the remainder is 1.

10. (A)

Since
$2^{x} + 2^{x} + 2^{x} + 2^{x} = 4 \times 2^{x} = 2^{2} \times 2^{x} = 2^{x+2}$,
then $2^{x+2} = 8^{x-2} = 2^{3x-6}$. Therefore,
$x + 2 = 3x - 6 \Rightarrow x = 4$.

11. (B)

$x^{2} - y^{2} = (x+y)(x-y) = 24$ and $x - y = 4$.
Therefore, $x + y = 6$.

12. (E)

The area of the rectangle $= x(x+7) = 60$.
Thus, $x^{2} + 7x - 60 = (x+12)(x-5) = 0$.
$x = 5$.
Therefore, the radius of the
circle$= \sqrt{5^{2} + 12^{2}} = 13$.
The area of the circle$= \pi(13)^{2} = 169\pi$.

13. (B)

Since $x = \dfrac{2}{5}y$ and $z = \dfrac{12}{5}y$, then

$\dfrac{x}{z} = \dfrac{\dfrac{2}{5}y}{\dfrac{12}{5}y} = \dfrac{1}{6}$.

14. (C)

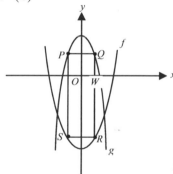

Since $PQ = 4$, $OW = 2$. Thus,
$QW = g(2) = -(2)^{2} + 9 = 5$ and
$RW = f(2) = 2(2)^{2} - 32 = -24$. Therefore,
$QR = 5 + |-24| = 29$. The perimeter is
$2(29+4) = 66$.

15. (C)

Linear means " constant slope."

Since $\dfrac{16-4}{3-a} = \dfrac{22-16}{b-3} = $ Constant slope.

Using cross-multiplication, $a + b = 6$.

16. (C)

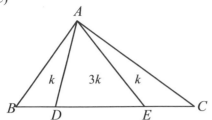

With the same height, the ratio of the bases must be equal to the ratio of the areas. Thus, the ratio of the areas is also $k : 3k : k$.
$3k = 12 \rightarrow k = 4$. Therefore, $k + k = 2k = 6$.

---

17. (A)

The slope of line ℓ is $\dfrac{1}{3}$. Therefore,

$\dfrac{3 - k}{6 - 0} = \dfrac{1}{3} \rightarrow k = 1$.

---

18. (C)
Choose $p = 3, 6$ and $q = 5, 10$.
I. $p + q = 3 + 5 = 8$ or $3 + 10 = 13$. Not always even.
II. $pq = 3 \times 5 = 15$ or $3 \times 10 = 30$. Not always odd.
III. $5p + 3q = 5(3) + 3(5) = 30$ or $5(3) + 3(10) = 45$.
     Always multiple of 15.

---

19. (E)
The worst case is as follows to have at least one pair of each color (RR)(BB)(GG),

RRRRRRRRRR BBBBBBBBBB GG
The 22nd sock is the correct answer.

---

20. (D)
Since $g(a) = 3f(a) - 10 = 14$, then $f(a) = 8$.
Therefore, for three values of a, the value of y is 8. $f(6) = 8$.

TEST 9   SECTION 5

1. (A)
If $b = 9$, $a = 5(9) + 3 = 48$, and if $b = 7$,
$a = 5(7) + 3 = 38$. Therefore, $48 - 38 = 10$.

2. (C)

The number of books $= \dfrac{600 + 750}{2} \simeq 675$.
Therefore 680 is best approximation.

---

3. (D)

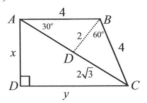

In the figure above, $\triangle ABD$ is a special triangle. Therefore, $AD = 4\sqrt{3}$.
$\rightarrow x^2 + y^2 = \left(4\sqrt{3}\right)^2 = 48$.

---

4. (D)
If $x = 8$, $2 < |8 - 4| = 4 < 5$. True.

---

5. (C)

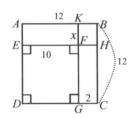

In the figure above, the area of $\square BCGK$ is 24. Thus, the area of $\square AEFK$ is, 54-24=30.
$AE = \dfrac{30}{10} = 3$. Therefore,
$FG = 12 - 3 = 9$ and the area of $\square EFGD$ is, $10 \times 9 = 90$.
Or, let $KF = x$. Then $FG = 12 - x$.
Therefore,
$(12 \times 12) - 10(12 - x) = 54 \rightarrow x = 3$.
$FG = 12 - 3 = 9$. The area is $9 \times 10 = 90$.

---

6. (A)

Since $\dfrac{a}{b} > 2$, a must be greater than 2.

If $a = 3$, then $b = 1. \rightarrow (3, 1)$
If $a = 4$, then $b = 1. \rightarrow (4, 1)$
If $a = 5$, then $b = 1, 2. \rightarrow (5, 1), (5, 2)$
If $a = 6$, then $b = 1, 2. \rightarrow (6, 1), (6, 2)$

Therefore, $P = \dfrac{6}{36} = \dfrac{1}{6}$.

7. (C)

Direct proportion: $y = kx$, where k is a constant.

Only (C) has a constant of -2.

---

8. (D)

The final price

$$= \frac{p}{3}\left(1 + \frac{s}{100}\right) = \frac{p}{3}\left(\frac{100+s}{100}\right) = \frac{p(100+s)}{300}.$$

---

9. 13

Since $k^2 \geq 165$, the smallest value of k is 13.

---

10. 2

Since $p = 2q$, $4q^2 + q^2 = 5q^2 = 20 \rightarrow q = 2$.

---

11. 9

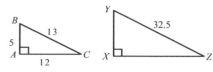

In the figure, $2a = 6 \rightarrow a = 3$. Therefore, $b = 3a = 3(3) = 9$.

---

12. 51

The average = median = $\frac{315}{7} = 45$.

▢ ▢ ▢ ▢ ▢ ▢ ▢

45 47 49 51

---

13. 75

In the figure above, $\dfrac{5}{XY} = \dfrac{12}{XZ} = \dfrac{13}{32.5} = \dfrac{2}{5}$.

Therefore, $XY = 12.5$ and $XZ = 30$.

The perimeter $= 32.5 + 12.5 + 30 = 75$.

---

14. 11

$g(6) = 2f\left(\frac{6}{2}\right) + 5 = 2f(3) + 5$.

Since $f(3) = 3$, $g(6) = 2 \times 3 + 5 = 11$.

---

15. 20

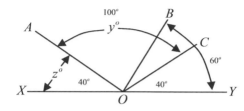

In the figure above,

$\angle BOC = 60^{\circ} - 40^{\circ} = 20^{\circ}$.

---

16. 5

Since $3(k^2 - 5) = 2(k^2 + 5)$,

$k^2 = 25 \rightarrow k = 5$.

---

17. 8

Point $P \rightarrow$

$0 = a^2 - 9 \rightarrow a = \pm 3 \rightarrow (3,0),(-3,0)$.

Point $Q \rightarrow$

$16 = b^2 - 9 \rightarrow b = \pm 5 \rightarrow (5,16),(-5,16)$.

Therefore, the slope between (5, 16) and (3, 0) is the greatest one. That is, $\dfrac{16-0}{5-3} = 8$.

---

18. 1440

If two books are kept together, the possible arrangements will be,

$2(6 \times 5 \times 4 \times 3 \times 2 \times 1) = 1440$.

TEST 9   SECTION 7

1. (D)

Since $q = 3$, then $3p = 1 \rightarrow p = \dfrac{1}{3}$.

---

2. (E)

Since $\dfrac{kn}{100} = 56$, $n = \dfrac{5600}{k}$.

---

3. (C)

Since $a + 2b = 9$, $5(a + 2b) = 5 \times 9 = 45$.

---

4. (D)

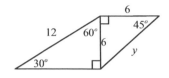

In the figure above, $y = 6\sqrt{2}$.

---

5. (C)

$\dfrac{k-0}{0-m} = -\dfrac{2}{3} \;\rightarrow\; 3k = 2m \;\rightarrow\; k = \dfrac{2m}{3}$.

---

6. (D)

Since $a + \dfrac{1}{a} = 2 + \dfrac{1}{2}$, then $a = 2$ or $\dfrac{1}{2}$.

---

7. (D)

Mean $= 45.5$, Median $= 16$, and Mode $= 16$

---

8. (B)

Since $t = 60n + 400$, $n = \dfrac{t - 400}{60}$.

---

9. (B)

$(x-y)^2 = (x+y)^2 - 4xy = 30 - 28 = 2.$

---

10. (D)

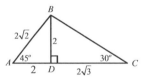

In the figure above, the area of

$\triangle BCD = \dfrac{2 \times 2\sqrt{3}}{2} = 2\sqrt{3}$.

---

11. (B)

Since $\dfrac{\frac{4}{p}}{100} \times n = \dfrac{p}{100} \times n$,

$\dfrac{4}{p} = p \;\rightarrow\; p^2 = 4 \;\rightarrow\; p = 2.$

---

12. (A)

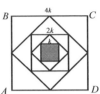

In the figure above, the ratio of the lengths of the largest square to the smallest square is 4:1.

Thus, the ratio of the areas $=16{:}1$. Therefore, the area of smallest square $= 48 \times \dfrac{1}{16} = 3$.

---

13. (A)

$\dfrac{p}{100} \times \dfrac{q}{100} \times r = \dfrac{pqr}{10000} = 0.0001\,pqr.$

---

14. (D)

$(-9, 4)$: The distance to the origin

$= \sqrt{(-9)^2 + 4^2}$

$= \sqrt{97} < \sqrt{100}$

---

15. (E)

Since $p < q < 0$, when divide by q

$\rightarrow \dfrac{p}{q} > \dfrac{q}{q} = 1$

Or, let $p = -11$ and $q = -10$. Then $\dfrac{p}{q} = 1.1$

(A) $-p = 11$ (B) $-q = 10$ (C) $pq = 110$

(D) 2   They are all greater than 1.1.

---

16. (A)

$g(3) = f(0) = h^2 + k = 5. \;\rightarrow\; k = 5 - h^2$.

$g(4) = f(1) = (1-h)^2 + k = 2$. Therefore,

$(1-h)^2 + 5 - h^2 = 2 \;\rightarrow\; -1 + 2h = 3 \;\rightarrow\; h = 2.$

$\boxed{END}$

Dr. John Chung's SAT Math

TEST
10

ANSWER SHEET

TEST #:

Math Scoring Worksheet

A. Section 3

    _____   _____
    numer of correct   *number of incorrect*

    +   +

B. Section 5 (1-8)

    _____   _____
    numer of correct   *number of incorrect*

    +

C. Section 5 (9-18)

    _____
    numer of correct

    +   +

D. Section 7

    _____   _____
    numer of correct   *number of incorrect*

    =   =

E. Total Unrounded Raw Score

    _____ − _____ ÷4 = _____
    numer of correct   *number of incorrect*

F. Total Rounded Raw Score   _____ (See table)

    Math Score Range = | _____ — _____ |

Math Conversion Table

Raw Score	Scaled Score	Raw Score	Scaled Score
54	800	23	490-550
53	780-800	22	480-540
52	760-800	21	470-530
51	740-800	20	460-520
50	720-780	19	450-510
49	700-760	18	450-510
48	690-750	17	440-500
47	680-740	16	430-490
46	670-730	15	420-480
45	660-720	14	420-480
44	650-710	13	410-470
43	650-710	12	400-460
42	640-700	11	390-450
41	630-690	10	380-440
40	620-680	9	390-430
39	610-670	8	380-420
38	610-670	7	370-410
37	600-660	6	360-400
36	590-650	5	340-380
35	580-640	4	320-370
34	570-630	3	310-360
33	560-620	2	300-350
32	560-620	1	270-320
31	550-610	0	240-300
30	540-600	-1	200-290
29	530-590	-2	200-270
28	530-590	-3	200-260
27	520-580	-4	200-240
26	510-570	-5	200-220
25	500-560	-6 and below	200
24	500-560		

SECTION 3
Time- 25 minutes
20 Questions

Turn to Section 3 (Page 1) of your answer sheet to answer the questions in this section.

Directions: For this section, solve each problem and decide which is the best of the choices given. Fill in the corresponding circle on the answer sheet. You may use any available space for scratchwork.

Notes

1. The use of a calculator is permitted.
2. All numbers used are real numbers.
3. Figures that accompany problems in this test are intended to provide information useful in solving the problems. They are drawn as accurately as possible EXCEPT when it is stated in a specific problem that the figure is not drawn to scale. All figure lie in a plane unless other indicated.
4. Unless otherwise specified, the domain of any function f is assumed to be set of all real numbers x for which $f(x)$ is a real number.

Reference Informatiom

$A = \pi r^2$
$C = 2\pi r$ $A = \ell w$ $A = \dfrac{1}{2}bh$ $V = \ell wh$ $V = \pi r^2 h$ $c^2 = a^2 + b^2$ Special Right Triangles

The numbers of degrees of arc in a circle is 360°.

The sum of the measures in degrees of the angles is 180°.

---

1. If x and y are integers and $x^2 - y^2 = 16$, which of the following cannot be the values of x?

   (A) -5
   (B) -4
   (C) 0
   (D) 4
   (E) 5

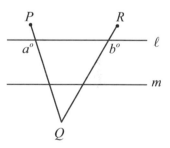

2. In the figure above, if $\angle PQR = 25^o$, what is the value of $a + b$?

   (A) 155
   (B) 205
   (C) 215
   (D) 235
   (E) 245

GO ON TO THE NEXT PAGE

3. If k is an integer and $k = \dfrac{m}{3}$, which of the following must be true?

      I. m is an even number.
     II. m is a multiple of 3.
    III. m is an odd number.

(A) I only
(B) II only
(C) III only
(D) I and II only
(E) I, II, and III

4. For what value of x does $(x-1)^3 = x^3 - 1$?

(A) $x = 1$ only
(B) $x = -1$ only
(C) $x = 0$ or $x = 1$
(D) $x = 1$ or $x = -1$
(E) $x = 0$ or $x = -1$

5. In a regular n-sided polygon, the measure of each interior angle d, in degrees, is given by the function $d = \dfrac{180(n-2)}{n}$. If the regular polygon has an interior angle of 135°, what is the number of sides of the polygon?

(A) 5
(B) 6
(C) 8
(D) 10
(E) 12

6. If $ab = k$, $b = ka$, and $k \neq 0$, which of the following could be equal to a ?

(A) 0
(B) $\dfrac{1}{2}$
(C) 1
(D) b
(E) $\dfrac{1}{b}$

GO ON TO THE NEXT PAGE

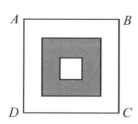

7. The figure above shows three squares of different sizes. The lengths of the edge of each square are 5, 3, and 1 respectively. If a point is chosen from the square $ABCD$, what is the probability that the point will be chosen from the shaded region?

(A) $\dfrac{2}{5}$

(B) $\dfrac{8}{25}$

(C) $\dfrac{9}{25}$

(D) $\dfrac{3}{5}$

(E) $\dfrac{12}{25}$

8. If ab is an odd integer, which of the following must also be an odd integer?

      I. $a+b$
      II. $a-b$
      III. $(a+b)(a-b)-a$

(A) I only
(B) II only
(C) III only
(D) I and II only
(E) II and III only

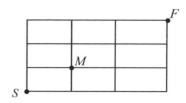

9. In the figure above, a path from point S to point F is determined by moving upward or to the right along the grid line. How many different paths can be drawn from S to F that includes point M ?

(A) 8
(B) 10
(C) 12
(D) 24
(E) 36

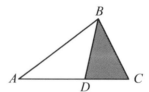

Note: Figure not drawn to scale.

10. In the figure above, the area of the shaded region is $70cm^2$. If the ratio of the lengths of $\overline{AD}$ to $\overline{DC}$ is 7:5, what is the area of $\triangle ABC$?

(A) $98cm^2$
(B) $128cm^2$
(C) $134cm^2$
(D) $156cm^2$
(E) $168cm^2$

GO ON TO THE NEXT PAGE

11. If $k^{-4} = 16a^2$, then $k^2 =$

(A) $4a^2$

(B) $\dfrac{1}{4a}$

(C) $\dfrac{a}{4}$

(D) $4a$

(E) $\dfrac{a^2}{4}$

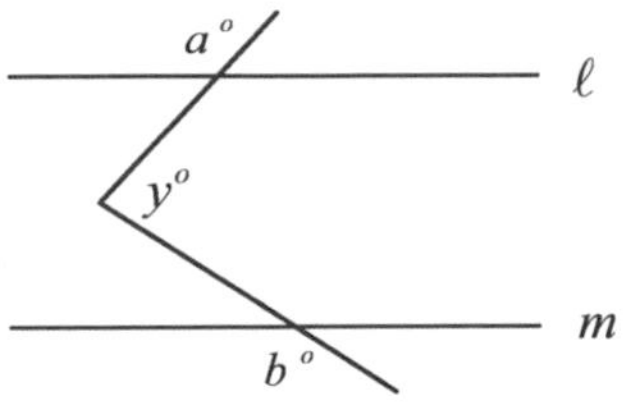

Note: Figure not drawn to scale.

13. In the figure above, line ℓ and m are parallel. What is the value of y ?

(A) 90
(B) $a + b$
(C) $90 + a + b$
(D) $180 - a - b$
(E) $360 - a - b$

12. The function h is defined by $h(x) = 3x - 2$ and the function g is defined by $g(x) = 3h(x) - 1$. If $g(k) = 11$, what is the value of k ?

(A) 2
(B) 3
(C) 4
(D) 5
(E) 6

14. If the median of 12 consecutive odd integers is 30, what is the smallest integer among these integers?

(A) 13
(B) 17
(C) 19
(D) 21
(E) 23

GO ON TO THE NEXT PAGE

MATH TEST RESULTS

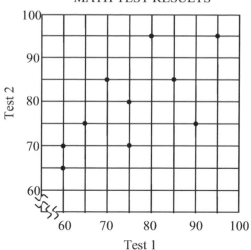

15. The scatter plot above shows the results of ten students on their two math tests. What is the median score on test 1?

(A) 70
(B) 72.5
(C) 75
(D) 77.5
(E) 80

16. The equation of line ℓ is given by $y = -2x - 5$. If line m is perpendicular to line ℓ, which of the following must be true?

(A) Line m has a negative slope.
(B) Line m has a positive slope.
(C) Line m has a positive x-intercept.
(D) Line m has a negative x-intercept.
(E) Line m has a positive y-intercept.

17. Let the operation $\odot$ be defined by

$a \odot b = \dfrac{ab}{a+b}$ for all positive values of a and

b. If $\dfrac{1}{a} \odot \dfrac{1}{b} = \dfrac{1}{2}$ and $a \odot \dfrac{1}{a} = \dfrac{1}{2}$, what is the

value of b ?

(A) 1
(B) 2
(C) 3
(D) 4
(E) 5

18. 10 men take x days to build a bridge. If n more men are hired and all men work at the same rate, how many fewer days will it take?

(A) $\dfrac{10x}{10+n}$

(B) $\dfrac{10x(10+n)}{x+n}$

(C) $\dfrac{10+n}{10x}$

(D) $\dfrac{nx}{10+n}$

(E) $\dfrac{nx}{10-n}$

GO ON TO THE NEXT PAGE ▷

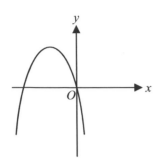

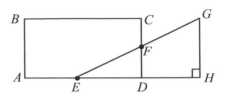

Note: Figure not drawn to scale.

19. The figure above shows the graph of the function f defined by $f(x) = ax^2 + bx + c$. Which of the following could be true?

(A) $a > 0$ and $b > 0$
(B) $a > 0$ and $b < 0$
(C) $a < 0$ and $b > 0$
(D) $a < 0$ and $b < 0$
(E) $a < 0$, $b < 0$, and $c > 0$

20. In the figure above, quadrilateral $ABCD$ is a rectangle such that E is the midpoint of $\overline{AD}$, F is the midpoint of $\overline{CD}$ and $\overline{EG}$, $AB = 10$, and $BC = 24$. What is the perimeter of quadrilateral $DFGH$?

(A) 35
(B) 40
(C) 45
(D) 50
(E) 55

STOP

If you finish before time is called, you may check your work on this section only.
Do not turn to any other section in the test.

SECTION 5
Time- 25 minutes
18 Questions

Turn to Section 5 (Page 1) of your answer sheet to answer the questions in this section.

Directions: For this section, solve each problem and decide which is the best of the choices given. Fill in the corresponding circle on the answer sheet. You may use any available space for scratchwork.

Reference Informatiom

$A = \pi r^2$
$C = 2\pi r$
$A = \ell w$
$A = \dfrac{1}{2}bh$
$V = \ell wh$
$V = \pi r^2 h$
$c^2 = a^2 + b^2$
Special Right Triangles

The numbers of degrees of arc in a circle is $360°$.

The sum of the measures in degrees of the angles is $180°$.

---

1. If $\left|4 - k^2\right| < 3$, which of the following could be the value of k?

(A) 0
(B) 1
(C) 2
(D) 3
(E) 4

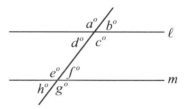

Note: Figure not drawn to scale.

2. In the figure above, if $\ell \parallel m$, which of the following cannot be true?

(A) $a = c$
(B) $c + f = 180$
(C) $d = f$
(D) $a + c = e + g$
(E) $c + e = 270$

GO ON TO THE NEXT PAGE

STUDENTS AT J.C. SCHOOL

	Boys	Girls	Total
Seniors			95
Juniors			
Total	100		160

3. The table above shows the number of students at J.C. School, and the numbers are partially filled. If $\frac{2}{3}$ of girls are seniors, how many boys are juniors?

(A) 20
(B) 40
(C) 45
(D) 60
(E) 65

4. If a six- pack of soda costs k cents, how many individual cans can be purchased for d dollars at the same rate?

(A) $\frac{100d}{k}$

(B) $\frac{6d}{k}$

(C) $\frac{100}{kd}$

(D) $\frac{600k}{d}$

(E) $\frac{600d}{k}$

5. If $(k-a)x+k=5$ for all value of x, where a and k are constants, which of the following is equal to a?

(A) 4
(B) 5
(C) 6
(D) 12
(E) 15

6. If the ratio of the numbers of boys to the number of girls in a certain group is $\frac{5}{7}$, which of the following could be the number of students in the group?

(A) 100
(B) 150
(C) 200
(D) 250
(E) 300

GO ON TO THE NEXT PAGE

7. If $\sqrt{100} = a\sqrt{b}$, where a and b are positive integers, which of the following could not be the value of a?

(A) 1
(B) 2
(C) 5
(D) 10
(E) 100

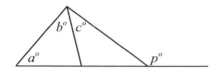

8. In the figure above, what is the value of p in terms of a, b, and c?

(A) $a + c$
(B) $180 - a - b - c$
(C) $360 - a - b - c$
(D) $a + b + c$
(E) $a + b + c - 180$

GO ON TO THE NEXT PAGE

5 ☐☐☐ 5 ☐☐☐ 5 ☐☐☐ 5

Directions: For Students-Produced Response questions 9-18, use the grid at the bottom of the answer sheet page on which you have answered questions 1-8.

Each of the remaining 10 questions requires you to solve the problem and enter your answer by making the circles in the special grid, as shown in the examples below. You may use any available space for scratchwork.

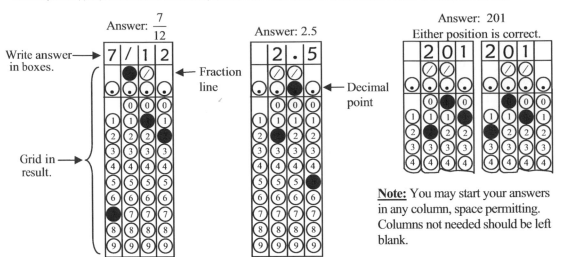

Answer: $\frac{7}{12}$

Write answer in boxes. → Fraction line

Grid in result.

Answer: 2.5 → Decimal point

Answer: 201
Either position is correct.

Note: You may start your answers in any column, space permitting. Columns not needed should be left blank.

- Mark no more than one circle in any column.

- Because the answer sheet will be machine-scored, **you will receive credit only if the circles are filled in correctly.**

- Although not required, it is suggested that you write your answer in the boxes at the top of the columns to help you fill in the circles accurately.

- Some problems may have more than one correct answer. In such cases, grid only one answer.

- No question has a negative answer.

- **Mixed numbers** such as $3\frac{1}{2}$ must be gridded as 3.5 or 7/2. (If ☐ is gridded, it will be interpreted as $\frac{31}{2}$, not $3\frac{1}{2}$.)

- **Decimal Answers:** If you obtain a decimal answer with more digits than the grid can accommodate, it may be either rounded or truncated, but it must fill the entire grid. For example, if you obtain an answer such as 0.6666…, you should record your result as .666 or .667. **A less accurate value such as .66 or .67 will be scored as incorrect.**

Acceptable ways to grid $\frac{2}{3}$ are:

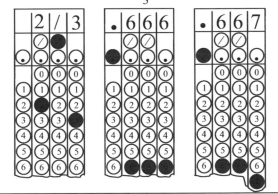

9. If $a^3 = 64$, what is the value of $a^{\frac{3}{2}}$?

x^2

10. On the number line above, there are 4 equal intervals between 2 and 3. What is the positive value of x?

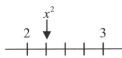

GO ON TO THE NEXT PAGE

11. The average of two angles of an isosceles triangle, in degrees, is 70. What is the greatest possible measure of the angles? (Discard degree unit)

12. If $x^2 - y^2 = 15$, where x and y are positive integers, what is one possible value of x?

13. The first term in a sequence is a, and each term after the first is r times the preceding term. If the third term is 18 and the sixth term is 486, what is the value of $a + r$?

14. Let the function f be defined by $f(x) = \dfrac{x^2}{3} + 6$. If $f(3m) = 9m$, what is one possible value of m?

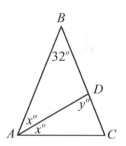

15. In the figure above, If $AB = BC$, what is the value of y?

GO ON TO THE NEXT PAGE

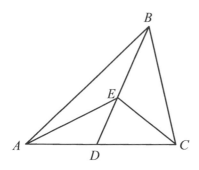

Note: Figure not drawn to scale.

16. In the figure above, $\dfrac{BE}{BD} = \dfrac{2}{3}$. What is the value

of the fraction $\dfrac{\text{area } \triangle ABE}{\text{area } \triangle AED}$?

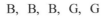

B, B, B, G, G

17. Three boys and two girls shown above are lined up side by side in a single row so that boys and girls are lined up in groups respectively. How many different arrangements are possible?

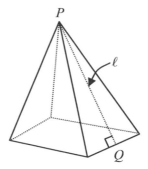

18. In the figure above, the formula for the surface area S of a regular pyramid is $S = \dfrac{1}{2}P\ell + B$, where P is the perimeter of the base, ℓ is the slant height ($\overline{PQ}$ shown in the figure), and B is the area of the base. If the length of a side of the square is 10 and the height of the pyramid is 12, what is the value of S ?

STOP

If you finish before time is called, you may check your work on this section only.
Do not turn to any other section in the test.

SECTION 7
Time- 20 minutes
16 Questions

Turn to Section 7 (Page 2) of your answer sheet to answer the questions in this section.

Directions: For this section, solve each problem and decide which is the best of the choices given. Fill in the corresponding circle on the answer sheet. You may use any available space for scratchwork.

1. The ratio of boys to girls is 3:5 in a certain class. $\frac{1}{5}$ of the boys are awarded for an excellent accomplishment. If the number of awarded students is 6, how many students are in the class?

   (A) 25
   (B) 40
   (C) 60
   (D) 80
   (E) 100

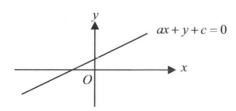

2. The figure above shows the graph of $ax + y + c = 0$. Which of the following is true?

   (A) $a > 0$
   (B) $a = 0$
   (C) $a < 0$
   (D) $c > 0$
   (E) $c = 0$

GO ON TO THE NEXT PAGE

PETER'S HOURS

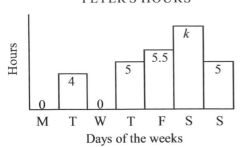

Days of the weeks

3. Peter works in a restaurant and is paid $12 an hour. If his average (arithmetic mean) amount of earning per day is $48.00, what is the value of k?

(A) 6
(B) 7
(C) 7.5
(D) 8
(E) 8.5

4. Positive integers a, b, and c satisfy the equations $a^b = 8$ and $b^c = 81$. If $b > a$, what is the value of c?

(A) 1
(B) 2
(C) 3
(D) 4
(E) 9

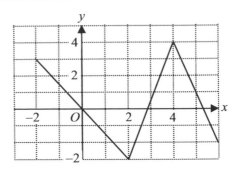

5. The figure above shows the graph of the function g. If $g(k) = 4$, what is the value of $g\left(\dfrac{k}{2}\right)$?

(A) −2
(B) −1
(C) 0
(D) 1
(E) 2

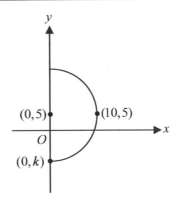

Note: Figure not drawn to scale.

6. In the semicircle above, the center is at $(0,5)$. What is the value of k?

(A) −3
(B) −5
(C) −6
(D) −8
(E) −10

GO ON TO THE NEXT PAGE

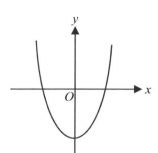

7. The figure above shows the graph of $y = ax^2 + b$. Which of the following graphs could represent the graph of $y = ax + b$?

(A)

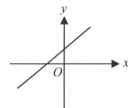

(B)

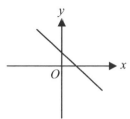

(C)

(D)

(E)

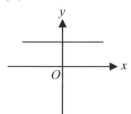

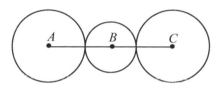

Note: Figure not drawn to scale.

8. In the figure above, the area of circle A : the area of circle B : the area of circle C is 9:1:9. If the radius of circle B is 2, what is the length of $\overline{AC}$?

(A) 12
(B) 14
(C) 16
(D) 18
(E) 20

9. The price of a postage-stamp was first increased by 10 percent and then the new price was increased again by 5 percent. By what percent was the final price increased from the initial price?

(A) 15.5%
(B) 15%
(C) 14.5%
(D) 14%
(E) 12.5%

GO ON TO THE NEXT PAGE

10. When the number k, where $k > 0$, is cubed and multiplied by 4, the result is the same as when the number is divided by 9. What is the value of k?

(A) 36

(B) 6

(C) $\dfrac{1}{6}$

(D) $\dfrac{1}{18}$

(E) $\dfrac{1}{36}$

12. If $\dfrac{p}{q}$ is a positive integer, which of the following must be true?

      I. p is an integer.

      II. q is an integer.

      III. $\dfrac{3p}{q}$ is an integer.

(A) I only
(B) II only
(C) III only
(D) I and II only
(E) 1, II, and III

11. In how many integers from 1 to 200 does the digit 9 appear at least once?

(A) 38
(B) 37
(C) 25
(D) 20
(E) 18

13. If $k > 10$ or $k < -20$, which of the following is equivalent to the interval of k?

(A) $|k| > 20$
(B) $|k - 5| < 15$
(C) $|k + 5| < 15$
(D) $|k + 5| > 15$
(E) $|k - 5| > 15$

GO ON TO THE NEXT PAGE

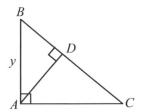

Note: Figure not drawn to scale.

14. The figure above shows the graph of $y = (1 + x)(5 - x)$. If $\triangle ABC$ is an isosceles, what is the area of $\triangle ABC$?

(A) 9
(B) 12
(C) 18
(D) 27
(E) 54

B
y
D
A
C

Note: Figure not drawn to scale.

15. In the figure above, if $BD = a$ and $CD = b$, which of the following could be used to find y ?

(A) $y = ab$
(B) $y^2 = ab$
(C) $y^2 = b(a + b)$
(D) $y^2 = a(a + b)$
(E) $y^2 = \sqrt{a(a + b)}$

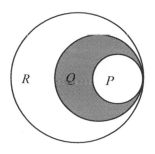

Note: Figure not drawn to scale.

16. In the figure above, circle P, Q, and R are internally tangent. The radius of circle P is $\dfrac{1}{3}$ of the radius of circle Q, and the radius of circle Q is $\dfrac{3}{4}$ of the radius of circle R. If the area of circle R is 48π, what is the area of the shaded region?

(A) 12π
(B) 24
(C) 24π
(D) 32
(E) 32π

STOP

If you finish before time is called, you may check your work on this section only.
Do not turn to any other section in the test.

TEST 10

#	SECTION 3	SECTION 5	SECTION 7
1	C	C	D
2	B	E	C
3	B	C	E
4	C	E	D
5	C	B	A
6	C	E	B
7	B	E	E
8	C	D	C
9	C	8	A
10	E	1.5	C
11	B	100	A
12	A	4 or 8	C
13	E	5	D
14	C	1 or 2	D
15	C	69	D
16	B	2	C
17	A	24	
18	D	360	
19	D		
20	B		

ANSWER KEY heading: **TEST 10   ANSWER KEY**

TEST 10   SECTION 3

1. (C)

If $x = 0$, $y^2 = -16$. y^2 cannot be negative.

2. (B)

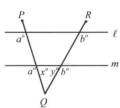

In the figure above, $x + y = 180 - 25 = 155$.

Since $a + x + y + b = 360$,

$a + b = 360 - (x + y) = 360 - 155 = 205$.

3. (B)

Since $m = 3k$, m is a multiple of 3.

I. m can be negative integer. $k = \dfrac{-6}{2} = -3$

m is not always positive.

III. m can be even number. $k = \dfrac{6}{2} = 3$

4. (C)

When $x = 0$, $-1 = -1$ (true)

When $x = 1$, $0 = 0$ (true)

Or, algebraically

$(x - 1)^3 = x^3 - 3x^2 + 3x - 1 = x^3 - 1$

The equation is simplified.

$-3x^2 + 3x = -3x(x - 1) = 0$

Therefore $x = 0$ or 1.

5. (C)

$\dfrac{180(n - 2)}{n} = 135 \rightarrow 180(n - 2) = 135n$

$45n = 360 \rightarrow n = 8$

6. (C)

$ab = k$

$\rightarrow a(ka) = k \rightarrow a^2 k = k \rightarrow a^2 = 1$

Therefore, $a = \pm 1$.

---

7. (B)

Since the ratio of the lengths is $5:3:1$, the ratio of the areas is $25:9:1 = 25k:9k:k$. The area of the shaded region = $9k - k = 8k$.

Therefore, the probability $P = \dfrac{8k}{25k} = \dfrac{8}{25}$.

Or, just use the number.

$P = \dfrac{9-1}{25} = \dfrac{8}{25}$

---

8. (C)

Since ab is odd, a is odd and b is odd. Let $a = 3$ and $b = 1$. Then $a + b = 4$ (even), $a - b = 2$ (even), and $(a+b)(a-b) - a = 4 \times 2 - 3 = 5$ (odd).

---

9. (C)

Use addition at intersection.

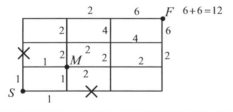

---

10. (E)

Since $AD:DC = 7:5$, the ratio of the areas is $7:5 = 7k:5k$. Therefore, $5k = 70$, $k = 14$. The area of $\triangle ABC = 12k = 12 \times 14 = 168$.

Or, use proportion.

$\dfrac{5}{70} = \dfrac{12}{x} \rightarrow x = 168$

---

11. (B)

$k^{-4} = 4^2 a^2 \rightarrow \left(k^{-4}\right)^{-\frac{1}{2}} = \left(4^2 a^2\right)^{-\frac{1}{2}} \rightarrow k^2 = 4^{-1} a^{-1}$

Therefore, $4^{-1} a^{-1} = \dfrac{1}{4a}$.

---

12. (A)

Since $g(k) = 3h(k) - 1$ and $h(k) = 3k - 2$, then $g(k) = 3(3k - 2) - 1 = 9k - 7$. Therefore, $9k - 7 = 11 \rightarrow k = 2$.

---

13. (E)

In the figure above, $a + b + y = 360$. Therefore, $y = 360 - (a + b)$.

---

14. (C)

The median is between 6 and 7th term. Since the sixth term is 29, the first term is $29 - (5 \times 2) = 19$.

---

15. (C)

The scores on Test 1 are as follows.

$60, 60, 65, 70, 75, 75, 80, 85, 90, 95$

Since the two terms are in the middle, the average is 75.

---

16. (B)

The equation of line m is $y = \dfrac{1}{2}x + b$. The x- and y-intercepts can be positive or negative. But the slope is always positive.

---

17. (A)

$\dfrac{1}{a} \odot \dfrac{1}{b} = \dfrac{\dfrac{1}{a} \times \dfrac{1}{b}}{\dfrac{1}{a} + \dfrac{1}{b}} = \dfrac{1}{a+b} = \dfrac{1}{2} \rightarrow a + b = 2.$ And

$a \odot \dfrac{1}{a} = \dfrac{a \times \dfrac{1}{a}}{a + \dfrac{1}{a}} = \dfrac{1}{a^2 + 1} = \dfrac{1}{2} \rightarrow a = 1.$

Therefore, $b = 1$.

---

18. (D)

Let y be the days for $(10 + n)$ workers.

Inverse proportion: $10 \times x = (10 + n) \times y$.

Since $y = \dfrac{10x}{10 + n}$, the difference $(x - y)$ is

$$x - \frac{10x}{10+n} = \frac{(10x+nx)-10x}{10+n} = \frac{nx}{10+n}.$$

19. (D)

Axis of symmetry $= \dfrac{-b}{2a} < 0$, $a < 0$, and

$c = 0$. Therefore, $a < 0$, and $b < 0$, and

$c = 0$.

20. (B)

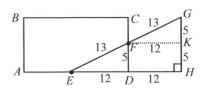

In the figure above, $\triangle DEF \simeq \triangle KFG$.
Therefore, the perimeter of $DFGH$ is
$5 + 13 + 10 + 12 = 40$.

TEST 10   SECTION 5

1. (C)
When $k = 2$, $0 < 3$ (ok)

2. (E)
The value of $c + e$ cannot be determined from
the information given.

3. (C)

STUDENTS AT J.C. SCHOOL

	Boys	Girls	Total
Seniors		40	95
Juniors	45	20	65
Total	100	60	160

$\dfrac{2}{3}$ of $60 = 40$.

4. (E)
Direct proportion: Let x be the number of
cans.

$$\frac{k}{6} = \frac{100d}{x} \;\rightarrow\; x = \frac{600d}{k}.$$

5. (B)
Identical equations: $k - a = 0$ and
$k = 5$. Therefore, $a = 5$.

---

6. (E)
Since boys $= 5k$ and girls $= 7k$, the total is
$12k$, where k is a positive integer. Therefore,
the number of students must be a multiple of
12.

7. (E)
$1\sqrt{100} = 2\sqrt{25} = 5\sqrt{4} = 10\sqrt{1}$. Therefore,
a cannot be 100.

8. (D)
Exterior angle = sum of the two non-adjacent
angles.

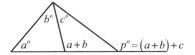

9. 8
Since $a^3 = 64$, $a = 4$. Therefore,
$$(4)^{\frac{3}{2}} = \left(2^2\right)^{\frac{3}{2}} = 2^3 = 8.$$

10. 1.5
Since $x^2 = 2.25$, $x = 1.5$.

11. 100
Two possibilities.

(1) $\dfrac{a+a}{2} = 70 \rightarrow a = 70$ and $b = 40$.

(2) $\dfrac{a+b}{2} = 70$ and

$2a + b = 180 \rightarrow a = 40$ and $b = 100$.

Therefore, the greatest possible one is 100.

---

12. 4 or 8
Since $x^2 - y^2 = (x+y)(x-y) = 15$,
$x + y = 15,\ 5$
$x - y = 1\ ,\ 3$
$2x = 16$ or $8\ \rightarrow x = 4$ or 8.

---

13. 5

Since $\dfrac{ar^5 = 486}{ar^2 = 18} \rightarrow r^3 = 27$. Therefore,

$r = 3$. From $ar^2 = 18$, $a = 2 \Rightarrow a + r = 5$

---

14. 1 or 2

Since

$f(3m) = \dfrac{9m^2}{3} + 6 = 3m^2 + 6,\ 3m^2 + 6 = 9m.$

Thus, $m^2 - 3m + 2 = (m-1)(m-2) = 0.$

Therefore $m = 1$ or 2.

---

15. 69

Since $\angle A = \angle C,\ 4x = 148 \rightarrow x = 37$.

Therefore, $y = 180 - 3x = 180 - 111 = 69.$

---

16. 2

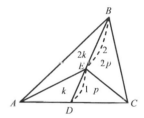

In the figure above, since the ratio of the lengths of $\triangle ABE$ to $\triangle AED$ is 2:1, then the ratio of their areas is also 2:1. Therefore, the

ratio is $\dfrac{2}{1}$. The two triangles have the same

height.

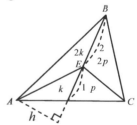

---

17. 24

Two possible arrangements as follows.
        BBBGG or GGBBB
Since each has

$(3 \times 2 \times 1)(2 \times 1) = 12$ arrangements, there are

$2 \times 12 = 24$ arrangements.

18. 360

$P = 10 \times 4 = 40,\ B = 10 \times 10 = 100$,and

$\ell = \sqrt{5^2 + 12^2} = 13.$ Therefore,

$S = \dfrac{1}{2}(40)(13) + 100 = 360.$

---

1. (D)

Let the number of students $= x.$

Since $\dfrac{1}{5}$ of $\dfrac{3}{8}x = \dfrac{3x}{40} = 6,\ x = 6 \times \dfrac{40}{3} = 80.$

---

2. (C)

$y = -ax - c.$ Since slope$= -a > 0$, and y-intercept $= -c > 0$, then $a < 0$ and $c < 0.$

---

3. (E)

Since total working hours
$= 4 + 5 + 5.5 + k + 5 = 19.5 + k$, then

$\dfrac{12(19.5 + k)}{7} = 48.$ Therefore, $k = 8.5.$

---

4. (D)

Since $a^b = 2^3$, b is 3. Thus, $3^c = 81 = 3^4.$

$c = 4.$

---

5. (A)

$g(k) = 4 \rightarrow k = 4.\ g\left(\dfrac{k}{2}\right) = g(2) = -2.$

---

6. (B)

Since the radius is 10, $k = 5 - 10 = -5.$

---

7. (E)

From the graph, $a > 0$ and $c < 0.$ Therefore the graph has a positive slope and a negative y-intercept.

---

8. (C)

Since the ratio of the lengths is $3 : 1 : 3$, let their radius be $3k,\ k$, and $3k.$ Because $k = 2,\ AC = 8k = 8 \times 2 = 16.$

---

9. (A)

If the original price is n, the final price is $(1.05)(1.1)n = 1.155n.$ Since

$1.155n = (1 + 0.155)n$, 15.5% was increased.

---

10. (C)

Since $4k^3 = \dfrac{k}{9}$ and $k > 0$,

$k^2 = \dfrac{1}{36} \rightarrow k = \dfrac{1}{6}$.

---

11. (A)

For only one 9,

One digit number $\rightarrow \boxed{9}$ --------- 1

Two digit-numbers $\rightarrow \boxed{}\,\boxed{9}$ --------- 8

$\rightarrow \boxed{9}\,\boxed{}$ --------- 9

Three digit-numbers $\rightarrow \boxed{1}\,\boxed{}\,\boxed{9}$ --------- 9

$\rightarrow \boxed{1}\,\boxed{9}\,\boxed{}$ --------- 9

For two 9's,

$\underset{}{99, \ 199}$ --------- 2

$\overline{38}$

Therefore, there are 38 integers.

Or, find all the numbers which do not contain digit 9 as follows.

One-digit number ---- 8 numbers
Two-digit number ---- $8 \times 9 = 72$ numbers
Three-digit number (100~199) ----
$1 \times 9 \times 9 = 81$ numbers
and 200---- 1 number
There are 162 numbers.
Therefore, $200 - 162 = 38$ numbers contain digit 9 at least one.

---

12. (C)

Think counterexamples. When $p = \dfrac{1}{2}$ and

$q = \dfrac{1}{2}$, $\dfrac{p}{q}$ is integer.

$3\left(\dfrac{p}{q}\right) = $ integer $\times$ integer $=$ integer.

---

13. (D)

The middle number $= \dfrac{10 + (-20)}{2} = -5$ and

difference is $10 - (-5) = 15$. Therefore,

$|k - (-5)| > 15 \rightarrow |k + 5| > 15$.

---

14. (D)

Since the zeros are $x = -1$ and $x = 5$,

$AC = 5 - (-1) = 6$. The axis of symmetry is

$x = \dfrac{5 + (-1)}{2} = 2$.

The maximum height (the height of the

triangle) $= f(2) = (1 + 2)(5 - 2) = 9$.

Therefore, the area of $\triangle ABC = \dfrac{6 \times 9}{2} = 27$.

---

15. (D)

It's the formula. $y^2 = BD \times (BD + BC)$.

---

16. (C)

The ratio of the lengths $= 1 : \dfrac{3}{4} : \dfrac{1}{4} = 4 : 3 : 1$.

The ratio of the areas $= 16 : 9 : 1 = 16k : 9k : k$.

Since $16k = 48\pi$, $k = 3\pi$. Therefore, the area

of the unshaded region $= 8k = 24\pi$.

$\boxed{END}$

Dr. John Chung's SAT Math

TEST

11

Math Scoring Worksheet

A. Section 3

_____   _____
numer of correct   *number of incorrect*

+   +

B. Section 5 (1-8)

_____   _____
numer of correct   *number of incorrect*

+

C. Section 5 (9-18)

_____
numer of correct

+   +

D. Section 7

_____   _____
numer of correct   *number of incorrect*

=   =

E. Total Unrounded Raw Score

_____   −   _____   ÷4 = _____
numer of correct   *number of incorrect*

F. Total Rounded Raw Score

_____ (See table)

Math Score Range =   [_____]

Math Conversion Table

Raw Score	Scaled Score	Raw Score	Scaled Score
54	800	23	490-550
53	780-800	22	480-540
52	760-800	21	470-530
51	740-800	20	460-520
50	720-780	19	450-510
49	700-760	18	450-510
48	690-750	17	440-500
47	680-740	16	430-490
46	670-730	15	420-480
45	660-720	14	420-480
44	650-710	13	410-470
43	650-710	12	400-460
42	640-700	11	390-450
41	630-690	10	380-440
40	620-680	9	390-430
39	610-670	8	380-420
38	610-670	7	370-410
37	600-660	6	360-400
36	590-650	5	340-380
35	580-640	4	320-370
34	570-630	3	310-360
33	560-620	2	300-350
32	560-620	1	270-320
31	550-610	0	240-300
30	540-600	-1	200-290
29	530-590	-2	200-270
28	530-590	-3	200-260
27	520-580	-4	200-240
26	510-570	-5	200-220
25	500-560	-6 and below	200
24	500-560		

SECTION 3
Time- 25 minutes
20 Questions

Turn to Section 3 (Page 1) of your answer sheet to answer the questions in this section.

Directions: For this section, solve each problem and decide which is the best of the choices given. Fill in the corresponding circle on the answer sheet. You may use any available space for scratchwork.

Notes

1. The use of a calculator is permitted.
2. All numbers used are real numbers.
3. Figures that accompany problems in this test are intended to provide information useful in solving the problems. They are drawn as accurately as possible EXCEPT when it is stated in a specific problem that the figure is not drawn to scale. All figure lie in a plane unless other indicated.
4. Unless otherwise specified, the domain of any function f is assumed to be set of all real numbers x for which $f(x)$ is a real number.

Reference Informatiom

$A = \pi r^2$
$C = 2\pi r$
$A = \ell w$
$A = \frac{1}{2}bh$
$V = \ell w h$
$V = \pi r^2 h$
$c^2 = a^2 + b^2$
Special Right Triangles

The numbers of degrees of arc in a circle is $360°$.

The sum of the measures in degrees of the angles is $180°$.

---

1. If $x^3 = y^4$ and $x = 16$, what is the value of y ?

(A) 4
(B) 8
(C) 16
(D) 27
(E) 36

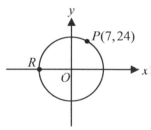

Note: Figure not drawn to scale.

2. In the figure above, the coordinates of point R are $(k, 0)$. If the circle is at the center $(0, 0)$, what is the value of k ?

(A) 24
(B) −24
(C) 25
(D) −25
(E) 32

GO ON TO THE NEXT PAGE

3. The Sporting Goods store offers 5 different types of bicycles and 3 different wheel sizes. How many bicycle choices does the store offer?

   (A) 8
   (B) 10
   (C) 12
   (D) 15
   (E) 30

4. For which of the following function does $f(3) = -f(-3)$?

   (A) $f(x) = x^2$

   (B) $f(x) = 4$

   (C) $f(x) = x + 2$

   (D) $f(x) = x^3 + 4$

   (E) $f(x) = \dfrac{3}{x}$

5. If the proportion $a : b : c = 3 : 1 : 5$, what is the value of $\dfrac{2a + 3b}{4b + 3c}$?

   (A) $\dfrac{1}{2}$

   (B) $\dfrac{9}{19}$

   (C) $\dfrac{5}{7}$

   (D) $\dfrac{6}{7}$

   (E) $\dfrac{11}{19}$

6. If y is 10% of a number, what is k percent of 50% of the number?

   (A) $\dfrac{k}{y}$

   (B) $\dfrac{y}{k}$

   (C) $\dfrac{k}{10y}$

   (D) $\dfrac{ky}{20}$

   (E) $\dfrac{ky}{100}$

GO ON TO THE NEXT PAGE

7. In a plane, M is the midpoint of $\overline{AB}$, the length of $\overline{PM}$ is 8, and the length of $\overline{AM}$ is 15. Which of the following could not be the length of $\overline{AP}$?

(A) 7
(B) 13
(C) 19
(D) 23
(E) 24

8. If $(a+b)^2 = (a-b)^2$, which of the following must be true?

(A) $a = 0$
(B) $b < 0$
(C) $a = b$
(D) $ab = 0$
(E) $a + b = 0$

9. For a linear function f, $f(1) = 8$ and $f(7) = -10$. If $f(k) = 5$, what is the value of k?

(A) 0
(B) 2
(C) 3
(D) 5
(E) 6

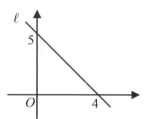

10. In the figure above, line m (not shown) is parallel to line ℓ, and passes through $(0, 0)$. Which of the following equations represents the m?

(A) $4x - 5y = 4$
(B) $4x + 5y = 0$
(C) $5x - 4y = 0$
(D) $5x + 4y = 0$
(E) $5x + 4y = 4$

GO ON TO THE NEXT PAGE

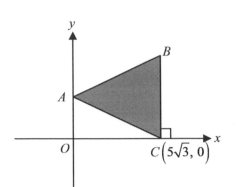

11. In the figure above, $\triangle ABC$ is an equilateral triangle. What is the area of $\triangle ABC$?

(A) 20
(B) 25
(C) $20\sqrt{3}$
(D) $25\sqrt{3}$
(E) $50\sqrt{3}$

12. If $-1 < x < 0$, which of the following is true?

(A) $x < x^2 < x^3$
(B) $x^2 < x < x^3$
(C) $x < x^3 < x^2$
(D) $x^3 < x^2 < x$
(E) $x^3 < x < x^2$

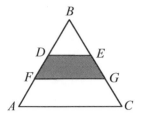

Note: Figure not drawn to scale.

13. In the figure above, $\overline{DE} \parallel \overline{FG} \parallel \overline{AC}$ and $AF = FD = DB$. If the area of $\square AFGC$ is 20, what is the area of the shaded region?

(A) 8
(B) 10
(C) 12
(D) 14
(E) 16

Tommy, Sonia, Matt, Roger

14. The students above are the four candidates for student council president. If their names are placed in random order on the ballot, what is the probability that Tommy's name will be first on the ballot followed by Sonia's name second?

(A) $\dfrac{1}{3}$

(B) $\dfrac{1}{6}$

(C) $\dfrac{1}{12}$

(D) $\dfrac{1}{18}$

(E) $\dfrac{1}{24}$

GO ON TO THE NEXT PAGE

15. For how many ordered pairs of positive integers (x, y) is $4x + 5y < 15$?

(A) None
(B) One
(C) Two
(D) Three
(E) Four

Note: Figure not drawn to scale.

16. In the figure above, the line m is perpendicular to line ℓ and passes through point P. What is the value of t ?

(A) 3
(D) 3.5
(C) 4
(D) 4.5
(E) 5

17. The function f is defined by

$f(x) = a(x - b)^2 + k$ for all real x, where a, b, and k are constants. If a and k are negative, which of the following CANNOT be true?

(A) $f(5) = -3$
(B) $f(0) = -5$
(C) $f(2) = 4$
(D) $f(1) = k$
(E) $f(-2) = b$

GO ON TO THE NEXT PAGE

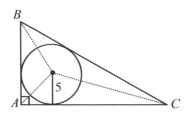

Note: Figure not drawn to scale.

18. In the figure above, the radius of a circle is 5 and the length of $\overline{BC}$ is 20. What is the area of $\triangle ABC$?

(A) 60
(B) 75
(C) 90
(D) 125
(E) 150

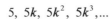

$$5, \; 5k, \; 5k^2, \; 5k^3, \ldots$$

19. The first term in the sequence above is 5, and each term after the first is k times the preceding term. If the sum of the first three terms is 15, which of the following could be the value of k ?

(A) −2
(B) −1
(C) 0
(D) 2
(E) 3

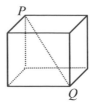

20. In the cube above, the length of a diagonal $\overline{PQ}$ is 12. What is the surface area of the cube?

(A) 27
(B) 64
(C) 125
(D) 288
(E) 316

STOP

If you finish before time is called, you may check your work on this section only.
Do not turn to any other section in the test.

SECTION 5
Time- 25 minutes
18 Questions

Turn to Section 5 (Page 1) of your answer sheet to answer the questions in this section.

Directions: For this section, solve each problem and decide which is the best of the choices given. Fill in the corresponding circle on the answer sheet. You may use any available space for scratchwork.

Notes

1. The use of a calculator is permitted.
2. All numbers used are real numbers.
3. Figures that accompany problems in this test are intended to provide information useful in solving the problems. They are drawn as accurately as possible EXCEPT when it is stated in a specific problem that the figure is not drawn to scale. All figure lie in a plane unless other indicated.
4. Unless otherwise specified, the domain of any function f is assumed to be set of all real numbers x for which $f(x)$ is a real number.

Reference Informatiom

$A = \pi r^2$ $A = \ell w$ $A = \frac{1}{2}bh$ $V = \ell wh$ $V = \pi r^2 h$ $c^2 = a^2 + b^2$ Special Right Triangles
$C = 2\pi r$

The numbers of degrees of arc in a circle is $360°$.

The sum of the measures in degrees of the angles is $180°$.

---

1. If $\dfrac{a+b}{a-b} = 1$, what is the value of b?

   (A) -1
   (B) 0
   (C) 1
   (D) 2
   (E) 3

2. At Camp David, there is one adult for every p campers. If there are c campers in each tent group, how many adults are needed for t tents groups?

   (A) $\dfrac{cp}{t}$

   (B) $\dfrac{pt}{c}$

   (C) $\dfrac{ct}{p}$

   (D) pct

   (E) $pc + t$

GO ON TO THE NEXT PAGE

VACATION TIME

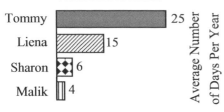

Average Number of Days Per Year

Tommy 25
Liena 15
Sharon 6
Malik 4

3. The bar graph above shows the average number of vacation days for workers per year. Which of the following circle graphs best describes the same data?

(A)

(B)

(C)

(D)

(E)

4. If p and q are multiples of 3, which of the following cannot be a multiple of 3?

(A) $p+q$

(B) $p-q$

(C) pq

(D) $\dfrac{p}{q}$

(E) $2p+3q$

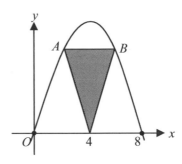

5. The figure above shows the graph of $y = x(8-x)$ and an isosceles triangle. If $AB = 4$, what is the area of the triangle?

(A) 48
(B) 36
(C) 32
(D) 28
(E) 24

GO ON TO THE NEXT PAGE

6. If $4^x = \left(\dfrac{1}{2}\right)^{y-3}$, what is the value of y in terms of x?

(A) $2x$
(B) $2x-3$
(C) $2x+3$
(D) $3-2x$
(E) $3+2x$

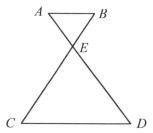

7. In the figure above, $\triangle ABE$ and $\triangle CDE$ are equilateral triangles. If the length of $\overline{AD}$ is equal to k, what is the perimeter of the figure?

(A) $2k$
(B) $2.5k$
(C) $3k$
(D) $3.5k$
(E) $4k$

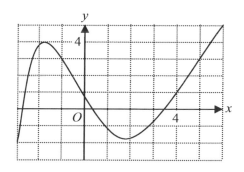

8. The figure above shows the graph of a function f. If $f\left(\dfrac{a}{2}\right) = f(5)$, which of the following could be the value of a?

(A) -3

(B) -2

(C) -1

(D) $-\dfrac{1}{2}$

(E) $-\dfrac{1}{4}$

GO ON TO THE NEXT PAGE

Dr. John Chung's SAT Math Test 11

Directions: For Students-Produced Response questions 9-18, use the grid at the bottom of the answer sheet page on which you have answered questions 1-8.

Each of the remaining 10 questions requires you to solve the problem and enter your answer by making the circles in the special grid, as shown in the examples below. You may use any available space for scratchwork.

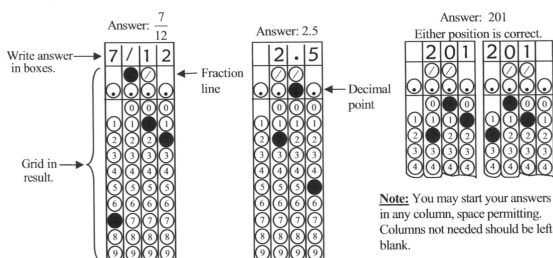

Answer: $\frac{7}{12}$

Write answer in boxes. → Fraction line

Grid in result.

Answer: 2.5 ← Decimal point

Answer: 201
Either position is correct.

Note: You may start your answers in any column, space permitting. Columns not needed should be left blank.

- Mark no more than one circle in any column.

- Because the answer sheet will be machine-scored, **you will receive credit only if the circles are filled in correctly.**

- Although not required, it is suggested that you write your answer in the boxes at the top of the columns to help you fill in the circles accurately.

- Some problems may have more than one correct answer. In such cases, grid only one answer.

- No question has a negative answer.

- **Mixed numbers** such as $3\frac{1}{2}$ must be gridded as 3.5 or 7/2. (If $\boxed{3\,1\,/\,2}$ is gridded, it will be interpreted as $\frac{31}{2}$, not $3\frac{1}{2}$.)

- **Decimal Answers:** If you obtain a decimal answer with more digits than the grid can accommodate, it may be either rounded or truncated, but it must fill the entire grid. For example, if you obtain an answer such as 0.6666..., you should record your result as .666 or .667. **A less accurate value such as .66 or .67 will be scored as incorrect.**

Acceptable ways to grid $\frac{2}{3}$ are:

9. The denominator of a certain fraction is 10 more than the numerator. If the fraction is equal to $\frac{1}{2}$, what is the denominator of the fraction?

10. The average (arithmetic mean) of a and b is 5, and the average of a^2 and b^2 is 30. What is the value of ab?

GO ON TO THE NEXT PAGE ⇨

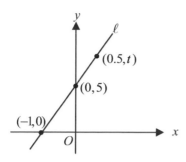

Note: Figure not drawn to scale.

11. In the figure above, three points lie on line ℓ.
    What is the value of t ?

12. The smallest integer in the set of consecutive
    even integers is -10. If the sum of these
    integers is 42, how many integers are in this set?

13. Let the functions f and g be defined by
    $g(x) = 2f(x) - 1$. If $g(1) = 3$, what is the
    value of $f(1)$?

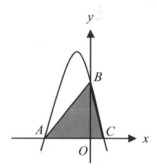

Note: Figure not drawn to scale.

14. The figure above shows the graph of
    $y = -(x-2)(x+5)$. What is the area of
    $\triangle ABC$?

GO ON TO THE NEXT PAGE

15. For all values of a and b, let $a \lozenge b$ be defined by $a \lozenge b = ab - a + 1$. If $k \lozenge (k-2) = 2 \lozenge 3$, what is the positive value of k?

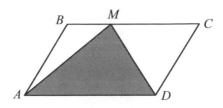

Note: Figure not drawn to scale.

17. In the figure above, $\square ABCD$ is a parallelogram. If $\dfrac{BM}{CM} = 2:3$ and the area of $\triangle ABM$ is 20, what is the area of the shaded region?

16. If $m + 5n$ is equal to 250 percent of $4n$, what is the value of $\dfrac{m+n}{m-n}$?

18. In the xy-coordinate plane, the distance between point $P(-1,6)$ and point $Q(3,a)$ is $4\sqrt{2}$. What is one possible value of a?

STOP

**If you finish before time is called, you may check your work on this section only.
Do not turn to any other section in the test.**

SECTION 7
Time- 20 minutes
16 Questions

Turn to Section 7 (Page 2) of your answer sheet to answer the questions in this section.

Directions: For this section, solve each problem and decide which is the best of the choices given. Fill in the corresponding circle on the answer sheet. You may use any available space for scratchwork.

Notes

1. The use of a calculator is permitted.
2. All numbers used are real numbers.
3. Figures that accompany problems in this test are intended to provide information useful in solving the problems. They are drawn as accurately as possible EXCEPT when it is stated in a specific problem that the figure is not drawn to scale. All figure lie in a plane unless other indicated.
4. Unless otherwise specified, the domain of any function f is assumed to be set of all real numbers x for which $f(x)$ is a real number.

Reference Informatiom

$A = \pi r^2$
$C = 2\pi r$ $A = \ell w$ $A = \dfrac{1}{2}bh$ $V = \ell wh$ $V = \pi r^2 h$ $c^2 = a^2 + b^2$ Special Right Triangles

The numbers of degrees of arc in a circle is $360°$.

The sum of the measures in degrees of the angles is $180°$.

---

1. If A is the set of positive odd integers less than 9, and B is the set of prime numbers, how many integers will be in both sets?

(A) 1
(B) 2
(C) 3
(D) 4
(E) 5

2. If $4 < \sqrt{6k} < 5$, then $k =$

(A) 1
(B) 3
(C) 5
(D) 7
(E) 9

GO ON TO THE NEXT PAGE

3. A bag contains two white marbles, four blue marbles, and an unknown number of black marbles. When one marble is drawn from the bag at random, the probability that the marble is black will be $\frac{1}{3}$. How many black marbles are in the bag?

(A) 3
(B) 4
(C) 5
(D) 6
(E) 8

PRICE OF MEAT

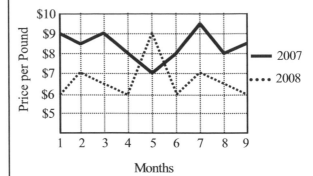

Months

5. The two graphs show the price per pound of meat for the first 9 months between 2007 and 2008. Which month had the greatest rate of change in price from year 2007 to 2008?

(A) Month 1
(B) Month 3
(C) Month 5
(D) Month 7
(E) Month 9

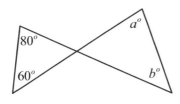

4. In the figure above, what is the value of $a + b$?

(A) 40
(B) 80
(C) 120
(D) 140
(E) 160

6. If a and b are positive integers, and $3^a \cdot 3^b = 27$, what is the value of $3^a + 3^b$?

(A) 9
(B) 12
(C) 18
(D) 27
(E) 36

GO ON TO THE NEXT PAGE

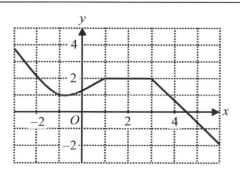

7. The graph of $y = f(x)$ is shown above. For how many values of k does $f(k) = 2$?

(A) One
(B) Two
(C) Three
(D) Four
(E) More than four

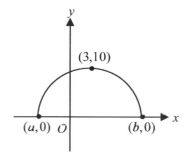

8. The figure above shows the graph of a semicircle with a radius of 10. What is the value of ab ?

(A) -80
(B) -83
(C) -91
(D) -95
(E) -97

9. A certain dancing group does not receive applicants who have a height of less than 5 feet or more than 6 feet. Which of the following inequalities can be used to determine the height h, in feet, that is not accepted in the group?

(A) $|h-5| > 6$
(B) $|h-6| > 5$
(C) $|h-5.5| < 0.5$
(D) $|h-5.5| > 0.5$
(E) $|h-5.5| > 6$

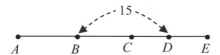

Note: Figure not drawn to scale.

10. In the figure above, $\dfrac{AC}{CE} = \dfrac{2}{1}$, B is the midpoint of $\overline{AC}$ and D is the midpoint of $\overline{CE}$. If $BD-15$, what is the length of $\overline{DE}$?

(A) 4
(B) 5
(C) 7
(D) 8
(E) 10

GO ON TO THE NEXT PAGE

$$y = ax + b$$

Note: Figure not drawn to scale.

11. The figure above shows the graph of $y = ax + b$. Which of the following graphs best represents the graph of $y = ax^2 + b$?

(A)

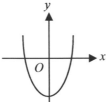

(B)

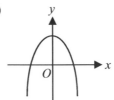

(C)

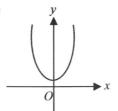

(D)

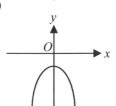

(E)

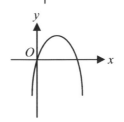

12. If $k = \left(\dfrac{1}{64}\right)^{\frac{1}{m}}$, where k is a positive integer, which of the following could be the value of m?

(A) 3
(B) 2
(C) 1
(D) 0
(E) −1

13. If a and b are positive numbers, and 125 percent of a^2 is equal to 5 percent of b^2, what is the value of $\dfrac{a}{b}$?

(A) 0.1
(B) 0.2
(C) 1
(D) 2
(E) 5

GO ON TO THE NEXT PAGE

14. If x and y are integers such that the value of $x^2 + y^2 + 2xy$ is odd, which of the following must be true?

(A) x is even.

(B) x is odd.

(C) y is even.

(D) y^2 is odd.

(E) xy is even.

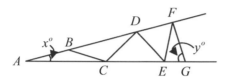

Note: Figure not drawn to scale.

15. In the figure above,
$AB = BC = CD = DE = EF = FG$,
$\angle BAG = x^o$, and $\angle EGF = y^o$. If $x = 15$,
what is the value of y?

(A) 70

(B) 75

(C) 78

(D) 80

(E) 82

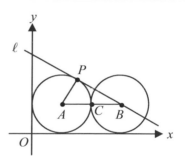

Note: Figure not drawn to scale.

16. In the figure above, the circle with center A and the circle with center B are tangent at point C. If the circles each have radius 10, and if line ℓ is tangent to the circle with center A at point P, what is the coordinates of point P?

(A) $(12, 13)$

(B) $(13, 15)$

(C) $(15, 10\sqrt{3})$

(D) $(15, 15\sqrt{3})$

(E) $(15, 10 + 5\sqrt{3})$

STOP

If you finish before time is called, you may check your work on this section only.
Do not turn to any other section in the test.

TEST 11 ANSWER KEY

#	SECTION 3	SECTION 5	SECTION 7
1	B	B	C
2	D	C	B
3	D	C	A
4	E	D	D
5	B	E	A
6	D	D	B
7	E	C	E
8	D	B	C
9	B	20	D
10	D	20	B
11	D	7.5 or $\frac{15}{2}$	C
12	C	14	E
13	C	2	B
14	C	35	E
15	D	4	B
16	A	1.5 or $\frac{3}{2}$	E
17	C	50	
18	D	2 or 10	
19	A		
20	D		

TEST 11   SECTION 3

1.   (B)
Since $(16)^3 = (2^4)^3 = (2^3)^4 = y^4$, then
$y = 2^3 = 8$.

2.   (D)
Since $OP = OR = \sqrt{7^2 + 24^2} = 25$, $k = -25$.

3.   (D)
$5 \times 3 = 15$.

4.   (E)
$f(x) = -f(-x)$. (E) is the odd function.

5.   (B)
Since $a:b:c = 3:1:5$, let $a = 3k$, $b = k$, and
$c = 5k$.

Therefore,
$$\frac{2a+3b}{4b+3c} = \frac{2(3k)+3(k)}{4(k)+3(5k)} = \frac{9k}{19k} = \frac{9}{19}.$$

---

6.   (D)
Since $y = 0.1x$, then $x = 10y$. Therefore,
$$\frac{k}{100} \times \frac{50}{100} \times 10y = \frac{ky}{20}.$$

7.   (E)

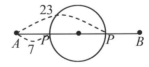

In the figure above, the minimum of
$AP = 7$, and the maximum of $AP = 23$.
Therefore, $7 \le AP \le 23$.

8. (D)

$(a+b)^2 = a^2 + 2ab + b^2$ and

$(a-b)^2 = a^2 - 2ab + b^2$.

Since $(a+b)^2 = (a-b)^2$, then $ab = 0$.

---

9. (B)

For the three point of $(1, 8)$, $(7, -10)$, and

$(k, 5)$, The slope is constant. Therefore,

$\dfrac{8-(-10)}{1-7} = \dfrac{5-8}{k-1}$.

$k = 2$.

---

10. (D)

Since the slope of line ℓ is $-\dfrac{5}{4}$, the slope of

line m is also $-\dfrac{5}{4}$. Therefore, the function of

line m is $y = -\dfrac{5}{4}x$, or $5x + 4y = 0$.

---

11. (D)

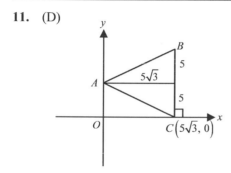

In the special triangle, the

area $= \dfrac{10 \times 5\sqrt{3}}{2} = 25\sqrt{3}$.

---

12. (C)

Select $x = -\dfrac{1}{2}$, then $x^2 = \dfrac{1}{4}$ and $x^3 = -\dfrac{1}{8}$.

Therefore, $x < x^3 < x^2$.

---

13. (C)

	$\triangle BDE$	$\triangle BFG$	$\triangle BAC$
(Length)	1 :	2 :	3
(Area)	1 :	4 :	9

Since the ratio of the areas is 1:4:9, let their

areas be k, $4k$, and $9k$. Therefore, the area

of $AFGC = 5k = 20$.

$k = 4$. The area of $DFGE = 3k = 12$.

---

14. (C)

All possible arrangements $= 4 \times 3 \times 2 \times 1 = 24$.

Desirable arrangements: T S M R or T S R M

$= 2$.

Therefore, probability is $P = \dfrac{2}{24} = \dfrac{1}{12}$.

---

15. (D)

Form the inequality $4x + 5y < 15$,

If $x = 1$, then $5y < 15 \to y = 1, 2$

If $x = 2$, then $5y < 7 \to y = 1$

If $x = 3$, then $5y < 3 \to$ no integer value of

y.

Therefore, the pairs will be $(1,1)$, $(1,2)$, and

$(2,1)$.

---

16. (A)

Since the slope of line ℓ is $-\dfrac{6}{10} = -\dfrac{3}{5}$, the

slope of line m is $\dfrac{5}{3}$. Therefore, $\dfrac{2t-1}{t-0} = \dfrac{5}{3}$.

$t = 3$.

---

17. (C)

Since $a < 0$ and $k < 0$, the graph of f is as

follows.

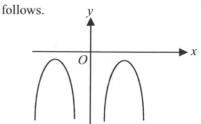

Therefore, for any x, $f(x) < 0$.

The value of b can be positive or negative.

---

18. (D)

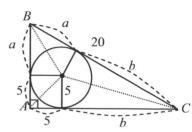

Since $a + b = 20$, the area (a square and 4

triangles) is

$$(5 \times 5) + 2\left(\frac{5a}{2}\right) + 2\left(\frac{5b}{2}\right) = 25 + 5(a+b) = 125.$$

Or (Three triangles)

$$\frac{20 \times 5}{2} + \frac{(a+5)5}{2} + \frac{(b+5)5}{2} = \frac{150 + 5(a+b)}{2}$$
$$= 125.$$

---

19. (A)

Since $5(1 + k + k^2) = 15$,

$k^2 + k - 2 = (k+2)(k-1) = 0$.

Therefore, $k = -2$ or 1.

---

20. (D)

Let x = the length of an edge of the cube.

Surface area $= 6x^2$.

$PQ = \sqrt{x^2 + x^2 + x^2} = x\sqrt{3} = 12$.

Since $x = \dfrac{12}{\sqrt{3}} = 4\sqrt{3}$, then

$6x^2 = 6\left(4\sqrt{3}\right)^2 = 288$.

TEST 11   SECTION 5

1. (B)

Since $a + b = a - b$, the $b = 0$.

---

2. (C)

The total campers$= ct$. Proportion:

$\dfrac{1}{p} = \dfrac{x}{ct}$ or $x = \dfrac{ct}{p}$ adults.

---

3. (C)

Tommy = 50%, Liena = 30%, Sharon = 12%, and Malik = 8%. (C) is correct.

---

4. (D)

Let $p = 6$ and $q = 3$. $\dfrac{p}{q} = \dfrac{6}{3} = 2$ (not

multiple of 3)

---

5. (E)

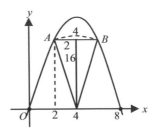

In the figure above, since the height of the triangle $= f(2) = 2(8-2) = 12$, then the area

$= \dfrac{4 \times 12}{2} = 24.$

---

6. (D)

$4^x = 2^{2x}$ and $\left(\dfrac{1}{2}\right)^{y-3} = \left(2^{-1}\right)^{y-3} = 2^{-y+3}$.

Therefore, $2^{2x} = 2^{-y+3}$

$\rightarrow 2x = -y + 3 \rightarrow y = 3 - 2x.$

---

7. (C)

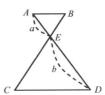

The perimeter $= 3a + 3b = 3(a+b) = 3k$.

---

8. (B)

Since $f(5) = 3$, $f\left(\dfrac{a}{2}\right) = 3$.

Therefore, $\dfrac{a}{2} = -1$ or 5.

$a = -2$ or 10.

---

9. 20

Since $\dfrac{n}{n+10} = \dfrac{1}{2}$, then $n = 10$. $n + 10 = 20$

---

10. 20

$\dfrac{a+b}{2} = 5$ and $\dfrac{a^2 + b^2}{2} = 30$. That is,

$a + b = 10$ and $a^2 + b^2 = 60$.

$(a+b)^2 = a^2 + b^2 + 2ab = 100$.

Therefore, $ab = 20$.

---

11. 7.5 or $^{15}/_{2}$

Since the slope is constant,

$$\frac{5-0}{0-(-1)} = \frac{t-5}{0.5-0} \rightarrow 2.5 = t-5 \rightarrow t = 7.5.$$

---

12. 14

$-10, -8, -6, \ldots 0, \ldots 6, 8, 10, \underline{12, 14, 16}$.

$\underline{\text{Sum} = 42}$

There are 14 even numbers.

---

13. 2

Since $g(1) = 2f(1) - 1 = 3$, then $f(2) = 2$.

---

14. 35

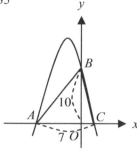

Since two zeros= $2, -5$,

then $AC = 2 - (-5) = 7$ and $OB = f(0) = 10$.

Therefore, the area $= \dfrac{10 \times 7}{2} = 35$.

---

15. 4

Since

$k \Diamond (k-2) = k(k-2) - k + 1 = k^2 - 3k + 1$ and

$2 \Diamond 3 = 2 \times 3 - 2 + 1 = 5$, then $k^2 - 3k + 1 = 5$.

Therefore, $k^2 - 3k - 4 = (k-4)(k+1) = 0$ or

$k = 4$ or -1.

---

16. 1.5 or $\dfrac{3}{2}$.

Since $m + 5n = 2.5 \times 4n = 10n$, then $m = 5n$.

Therefore, $\dfrac{m+n}{m-n} = \dfrac{6n}{4n} = \dfrac{3}{2}$.

---

17. 50

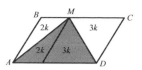

In the figure above, the ratio of

$\dfrac{\text{Area of } \triangle ABO}{\text{Area of } \triangle DMC} = \dfrac{2}{3} = \dfrac{2k}{3k}$. Since $2k = 20$,

then $5k(\text{Area of } \triangle AMD) = 50$.

---

18. 2 or 10

Distance $= \sqrt{(3-(-1))^2 + (a-6)^2}$

$= 4\sqrt{2}$.

Since $a^2 - 12a + 52 = 32$

then $a^2 - 12a + 20 = 0$.

Therefore, $(a-2)(a-10) = 0 \rightarrow a = 2$ or 10.

TEST 11   **SECTION 7**

1. (C)

$\cancel{1}, 3, 5, 7$ ----- they are all prime.

---

2. (B)

Since $16 < 6k < 25$, then $2,66.. < k < 4,16...$

Therefore, $k = 3$. Or substitute the choices

into the inequality and check.

---

3. (A)

If x is the number of black marbles,

$\dfrac{x}{2+4+x} = \dfrac{1}{3}$. Therefore,

$3x = x + 6 \rightarrow x = 3$.

---

4. (D)

$a + b = 80 + 60 = 140$.

---

5. (A)

Month 1 $\rightarrow \left| \dfrac{6-9}{9} \right| = 0.33..$

Month 3 $\rightarrow \left| \dfrac{2.5}{9} \right| = \dfrac{2.5}{9}$

Month 5 $\rightarrow \left| \dfrac{9-7}{7} \right| = \dfrac{2}{7}$

Month 7 $\rightarrow \left| \dfrac{7-9.5}{9.5} \right| = \dfrac{2.5}{9.5}$

Month 9 → $\left|\dfrac{6-8.5}{8.5}\right| = \dfrac{2.5}{8.5}$.

Therefore, month 1 has the greatest change.

---

6. (B)

$3^{a+b} = 3^3 \rightarrow a+b=3$. Since a and b are integers, $(a=1, b=2)$ or $(a=2, b=1)$.

Therefore, $3^1 + 3^2 = 12$.

---

7. (E)

When $x=-1$ and $1 \le x \le 3$, $f(x)=2$.

Therefore, there are infinite values of k.

---

8. (C)

In the figure below, the radius of the circle is 10. Therefore, $a = 3 - 10 = -7$.

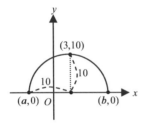

---

9. (D)

The middle value of $h = \dfrac{5+6}{2} = 5.5$ and the distance $= 6 - 5.5 = 0.5$.

Therefore, $|h - 5.5| > 0.5$.

---

10. (B)

If $AC = 2k$ and

$CE = k$, $BD = k + 0.5k = 1.5k$.

Since $1.5k = 15$, $k = 10$.

Therefore, $DE = 0.5k = 0.5 \times 10 = 5$.

---

11. (C)

Since $a > 0$ and $b > 0$, the graph of the parabola must be open-up and a positive y-intercept.

---

12. (E)

$k = \left(2^{-4}\right)^{\frac{1}{m}} = 2^{-\frac{4}{m}}$. When $m = -1$,

$k = 2^4 = 16$.

---

13. (B)

Since $1.25 \times a^2 = 0.05 \times b^2$,

$b^2 = 25a^2 \rightarrow b = 5a$.

Therefore, $\dfrac{a}{b} = \dfrac{a}{5a} = \dfrac{1}{5}$.

---

14. (E)

$x^2 + y^2 + 2xy = (x+y)^2$ is odd.

$\rightarrow (x+y) =$ odd. That is, one of x and y must be even (ex. $x=1$ and $y=2$, or $x=2$ and $y=3$). Therefore, their product must be even.

---

15. (B)

In the figure above, $y = 75$.

---

16. (E)

In the figure above, $AB = 20$, $AP = 10$, and $\angle APB = 90^o$. Thus $\angle A = 60^o$, $AH = 5$, and $PH = 5\sqrt{5}$. Therefore, the coordinates of point P are,

$\left(10+5, 10+5\sqrt{5}\right) \rightarrow \left(15, 10+5\sqrt{3}\right)$.

$\boxed{END}$

Dr. John Chung's SAT Math

TEST
12

ANSWER SHEET   TEST #:

SECTION 3

1 Ⓐ Ⓑ Ⓒ Ⓓ Ⓔ	11 Ⓐ Ⓑ Ⓒ Ⓓ Ⓔ	21 Ⓐ Ⓑ Ⓒ Ⓓ Ⓔ	31 Ⓐ Ⓑ Ⓒ Ⓓ Ⓔ
2 Ⓐ Ⓑ Ⓒ Ⓓ Ⓔ	12 Ⓐ Ⓑ Ⓒ Ⓓ Ⓔ	22 Ⓐ Ⓑ Ⓒ Ⓓ Ⓔ	32 Ⓐ Ⓑ Ⓒ Ⓓ Ⓔ
3 Ⓐ Ⓑ Ⓒ Ⓓ Ⓔ	13 Ⓐ Ⓑ Ⓒ Ⓓ Ⓔ	23 Ⓐ Ⓑ Ⓒ Ⓓ Ⓔ	33 Ⓐ Ⓑ Ⓒ Ⓓ Ⓔ
4 Ⓐ Ⓑ Ⓒ Ⓓ Ⓔ	14 Ⓐ Ⓑ Ⓒ Ⓓ Ⓔ	24 Ⓐ Ⓑ Ⓒ Ⓓ Ⓔ	34 Ⓐ Ⓑ Ⓒ Ⓓ Ⓔ
5 Ⓐ Ⓑ Ⓒ Ⓓ Ⓔ	15 Ⓐ Ⓑ Ⓒ Ⓓ Ⓔ	25 Ⓐ Ⓑ Ⓒ Ⓓ Ⓔ	35 Ⓐ Ⓑ Ⓒ Ⓓ Ⓔ
6 Ⓐ Ⓑ Ⓒ Ⓓ Ⓔ	16 Ⓐ Ⓑ Ⓒ Ⓓ Ⓔ	26 Ⓐ Ⓑ Ⓒ Ⓓ Ⓔ	36 Ⓐ Ⓑ Ⓒ Ⓓ Ⓔ
7 Ⓐ Ⓑ Ⓒ Ⓓ Ⓔ	17 Ⓐ Ⓑ Ⓒ Ⓓ Ⓔ	27 Ⓐ Ⓑ Ⓒ Ⓓ Ⓔ	37 Ⓐ Ⓑ Ⓒ Ⓓ Ⓔ
8 Ⓐ Ⓑ Ⓒ Ⓓ Ⓔ	18 Ⓐ Ⓑ Ⓒ Ⓓ Ⓔ	28 Ⓐ Ⓑ Ⓒ Ⓓ Ⓔ	38 Ⓐ Ⓑ Ⓒ Ⓓ Ⓔ
9 Ⓐ Ⓑ Ⓒ Ⓓ Ⓔ	19 Ⓐ Ⓑ Ⓒ Ⓓ Ⓔ	29 Ⓐ Ⓑ Ⓒ Ⓓ Ⓔ	39 Ⓐ Ⓑ Ⓒ Ⓓ Ⓔ
10 Ⓐ Ⓑ Ⓒ Ⓓ Ⓔ	20 Ⓐ Ⓑ Ⓒ Ⓓ Ⓔ	30 Ⓐ Ⓑ Ⓒ Ⓓ Ⓔ	40 Ⓐ Ⓑ Ⓒ Ⓓ Ⓔ

SECTION 5

1 Ⓐ Ⓑ Ⓒ Ⓓ Ⓔ	11 Ⓐ Ⓑ Ⓒ Ⓓ Ⓔ	21 Ⓐ Ⓑ Ⓒ Ⓓ Ⓔ	31 Ⓐ Ⓑ Ⓒ Ⓓ Ⓔ
2 Ⓐ Ⓑ Ⓒ Ⓓ Ⓔ	12 Ⓐ Ⓑ Ⓒ Ⓓ Ⓔ	22 Ⓐ Ⓑ Ⓒ Ⓓ Ⓔ	32 Ⓐ Ⓑ Ⓒ Ⓓ Ⓔ
3 Ⓐ Ⓑ Ⓒ Ⓓ Ⓔ	13 Ⓐ Ⓑ Ⓒ Ⓓ Ⓔ	23 Ⓐ Ⓑ Ⓒ Ⓓ Ⓔ	33 Ⓐ Ⓑ Ⓒ Ⓓ Ⓔ
4 Ⓐ Ⓑ Ⓒ Ⓓ Ⓔ	14 Ⓐ Ⓑ Ⓒ Ⓓ Ⓔ	24 Ⓐ Ⓑ Ⓒ Ⓓ Ⓔ	34 Ⓐ Ⓑ Ⓒ Ⓓ Ⓔ
5 Ⓐ Ⓑ Ⓒ Ⓓ Ⓔ	15 Ⓐ Ⓑ Ⓒ Ⓓ Ⓔ	25 Ⓐ Ⓑ Ⓒ Ⓓ Ⓔ	35 Ⓐ Ⓑ Ⓒ Ⓓ Ⓔ
6 Ⓐ Ⓑ Ⓒ Ⓓ Ⓔ	16 Ⓐ Ⓑ Ⓒ Ⓓ Ⓔ	26 Ⓐ Ⓑ Ⓒ Ⓓ Ⓔ	36 Ⓐ Ⓑ Ⓒ Ⓓ Ⓔ
7 Ⓐ Ⓑ Ⓒ Ⓓ Ⓔ	17 Ⓐ Ⓑ Ⓒ Ⓓ Ⓔ	27 Ⓐ Ⓑ Ⓒ Ⓓ Ⓔ	37 Ⓐ Ⓑ Ⓒ Ⓓ Ⓔ
8 Ⓐ Ⓑ Ⓒ Ⓓ Ⓔ	18 Ⓐ Ⓑ Ⓒ Ⓓ Ⓔ	28 Ⓐ Ⓑ Ⓒ Ⓓ Ⓔ	38 Ⓐ Ⓑ Ⓒ Ⓓ Ⓔ
9 Ⓐ Ⓑ Ⓒ Ⓓ Ⓔ	19 Ⓐ Ⓑ Ⓒ Ⓓ Ⓔ	29 Ⓐ Ⓑ Ⓒ Ⓓ Ⓔ	39 Ⓐ Ⓑ Ⓒ Ⓓ Ⓔ
10 Ⓐ Ⓑ Ⓒ Ⓓ Ⓔ	20 Ⓐ Ⓑ Ⓒ Ⓓ Ⓔ	30 Ⓐ Ⓑ Ⓒ Ⓓ Ⓔ	40 Ⓐ Ⓑ Ⓒ Ⓓ Ⓔ

Grid-in response boxes numbered: 9, 10, 11, 12, 13, 14, 15, 16, 17, 18

Each grid-in box contains four columns with fraction bars (/) and decimal points, and digit bubbles 0 through 9.

Math Scoring Worksheet

A. Section 3

_____   _____
numer of correct   number of incorrect

+   +

B. Section 5 (1-8)

_____   _____
numer of correct   number of incorrect

+

C. Section 5 (9-18)

_____
numer of correct

+   +

D. Section 7

_____   _____
numer of correct   number of incorrect

=   =

E. Total Unrounded Raw Score

_____   −   _____   ÷4 = _____
numer of correct   number of incorrect

F. Total Rounded Raw Score   _____   (See table)

Math Score Range =   [_____]

Math Conversion Table

Raw Score	Scaled Score	Raw Score	Scaled Score
54	800	23	490-550
53	780-800	22	480-540
52	760-800	21	470-530
51	740-800	20	460-520
50	720-780	19	450-510
49	700-760	18	450-510
48	690-750	17	440-500
47	680-740	16	430-490
46	670-730	15	420-480
45	660-720	14	420-480
44	650-710	13	410-470
43	650-710	12	400-460
42	640-700	11	390-450
41	630-690	10	380-440
40	620-680	9	390-430
39	610-670	8	380-420
38	610-670	7	370-410
37	600-660	6	360-400
36	590-650	5	340-380
35	580-640	4	320-370
34	570-630	3	310-360
33	560-620	2	300-350
32	560-620	1	270-320
31	550-610	0	240-300
30	540-600	-1	200-290
29	530-590	-2	200-270
28	530-590	-3	200-260
27	520-580	-4	200-240
26	510-570	-5	200-220
25	500-560	-6 and below	200
24	500-560		

SECTION 3
Time- 25 minutes
20 Questions

Turn to Section 3 (Page 1) of your answer sheet to answer the questions in this section.

Directions: For this section, solve each problem and decide which is the best of the choices given. Fill in the corresponding circle on the answer sheet. You may use any available space for scratchwork.

Notes

1. The use of a calculator is permitted.
2. All numbers used are real numbers.
3. Figures that accompany problems in this test are intended to provide information useful in solving the problems. They are drawn as accurately as possible EXCEPT when it is stated in a specific problem that the figure is not drawn to scale. All figure lie in a plane unless other indicated.
4. Unless otherwise specified, the domain of any function f is assumed to be set of all real numbers x for which $f(x)$ is a real number.

Reference Informatiom

$A = \pi r^2$
$C = 2\pi r$

$A = \ell w$

$A = \frac{1}{2}bh$

$V = \ell w h$

$V = \pi r^2 h$

$c^2 = a^2 + b^2$

Special Right Triangles

The numbers of degrees of arc in a circle is $360°$.

The sum of the measures in degrees of the angles is $180°$.

---

1. A certain flag has a length that is 2 times its width. What is the width of a flag that has an area of 100?

(A) 5
(B) $5\sqrt{2}$
(C) 10
(D) $10\sqrt{2}$
(E) 20

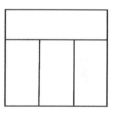

2. In the figure above, the mats are equal in size. If the floor covered with four mats shown above has an area of 108, what is the perimeter of one mat?

(A) 18
(B) 20
(C) 24
(D) 27
(E) 28

GO ON TO THE NEXT PAGE

3. If $2a+4b=\sqrt{k}$, then $(a+2b)\sqrt{k}=$

(A) $3k$

(B) $2k$

(C) k

(D) $\dfrac{k}{2}$

(E) $\dfrac{k}{4}$

Questions 4-6 refer to the following graph.

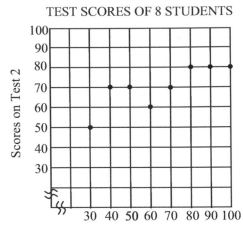

TEST SCORES OF 8 STUDENTS

4. What was the greatest change in scores from test 1 to test 2?

(A) 10
(B) 20
(C) 30
(D) 40
(E) 50

5. What was the median of the scores of the 8 students on test 1?

(A) 60
(B) 65
(C) 70
(D) 75
(E) 80

6. What was the average (arithmetic mean) of the scores of 8 students on test 2?

(A) 60
(B) 65
(C) 68
(D) 70
(E) 72

GO ON TO THE NEXT PAGE

7. If $x^2 = 3$, what is the value of $\dfrac{1}{x+1} - \dfrac{1}{x-1}$?

(A) -2
(B) -1
(C) 0
(D) 1
(E) 2

9. If it takes 10 people 18 hours to do a certain job, how many hours would it take 15 people, working at the same rate, to do $\dfrac{2}{3}$ of the same job?

(A) 6
(B) 8
(C) 9
(D) 10
(E) 12

8. In triangle XYZ , the length of $\overline{XY}$ is 5 and the length of $\overline{YZ}$ is 18. Which of the following could not be the length of $\overline{ZX}$?

(A) 12.5
(B) 13.5
(C) 15
(D) 16
(E) 16.5

x	-1	0	1	2
$g(x)$	0	-3	0	9

10. The table above shows values of function g for selected values of x. Which of the following defines g ?

(A) $g(x) = x^2 - 3$
(B) $g(x) = 2x^2 - 3$
(C) $g(x) = 3x^2 - 3$
(D) $g(x) = x + 1$
(E) $g(x) = 3x + 3$

GO ON TO THE NEXT PAGE

11. If a child was born exactly k years ago, how old was the child m years ago?

(A) m
(B) $-m$
(C) $m-k$
(D) $k-m$
(E) $m+k$

$$S = 1+2+3+4+...+49+50$$
$$T = 51+52+53+...+99+100$$

12. If the sum of the positive integers from 1 to 50 is S, and the sum of the positive integers from 51 to 100, what is T in terms of S ?

(A) $2S$
(B) $S+50$
(C) $S+100$
(D) $S+2500$
(E) $S+10000$

13. If $f = 1+2+4+8+16$ and $g = \dfrac{1}{2}f + \dfrac{1}{2}$, then f exceeds g by

(A) 2
(B) 4
(C) 8
(D) 15
(E) 16

14. The figure above is composed of rectangles with dimension x by y. What is the area of the figure in terms of x ?

(A) $2x^2$
(B) $4x^2$
(C) $6x^2$
(D) $8x^2$
(E) $9x^2$

GO ON TO THE NEXT PAGE

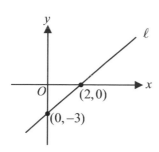

Note: Figure not drawn to scale.

15. In the figure above, which of the following points lies on the line ℓ ?

(A) $(4, 6)$
(B) $(3, 4)$
(C) $(1, 2)$
(D) $(-1, 3)$
(E) $(-2, -6)$

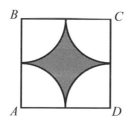

16. In square $ABCD$ above, four quarter circles are at the center $A, B, C,$ and D, respectively. If the radius of each quarter circle is 1, what is the area of the shaded region?

(A) $1 - \pi$

(B) $2 - \dfrac{\pi}{2}$

(C) $4 - \pi$

(D) $4 - \dfrac{\pi}{2}$

(E) $4 - \dfrac{\pi}{3}$

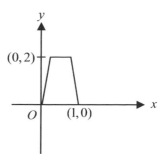

Note: Figure not drawn to scale.

17. The figure above shows the graph of $y = f(x)$. Which of the following could be the graph of

$$y = \frac{1}{2} f(x+1) \, ?$$

(A)   (B)

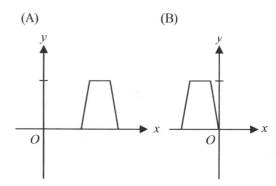

(C)   (D)

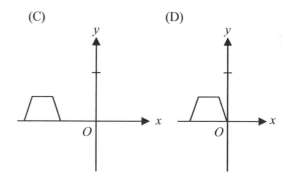

(E)

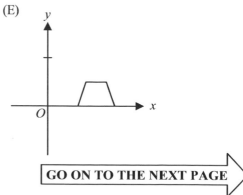

GO ON TO THE NEXT PAGE

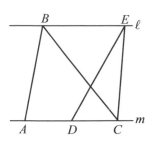

Note: Figure not drawn to scale.

18. In the figure above, $\ell \parallel m$ and D is the midpoint of $\overline{AC}$. If the area of $\triangle ABC$ is 40, what is the area of $\triangle CDE$?

(A) 10
(B) 15
(C) 20
(D) 25
(E) 40

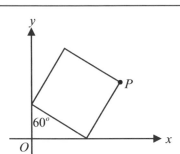

Note: Figure not drawn to scale.

19. In the figure above, the length of a side of the square is 10. Which of the following is the coordinates of point P?

(A) $(15, 8)$
(B) $(10, \ 5\sqrt{3})$
(C) $(15, \ 5\sqrt{3})$
(D) $(5+5\sqrt{3}, \ 5\sqrt{3})$
(E) $\left(10+5\sqrt{3}, 5\sqrt{3}\right)$

20. For all positive numbers p and q, let the operation $\Diamond$ be defined by $p \Diamond q = p^2 - q^2$. If a and b are positive integers, which of the following can be equal to zero?

    I. $a \Diamond b$
   II. $(a+b) \Diamond (a-b)$
  III. $\dfrac{1}{a} \Diamond \dfrac{1}{b}$

(A) I only
(B) II only
(C) III only
(D) I and II only
(E) I and III only

STOP

If you finish before time is called, you may check your work on this section only.
Do not turn to any other section in the test.

SECTION 5
Time- 25 minutes
18 Questions

Turn to Section 5 (Page 1) of your answer sheet to answer the questions in this section.

Directions: For this section, solve each problem and decide which is the best of the choices given. Fill in the corresponding circle on the answer sheet. You may use any available space for scratchwork.

Notes

1. The use of a calculator is permitted.
2. All numbers used are real numbers.
3. Figures that accompany problems in this test are intended to provide information useful in solving the problems. They are drawn as accurately as possible EXCEPT when it is stated in a specific problem that the figure is not drawn to scale. All figure lie in a plane unless other indicated.
4. Unless otherwise specified, the domain of any function f is assumed to be set of all real numbers x for which $f(x)$ is a real number.

Reference Informatiom

$A = \pi r^2$
$C = 2\pi r$
$A = \ell w$
$A = \frac{1}{2}bh$
$V = \ell wh$
$V = \pi r^2 h$
$c^2 = a^2 + b^2$
Special Right Triangles

The numbers of degrees of arc in a circle is $360°$.

The sum of the measures in degrees of the angles is $180°$.

1. If $3\sqrt{k^2} + 8 = 20$, which of the following could be the value of k ?

   (A) -4
   (B) -1
   (C) 6
   (D) 8
   (E) 36

2. If a is three more than $\frac{1}{3}$ of b, then what is b in terms of a ?

   (A) $a - 9$
   (B) $a - 3$
   (C) $3(a - 3)$
   (D) $3(a + 3)$
   (E) $3a - 3$

GO ON TO THE NEXT PAGE

TEST SCORES

Scores	Number of Students
90	2
85	3
80	3
70	1
60	x

3. The table above shows the test scores for students on an European History test. If the median for the test scores is 80, what is the least possible value of x ?

(A) 1
(B) 2
(C) 3
(D) 4
(E) 5

4. Let Q be the set of all integers that can be written as m^k, where m and k are positive integers. If $k > m$, which of the following integers is in Q ?

(A) 9
(B) 25
(C) 27
(D) 64
(E) 100

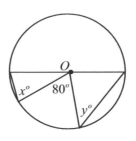

Note: Figure not drawn to scale.

5. In the figure above, point O is the center of the circle. What is the value of $x + y$?

(A) 100
(B) 110
(C) 120
(D) 130
(E) 140

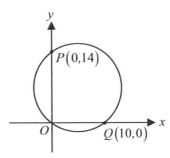

6. The circle in the xy-plane above intersects the x- and y-axis at the origin and at points P and Q. What is the area of the circle?

(A) 49π
(B) 74π
(C) 100π
(D) 125π
(E) 169π

GO ON TO THE NEXT PAGE

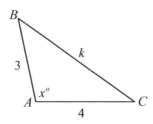

Note: Figure not drawn to scale.

7. In $\triangle ABC$ above, the lengths of each side are 3, 4, and k. If $x > 90$, which of the following could be the value of k?

(A) 4
(B) 5
(C) 6
(D) 7
(E) 8

8. If a and b are positive distinct integers and $(a^{-b})^3 = \dfrac{1}{64}$, what is the value of $a - b$?

(A) 2
(B) 3
(C) 4
(D) 8
(E) 12

GO ON TO THE NEXT PAGE

5 □□□ 5 □□□ 5 □□□ 5

Directions: For Students-Produced Response questions 9-18, use the grid at the bottom of the answer sheet page on which you have answered questions 1-8.

Each of the remaining 10 questions requires you to solve the problem and enter your answer by making the circles in the special grid, as shown in the examples below. You may use any available space for scratchwork.

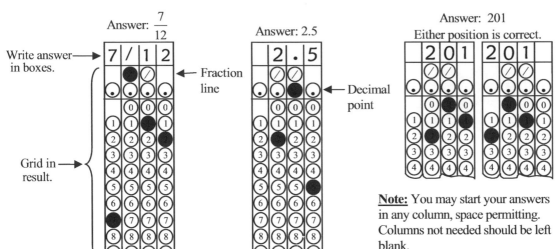

Note: You may start your answers in any column, space permitting. Columns not needed should be left blank.

- Mark no more than one circle in any column.

- Because the answer sheet will be machine-scored, **you will receive credit only if the circles are filled in correctly.**

- Although not required, it is suggested that you write your answer in the boxes at the top of the columns to help you fill in the circles accurately.

- Some problems may have more than one correct answer. In such cases, grid only one answer.

- No question has a negative answer.

- **Mixed numbers** such as $3\frac{1}{2}$ must be gridded as 3.5 or 7/2. (If $\boxed{3\,1\,/\,2}$ is gridded, it will be interpreted as $\frac{31}{2}$, not $3\frac{1}{2}$.)

- **Decimal Answers:** If you obtain a decimal answer with more digits than the grid can accommodate, it may be either rounded or truncated, but it must fill the entire grid. For example, if you obtain an answer such as 0.6666..., you should record your result as .666 or .667. **A less accurate value such as .66 or .67 will be scored as incorrect.**

Acceptable ways to grid $\frac{2}{3}$ are:

9. What is the greatest two-digit integer that has a factor of 7 ?

10. A dog can run 80 yards in 10 seconds. At this rate, how many yards can the dog run in 25 seconds?

GO ON TO THE NEXT PAGE

11. In $\triangle ABC$, angle A has a value between 50° and 75°, and angle B has a value between 45° and 65°. What is the greatest possible integer value of angle C? (Disregard the degree sign when gridding your answer.)

14. If $\sqrt{(x+y)(x-y)} = 3$, where x and y are positive integers, what is the value of x?

12. What is the sum of 10 consecutive odd integers if the median of these integers is 40?

15. The price of a certain item was p dollars. During a sale, the price was decreased by 10 percent. A sales tax of 8 percent of the price of the item is added to the price to obtain the final cost. If p is an integer, then the final cost, in dollars, is p times what number?

13. In a rectangular coordinate system, the center of a circle has coordinates $(6,10)$, and the circle touches the y-axis at one point only. What is the radius of the circle?

GO ON TO THE NEXT PAGE

16. Let the function f be defined for all x by
$f(x) = ax + b$, where a and b are constants.
If $f(2) = 5$ and $f(4) = 9$, what is the value
of a?

18. If the 7 cards shown above are placed in a row
so that cards A and G are at either end, and
if $B, C,$ and D must be kept together, how
many different arrangements are possible?

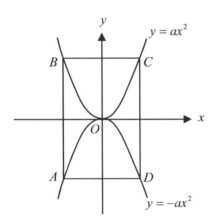

Note: Figure not drawn to scale.

17. In the figure above, $ABCD$ is a rectangle and
points $A, B, C,$ and D lie on the graphs of
$y = ax^2$ and $y = -ax^2$ respectively. If the
area of $ABCD$ is 72 and $BC = 6$, What is the
value of a?

STOP

**If you finish before time is called, you may check your work on this section only.
Do not turn to any other section in the test.**

SECTION 7
Time- 20 minutes
16 Questions

Turn to Section 7 (Page 2) of your answer sheet to answer the questions in this section.

Directions: For this section, solve each problem and decide which is the best of the choices given. Fill in the corresponding circle on the answer sheet. You may use any available space for scratchwork.

Notes

1. The use of a calculator is permitted.
2. All numbers used are real numbers.
3. Figures that accompany problems in this test are intended to provide information useful in solving the problems. They are drawn as accurately as possible EXCEPT when it is stated in a specific problem that the figure is not drawn to scale. All figure lie in a plane unless other indicated.
4. Unless otherwise specified, the domain of any function f is assumed to be set of all real numbers x for which $f(x)$ is a real number.

Reference Informatiom

  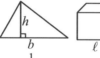 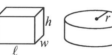

$A = \pi r^2$
$C = 2\pi r$ $A = \ell w$ $A = \dfrac{1}{2}bh$ $V = \ell wh$ $V = \pi r^2 h$ $c^2 = a^2 + b^2$ Special Right Triangles

The numbers of degrees of arc in a circle is $360°$.
The sum of the measures in degrees of the angles is $180°$.

---

1. A telephone call cost 2 dollars for the first three minutes and 20 cents each additional minute. If a call started at 10:20 a.m. and ended at 10:50 a.m., how much did the call cost?

(A) $7.00
(B) $7.20
(C) $7.40
(D) $8.00
(E) $8.20

2. If a linear function is given by $ax + by - 2 = 0$ with a negative slope and a negative y-intercept, which of the following must be true?

(A) $a > 0$
(B) $b > 0$
(C) $a = 0$
(D) $b = 0$
(E) $a < 0$

GO ON TO THE NEXT PAGE

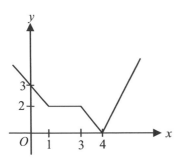

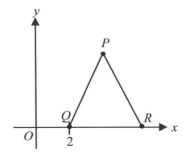

Note: Figure not drawn to scale.

3. The figure above shows the graph of the function g. If $g(k) = 2$, which of the following could be the value of k?

(A) 0
(B) 0.5
(C) 2.5
(D) 3.5
(E) 4

5. In the figure above, $\triangle PQR$ is equilateral and the length of $\overline{PR}$ is equal to 4. Which of the following are the coordinates of point P?

(A) (3, 3)
(B) (3, 4)
(C) $(2, \sqrt{3})$
(D) $(4, 2\sqrt{3})$
(E) $(4, 4\sqrt{3})$

4. If positive integers a and b satisfy the equation $a^b = 64$, which of the following cannot be the value of $a + b$?

(A) 7
(B) 8
(C) 10
(D) 15
(E) 65

6. In the xy-coordinate plane, the center of a circle is at (4, 0) and a point on the circle is at (7, 4). What is the area of the circle?

(A) 4π
(B) 9π
(C) 16π
(D) 25π
(E) 36π

GO ON TO THE NEXT PAGE

7. If k is an integer and 4 is the remainder when $k^2 + 3$ is divided by 7, which of the following could be the value of k ?

(A) 3
(B) 4
(C) 5
(D) 6
(E) 7

8. Benjamin and Jerry start with the same number of trading cards. After Benjamin gives 12 of his cards to Jerry, Jerry has two times as many cards as Benjamin does. How many cards did Benjamin have at the start?

(A) 18
(B) 24
(C) 36
(D) 40
(E) 42

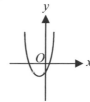

9. The figure above shows the graph of a function f. If a function g is defined by $g(x) = f(x-1) - 1$, which of the following could be the graph of g ?

(A)

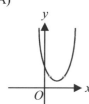

(B)

(C)

(D)

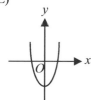

(E)

GO ON TO THE NEXT PAGE

10. The average rate of change of a function f between a and b is defined by $C(x) = \dfrac{f(b) - f(a)}{b - a}$. If $f(2) = 6$ and $f(5) = 15$, what is the average rate of change between 2 and 5?

(A) 3
(B) 4
(C) 5
(D) 7
(E) 9

Sequence I: 3, 6, 9,...

Sequence II: $\dfrac{1}{81}, \dfrac{1}{27}, \dfrac{1}{9},...$

11. In sequence I above, the first term is 3 and every term after the first is 3 more than the preceding term. In sequence II, the first term is $\dfrac{1}{81}$ and every term after the first is 3 times the preceding term. What is the first positive integer n for which the nth term of sequence II is greater than the nth term of sequence I ?

(A) 8
(B) 9
(C) 10
(D) 16
(E) 20

12. If a positive number a is 150 percent of $4p$, and if p is 25 percent of $8b$, then what is the value of $\dfrac{a}{b}$?

(A) 4
(B) 8
(C) 12
(D) 16
(E) 32

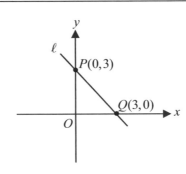

13. A line m (not shown) is perpendicular to line ℓ in the graph above, and passes through between point P and Q. Which of the following could be an equation of line m ?

(A) $y = -x$
(B) $y = x + 5$
(C) $y = x - 2$
(D) $y = x - 4$
(E) $y = -x + 2$

GO ON TO THE NEXT PAGE

14. If a is a real number, and if $a^2 - 1 = a + 1$, which of the following could be true?

      I. $a = -1$
      II. $a = 1$
      III. $a = 2$

(A) I only
(B) II only
(C) III only
(D) I and III only
(E) I, II, and III

15. Some boys and girls were having a car wash to raise money for a class trip. Initially, $\frac{2}{5}$ of the group were girls. Shortly thereafter two girls left and two boys entered, then $\frac{1}{3}$ of the group are girls. How many girls were initially in the group?

(A) 8
(B) 10
(C) 12
(D) 14
(E) 20

16. If $|p| > 10$, which of the following must be true?

      I. $p > 10$
      II. $p^2 > 10$
      III. $p^3 > 10$

(A) I only
(B) II only
(C) III only
(D) I and II only
(E) I, II, and III

STOP

If you finish before time is called, you may check your work on this section only.
Do not turn to any other section in the test.

TEST 12 ANSWER KEY

#	SECTION 3	SECTION 5	SECTION 7
1	B	A	C
2	C	C	E
3	D	B	C
4	C	D	D
5	B	D	D
6	D	B	D
7	B	C	D
8	A	B	C
9	B	98	C
10	C	200	A
11	D	84	A
12	D	400	C
13	D	6	C
14	D	5	D
15	E	.972	C
16	C	2	B
17	D	$\frac{2}{3}$ or .666, .667	
18	C	72	
19	D		
20	E		

TEST 12   SECTION 3

1. (B)
Let width $= x$ and length $= 2x$. Therefore,
$(2x)(x) = 100 \rightarrow x = \sqrt{50} = 5\sqrt{2}$.

2. (C)
Let width $= x$ and length $= y$. $y = 3x$. The
area of one mat is $\frac{108}{4} = 27$.

$x(3x) = 3x^2 = 27$. $x = 3$. Therefore, the
perimeter of one mat is $8x = 8(3) = 24$.

3. (D)
$2a + 4b = 2(a + 2b) = \sqrt{k} \Rightarrow a + 2b = \frac{\sqrt{k}}{2}$

Therefore, $(a + 2b)\sqrt{k} = \frac{\sqrt{k}}{2}\sqrt{k} = \frac{k}{2}$.

4. (C)

When test I $= 40$ and test II $= 70$,
$70 - 40 = 30$.

5. (B)
Test 1: 30, 40, 50, <u>60, 70,</u> 80, 90, 100
The median $= \frac{60 + 70}{2} = 65$.

6. (D)
Average $= \frac{50 + 60 + 3(70) + 3(80)}{8} = 70$.

7. (B)
$\frac{1}{x+1} - \frac{1}{x-1} = \frac{(x-1) - (x+1)}{x^2 - 1} = \frac{-2}{x^2 - 1}$.
Since $x^2 = 3$, $\rightarrow \frac{-2}{3-1} = -1$.

8. (A)
Triangle inequality: $18 - 5 < \overline{ZX} < 18 + 5$.
Therefore, $13 < \overline{ZX} < 23$.

- 408 -

9. (B)

Inverse proportion:
$10 \times 18 = 15 \times x \rightarrow x = 12$.

Therefore, to do $\dfrac{2}{3}$ of the job, they need

$12 \times \dfrac{2}{3} = 8$ hrs.

10. (C)

Substitute the values of x into the given functions.

11. (D)

Use numbers: If a child was born 10 years ago, how old was the child 5 years ago? The answer is 10-5 = 5. That is $(k-m)$ years old.

12. (D)

$$-\begin{vmatrix} T = 51 + 52 + 53 + ... + 99 + 100 \\ S = 1 + 2 + 3 + 4 + ... + 49 + 50 \end{vmatrix}$$

$T - S = \ 50 + 50 + ...50.$
$\rightarrow 50 \times 50 = 2500$
Therefore, $T = S + 2500$.

13. (D)

$g = \dfrac{1}{2}(1 + 2 + 4 + 8 + 16) + \dfrac{1}{2} \rightarrow g = 16$ and
$f = 31$.
Therefore, $31 - 16 = 15$.

14. (D)

In the figure above, $y = 2x$. Therefore, the area of the small rectangle is
$xy = x \cdot 2x = 2x^2$. The entire area is,
$4 \times 2x^2 = 8x^2$.

15. (E)

Slope $= \dfrac{0 - (-3)}{2 - 0} = \dfrac{3}{2}$ and y-intercept $= -3$.
Therefore, the equation of line ℓ is
$y = \dfrac{3}{2}x - 3$. (E): When $x = -2$,

$y = \dfrac{3}{2}(-2) - 3 = -6$.

16. (C)

The area of the square$= 2 \times 2 = 4$ and the area of the quarter circle$= \dfrac{\pi}{4}$. The area of four quarter circles$= \pi$. Therefore, $4 - \pi$ is the area of the shaded region.

17. (D)

Translate to the left by 1 and reduce y by half.

18. (C)

With the same height, the ratio of
$\dfrac{\text{Area of } \triangle ABC}{\text{Area of } \triangle CDE} = \dfrac{2}{1} = \dfrac{2k}{k}$.
If $2k = 40$, then $k = 20$.

19. (D)

Special triangles:

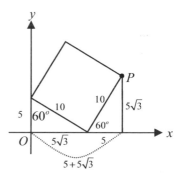

In the figure above, The coordinates of point P are $(5 + 5\sqrt{3}, 5\sqrt{3})$

20. (E)

I.
$a \Diamond b = a^2 - b^2 = (a+b)(a-b) = 0$
$\rightarrow (a+b) > 0$,
    If $a = b$, then it can be zero.

II.
$(a+b) \Diamond (a-b) = (a+b)^2 - (a-b)^2 = 4ab > 0$
    It cannot be zero.

III.
$\dfrac{1}{a} \Diamond \dfrac{1}{b} = \dfrac{1}{a^2} - \dfrac{1}{b^2} = \dfrac{b^2 - a^2}{a^2 b^2} = \dfrac{(b-a)(b+a)}{a^2 b^2}$.
    If $a = b$, it can be zero.

1. (A)

$$3\sqrt{(-4)^2} + 8 = 3\sqrt{16} + 8 = 12 + 8 = 20.$$

2. (C)

Since $a = \dfrac{1}{3}b + 3$, $\dfrac{1}{3}b = a - 3 \rightarrow b = 3(a-3)$.

3. (B)

Arrange the data in order:

60..   70,80,80,80,85,85,85,90,90
$\quad\quad\quad$ (1)(2)(3)

If 80 in (1) is the median, we need six 60s as follows.

60,60,60,60,60,60,70,80,80,80,85,85,85,90,90.

If 80 in (3) is the median, we need two 60s as follows.

60,60,70,80,80,80,85,85,85,90,90,

Therefore, the least number of x is 2.

4. (D)

$$64 = 8^2 = 4^3 = 2^6.$$

5. (D)

Supplementary of 80 is 100 and two triangles are isosceles. Therefore,

$$2(x+y) = 360 - 100 = 260 \rightarrow x+y = 130.$$

6. (B)

The coordinates of the center $= (5,7)$

The length of the radius $= \sqrt{5^2 + 7^2} = \sqrt{74}$

$\therefore$ Area $= \pi r^2 = 74\pi$

7. (C)

When $x = 90$ (right triangle), $k = 5$. From triangle inequality,

$(4-3) < k < (4+3) \rightarrow 1 < k < 7$. For obtuse triangle, $5 < k < 7$. Therefore, $k = 6$.

8. (B)

Since, $a^{-3b} = \left(a^b\right)^{-3} = 4^{-3}$, $a^b = 4$. Therefore,

$4 = 4^1 = \cancel{2}$. $a = 4$ and

$b = 1 \rightarrow a - b = 4 - 1 = 3$.

9. 98

The number must be a multiple of 7. The greatest number is 98.

10. 200

Proportion: $\dfrac{80}{10} = \dfrac{x}{25} \rightarrow x = 200.$

11. 84

Since $50 < A < 75$ and $45 < B < 65$,

$95 < A + B < 140$.

$A + B + C = 180$. Therefore, $40 < C < 85$.

The greatest possible integer of C is 84.

Or, algebraically,

$A + B = 180 - C \rightarrow 95 < 180 - C < 140$

$\rightarrow 40 < C < 85$

12. 400

Median = average. The sum is $40 \times 10 = 400$.

13. 6

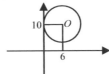

In the figure above, the radius of the circle is 6.

14. 5

$(x+y)(x-y) = 9$. Since x and y are positive integers,

$x + y = 9$ and $x - y = 1$. Therefore, $x = 5$ and $y = 4$.

If $x + y = 3$ and $x - y = 3$, then $x = 3$, $y = 0$.

This is not solution, because y is not positive integer.

15. .972

$P - 0.1p = 0.9p$. The final price

$= (1 + 0.08) \times 0.9p = 0.972p$.

16. 2

$f(2) = 2a + b = 5$ and $f(4) = 4a + b = 9$.

Since $f(4) - f(2) = 2a = 4$, $a = 2$.

Or, a is the slope of (2, 5) and (4, 9).

Therefore, $a = \dfrac{9-5}{4-2} = 2$.

17. $\frac{2}{3}$, .666 , or .667

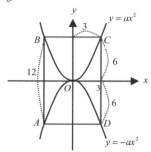

In the figure above, $AB = 12$ and point C has coordinates $(3,6)$. Substitute into the equation. $6 = a(3)^2$. Therefore, $a = \frac{2}{3}$.

18. 72

$A \square \square \square G \rightarrow 3 \cdot 2 \cdot 1 \cdot (3 \cdot 2 \cdot 1) = 36$

and

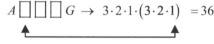

$G \square \square \square A \rightarrow 3 \cdot 2 \cdot 1 \cdot (3 \cdot 2 \cdot 1) = 36$

Therefore, $36 + 36 = 72$.

TEST 12   SECTION 7

1. (C)
Since it is a 30 minute phone call, the cost will be $^{\$}2.00 + {^{\$}}0.2(30 - 3) = {^{\$}}7.40$.

---

2. (E)
$ax + by - 2 = 0 \rightarrow y = -\frac{a}{b}x + \frac{2}{b}$. Since
$-\frac{a}{b} < 0$ and $\frac{2}{b} < 0$, b is negative and a is also negative.

---

3. (C)
For $1 \le x \le 3$, $f(x) = 2$. Therefore, $1 \le k \le 3$.

---

4. (D)

Since $a^b = 64^1 = 2^6 = 4^3 = 8^2$,
$a + b = 65, 8, 7,$ and 10.

---

5. (D)

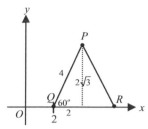

In the figure above, the coordinates of point P are $\left(4, 2\sqrt{3}\right)$.

---

6. (D)
The radius of the circle,
$r = \sqrt{(4-0)^2 + (7-4)^2} = 5$.
Therefore the area $= \pi r^2 = 25\pi$.

---

7. (D)
Backsolving: If $k = 6$, then 39 has remainder 4 when divided by 7.

---

8. (C)
Let the number of cards at the start be n.
Benjamin $n \rightarrow n - 12$
$\dfrac{\text{Jerry} \qquad n \rightarrow n + 12}{n + 12 = 2(n - 12)}$. Therefore, $n = 36$.

---

9. (C)
Translate: Move to the right by 1 and 1 down.

---

10. (A)
The average rate of change $= \dfrac{15 - 6}{5 - 2} = 3$.

---

11. (A)
$SQ1: \ 3, \ 6, \ 9, 12, 15, 18, 21, \underline{24}, 27$
$SQ2: \dfrac{1}{81}, \dfrac{1}{27}, \dfrac{1}{9}, \dfrac{1}{3}, \ 1, \ 3, \ 9, \ \underline{27} \ \rightarrow$ 8th term.

---

12. (C)

$a = 1.5(4p) = 6p$ and $p = 0.25(8b) = 2b$.

Therefore, $\dfrac{a}{b} = \dfrac{6p}{p/2} = 12$.

---

13. (C)

Since the equation of line ℓ is $y = -x + 3$, the equation of line m is $y = x + b$, where $-3 < b < 3$.

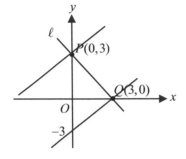

---

14. (D)

$(a+1)(a-1) = a + 1 \rightarrow (a+1)(a-2) = 0$.

Therefore, $a = -1$ or $a = 2$.

---

15. (C)

Since $\dfrac{2}{5}$ are girls and $\dfrac{3}{5}$ are boys, let girls be $2k$ and boys be $3k$.

Boys $\rightarrow 3k + 2$

Girls $\rightarrow 2k - 2$

Therefore, $\dfrac{2k-2}{5k} = \dfrac{1}{3} \rightarrow k = 6$.

Girls $= 2k = 2(6) = 12$.

---

16. (B)

$|p| > 10 \rightarrow p > 10$ or $p < -10$. Only II is correct.

$\boxed{END}$

Dr. John Chung's SAT Math

TEST
13

ANSWER SHEET

TEST #:

SECTION 3

1 Ⓐ Ⓑ Ⓒ Ⓓ Ⓔ	11 Ⓐ Ⓑ Ⓒ Ⓓ Ⓔ	21 Ⓐ Ⓑ Ⓒ Ⓓ Ⓔ	31 Ⓐ Ⓑ Ⓒ Ⓓ Ⓔ
2 Ⓐ Ⓑ Ⓒ Ⓓ Ⓔ	12 Ⓐ Ⓑ Ⓒ Ⓓ Ⓔ	22 Ⓐ Ⓑ Ⓒ Ⓓ Ⓔ	32 Ⓐ Ⓑ Ⓒ Ⓓ Ⓔ
3 Ⓐ Ⓑ Ⓒ Ⓓ Ⓔ	13 Ⓐ Ⓑ Ⓒ Ⓓ Ⓔ	23 Ⓐ Ⓑ Ⓒ Ⓓ Ⓔ	33 Ⓐ Ⓑ Ⓒ Ⓓ Ⓔ
4 Ⓐ Ⓑ Ⓒ Ⓓ Ⓔ	14 Ⓐ Ⓑ Ⓒ Ⓓ Ⓔ	24 Ⓐ Ⓑ Ⓒ Ⓓ Ⓔ	34 Ⓐ Ⓑ Ⓒ Ⓓ Ⓔ
5 Ⓐ Ⓑ Ⓒ Ⓓ Ⓔ	15 Ⓐ Ⓑ Ⓒ Ⓓ Ⓔ	25 Ⓐ Ⓑ Ⓒ Ⓓ Ⓔ	35 Ⓐ Ⓑ Ⓒ Ⓓ Ⓔ
6 Ⓐ Ⓑ Ⓒ Ⓓ Ⓔ	16 Ⓐ Ⓑ Ⓒ Ⓓ Ⓔ	26 Ⓐ Ⓑ Ⓒ Ⓓ Ⓔ	36 Ⓐ Ⓑ Ⓒ Ⓓ Ⓔ
7 Ⓐ Ⓑ Ⓒ Ⓓ Ⓔ	17 Ⓐ Ⓑ Ⓒ Ⓓ Ⓔ	27 Ⓐ Ⓑ Ⓒ Ⓓ Ⓔ	37 Ⓐ Ⓑ Ⓒ Ⓓ Ⓔ
8 Ⓐ Ⓑ Ⓒ Ⓓ Ⓔ	18 Ⓐ Ⓑ Ⓒ Ⓓ Ⓔ	28 Ⓐ Ⓑ Ⓒ Ⓓ Ⓔ	38 Ⓐ Ⓑ Ⓒ Ⓓ Ⓔ
9 Ⓐ Ⓑ Ⓒ Ⓓ Ⓔ	19 Ⓐ Ⓑ Ⓒ Ⓓ Ⓔ	29 Ⓐ Ⓑ Ⓒ Ⓓ Ⓔ	39 Ⓐ Ⓑ Ⓒ Ⓓ Ⓔ
10 Ⓐ Ⓑ Ⓒ Ⓓ Ⓔ	20 Ⓐ Ⓑ Ⓒ Ⓓ Ⓔ	30 Ⓐ Ⓑ Ⓒ Ⓓ Ⓔ	40 Ⓐ Ⓑ Ⓒ Ⓓ Ⓔ

SECTION 5

1 Ⓐ Ⓑ Ⓒ Ⓓ Ⓔ	11 Ⓐ Ⓑ Ⓒ Ⓓ Ⓔ	21 Ⓐ Ⓑ Ⓒ Ⓓ Ⓔ	31 Ⓐ Ⓑ Ⓒ Ⓓ Ⓔ
2 Ⓐ Ⓑ Ⓒ Ⓓ Ⓔ	12 Ⓐ Ⓑ Ⓒ Ⓓ Ⓔ	22 Ⓐ Ⓑ Ⓒ Ⓓ Ⓔ	32 Ⓐ Ⓑ Ⓒ Ⓓ Ⓔ
3 Ⓐ Ⓑ Ⓒ Ⓓ Ⓔ	13 Ⓐ Ⓑ Ⓒ Ⓓ Ⓔ	23 Ⓐ Ⓑ Ⓒ Ⓓ Ⓔ	33 Ⓐ Ⓑ Ⓒ Ⓓ Ⓔ
4 Ⓐ Ⓑ Ⓒ Ⓓ Ⓔ	14 Ⓐ Ⓑ Ⓒ Ⓓ Ⓔ	24 Ⓐ Ⓑ Ⓒ Ⓓ Ⓔ	34 Ⓐ Ⓑ Ⓒ Ⓓ Ⓔ
5 Ⓐ Ⓑ Ⓒ Ⓓ Ⓔ	15 Ⓐ Ⓑ Ⓒ Ⓓ Ⓔ	25 Ⓐ Ⓑ Ⓒ Ⓓ Ⓔ	35 Ⓐ Ⓑ Ⓒ Ⓓ Ⓔ
6 Ⓐ Ⓑ Ⓒ Ⓓ Ⓔ	16 Ⓐ Ⓑ Ⓒ Ⓓ Ⓔ	26 Ⓐ Ⓑ Ⓒ Ⓓ Ⓔ	36 Ⓐ Ⓑ Ⓒ Ⓓ Ⓔ
7 Ⓐ Ⓑ Ⓒ Ⓓ Ⓔ	17 Ⓐ Ⓑ Ⓒ Ⓓ Ⓔ	27 Ⓐ Ⓑ Ⓒ Ⓓ Ⓔ	37 Ⓐ Ⓑ Ⓒ Ⓓ Ⓔ
8 Ⓐ Ⓑ Ⓒ Ⓓ Ⓔ	18 Ⓐ Ⓑ Ⓒ Ⓓ Ⓔ	28 Ⓐ Ⓑ Ⓒ Ⓓ Ⓔ	38 Ⓐ Ⓑ Ⓒ Ⓓ Ⓔ
9 Ⓐ Ⓑ Ⓒ Ⓓ Ⓔ	19 Ⓐ Ⓑ Ⓒ Ⓓ Ⓔ	29 Ⓐ Ⓑ Ⓒ Ⓓ Ⓔ	39 Ⓐ Ⓑ Ⓒ Ⓓ Ⓔ
10 Ⓐ Ⓑ Ⓒ Ⓓ Ⓔ	20 Ⓐ Ⓑ Ⓒ Ⓓ Ⓔ	30 Ⓐ Ⓑ Ⓒ Ⓓ Ⓔ	40 Ⓐ Ⓑ Ⓒ Ⓓ Ⓔ

9   10   11   12   13

14   15   16   17   18

Math Scoring Worksheet

A. Section 3

_____   _____
numer of correct   *number of incorrect*

+   +

B. Section 5 (1-8)

_____   _____
numer of correct   *number of incorrect*

+

C. Section 5 (9-18)

_____
numer of correct

+   +

D. Section 7

_____   _____
numer of correct   *number of incorrect*

=   =

E. Total Unrounded Raw Score

_____ − _____ ÷4 = _____
numer of correct   *number of incorrect*

F. Total Rounded Raw Score

_____ (See table)

Math Score Range = [_____ — _____]

Math Conversion Table

Raw Score	Scaled Score	Raw Score	Scaled Score
54	800	23	490-550
53	780-800	22	480-540
52	760-800	21	470-530
51	740-800	20	460-520
50	720-780	19	450-510
49	700-760	18	450-510
48	690-750	17	440-500
47	680-740	16	430-490
46	670-730	15	420-480
45	660-720	14	420-480
44	650-710	13	410-470
43	650-710	12	400-460
42	640-700	11	390-450
41	630-690	10	380-440
40	620-680	9	390-430
39	610-670	8	380-420
38	610-670	7	370-410
37	600-660	6	360-400
36	590-650	5	340-380
35	580-640	4	320-370
34	570-630	3	310-360
33	560-620	2	300-350
32	560-620	1	270-320
31	550-610	0	240-300
30	540-600	-1	200-290
29	530-590	-2	200-270
28	530-590	-3	200-260
27	520-580	-4	200-240
26	510-570	-5	200-220
25	500-560	-6 and below	200
24	500-560		

SECTION 3
Time- 25 minutes
20 Questions

Turn to Section 3 (Page 1) of your answer sheet to answer the questions in this section.

Directions: For this section, solve each problem and decide which is the best of the choices given. Fill in the corresponding circle on the answer sheet. You may use any available space for scratchwork.

Notes

1. The use of a calculator is permitted.
2. All numbers used are real numbers.
3. Figures that accompany problems in this test are intended to provide information useful in solving the problems. They are drawn as accurately as possible EXCEPT when it is stated in a specific problem that the figure is not drawn to scale. All figure lie in a plane unless other indicated.
4. Unless otherwise specified, the domain of any function f is assumed to be set of all real numbers x for which $f(x)$ is a real number.

Reference Informatiom

$A = \pi r^2$
$C = 2\pi r$ $A = \ell w$ $A = \frac{1}{2}bh$ $V = \ell wh$ $V = \pi r^2 h$ $c^2 = a^2 + b^2$ Special Right Triangles

The numbers of degrees of arc in a circle is $360°$

The sum of the measures in degrees of the angles is $180°$

---

1. If $3(a + 2b) = 7$, what is the value of $9a + 18b$?

(A) 21
(B) 28
(C) 49
(D) 63
(E) 81

2. If $\left|5 - \sqrt{k}\right| = 4$, which of the following could be a value of k?

(A) -1
(B) 9
(C) 36
(D) 81
(E) 100

GO ON TO THE NEXT PAGE

3. If b is 6 more than $2a$, and b is 3 less than c, then $c =$

(A) $\dfrac{a}{3}$

(B) $\dfrac{a}{2} + 9$

(C) $\dfrac{2a}{3}$

(D) $2a + 9$

(E) $2a + 15$

4. For which of the following functions is $f(-2) > f(2)$?

(A) $f(x) = x^2$

(B) $f(x) = x + 2$

(C) $f(x) = x^3 + 2$

(D) $f(x) = \dfrac{2}{x}$

(E) $f(x) = -x^3 + 2$

5. A wooden cube is 3 inches long on each side and has its exterior faces painted red. If the cube is cut into smaller cubes 1 inch long each side, how many of the smaller cubes have exactly two red faces?

(A) 3
(B) 6
(C) 8
(D) 10
(E) 12

6. If Y is the midpoint of $\overline{XZ}$, and P is the midpoint of $\overline{XY}$, which of the following must be true?

I. $PY = \dfrac{1}{2}YZ$

II. $XZ = 4XP$

III. $\dfrac{1}{2}XZ = 2PY$

(A) I only
(B) I, II only
(C) I, III only
(D) II and III only
(E) I, II, and III

GO ON TO THE NEXT PAGE

7. If $p = 3^a$, $q = 3^b$, and $a - b = 3$, what is p in terms of q?

(A) $9q$

(B) $18q$

(C) $\dfrac{1}{27q}$

(D) $27q$

(E) $\dfrac{q}{27}$

8. Marty has now m more marbles than Jane has. After he gives n marbles to Jane, how many more marbles will Jane have than Marty?

(A) m
(B) $n + m$
(C) $2n - m$
(D) $3n - 2m$
(E) $m - n$

$$x^2 < \frac{1}{x^2}$$

9. For what value of x is the statement above true?

(A) 2
(B) 1
(C) −0.3
(D) −1
(E) −2

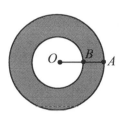

Note: Figure not drawn to scale.

10. The figure above shows two circles, each with center O. If $OB : AB = 3 : 2$ and the area of the shaded region is 64π, what is the radius of the smaller circle?

(A) 3
(B) 5
(C) 6
(D) 8
(E) 9

GO ON TO THE NEXT PAGE

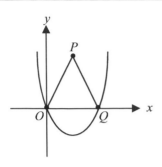

Note: Figure not drawn to scale.

11. The figure above shows the graph of f defined by $f(x) = x^2 - 4x$. If $\triangle OPQ$ is an equilateral triangle, what is the area of the triangle?

(A) 4
(B) $4\sqrt{3}$
(C) 8
(D) $8\sqrt{3}$
(E) 12

Note: Figure not drawn to scale.

12. In the figure above, Point P and Q are the midpoints of $\overline{AB}$ and $\overline{BC}$ respectively. If the length of $\overline{AB}$ is equal to $\frac{1}{3}$ of the length of $\overline{BC}$, and the length of $\overline{PQ}$ is 12, what is the length of $\overline{BC}$?

(A) 15
(B) 16
(C) 17
(D) 18
(E) 20

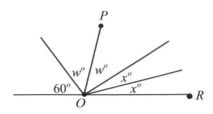

Note: Figure not drawn to scale.

13. In the figure above, If $\angle POR = 85$, what is the value of x?

(A) 15
(B) 20
(C) 22.5
(D) 25
(E) 27.5

14. A certain math club paid $80 for pizzas to be delivered to the club. The cost per person was calculated using the total number of members. However six members did not attend. The cost was recalculated which caused the cost per person attending to increase by $0.80. Which of the following equations can be used to find the number x of people who attended the meeting?

(A) $\dfrac{80}{x} - \dfrac{80}{x-6} = 0.8x$

(B) $\dfrac{80}{x} - \dfrac{80}{x+6} = 0.8$

(C) $(x-6)(80-6x) = 0.8$

(D) $\dfrac{80}{x-6} - \dfrac{80}{x} = 0.8$

(E) $\dfrac{x+0.8}{x-6} = \dfrac{x}{80}$

GO ON TO THE NEXT PAGE

15. The product of two numbers is k. If each of these two numbers is increased by 2, the new product is how much greater than twice the sum of the two original numbers?

(A) k

(B) $2k$

(C) $k+2$

(D) $2k+2$

(E) $k+4$

16. For how many ordered pairs of positive integers (x, y) is $x^2 + y^2 < 7$?

(A) One

(B) Two

(C) Three

(D) Four

(E) Five

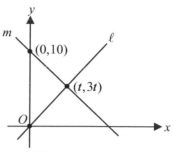

Note: Figure not drawn to scale.

17. In the xy-coordinate plane above, line ℓ passes through the origin and is perpendicular to line m. If the two lines intersect at point $(t, 3t)$, what is the value of t?

(A) 3

(B) 4

(C) 5

(D) 6

(E) 8

18. If there are m gallons of salt water that is $m\%$ salt, how many gallons of water must be added to make a solution that is 10% salt?

(A) $\dfrac{m-10}{10}$

(B) $\dfrac{m^2}{m-10}$

(C) $\dfrac{m^2-10m}{10}$

(D) $\dfrac{10}{m^2-10}$

(E) $\dfrac{m^2-10}{m+10}$

GO ON TO THE NEXT PAGE

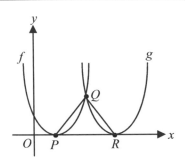

Note: Figure not drawn to scale.

19. The figure above shows the graphs of function f and g. If the function f is defined by $f(x) = (x-2)^2$, and the function g is defined by $g(x) = f(x-2)$. If the graphs intersect at point Q, what is the area of $\triangle PQR$?

(A) $\dfrac{1}{2}$

(B) 1

(C) $\dfrac{3}{2}$

(D) 2

(E) $\dfrac{9}{2}$

20. A ball is thrown straight up from the ground with an initial velocity of 100 feet per second, and height h, in t seconds, is given by $h = -10t^2 + 100t$. What is the maximum height that the ball can reach?

(A) 200 feet
(B) 250 feet
(C) 275 feet
(D) 300 feet
(E) 325 feet

STOP

If you finish before time is called, you may check your work on this section only.
Do not turn to any other section in the test.

SECTION 5
Time- 25 minutes
18 Questions

Turn to Section 5 (Page 1) of your answer sheet to answer the questions in this section.

Directions: For this section, solve each problem and decide which is the best of the choices given. Fill in the corresponding circle on the answer sheet. You may use any available space for scratchwork.

Reference Information

$A = \pi r^2$
$C = 2\pi r$ $A = \ell w$ $A = \dfrac{1}{2}bh$ $V = \ell wh$ $V = \pi r^2 h$ $c^2 = a^2 + b^2$ Special Right Triangles

The numbers of degrees of arc in a circle is 360°

The sum of the measures in degrees of the angles is 180°

---

1. If a machine can produce w widgets in s seconds, how many widgets will the same machine produce in 2 minutes at the same rate?

    (A) $2w$

    (B) $2ws$

    (C) $\dfrac{ws}{120}$

    (D) $\dfrac{120w}{s}$

    (E) $\dfrac{120}{sw}$

2. If $|1 - 2k| = k + 4$, which of the following could be the value of k?

    (A) -2
    (B) -1
    (C) 0
    (D) 2
    (E) 4

GO ON TO THE NEXT PAGE

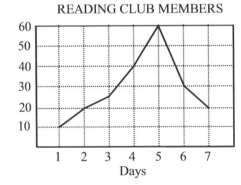

3. In the figure above, line ℓ, m, and n are parallel. If $b = 20$, what is the value of $a + c + d$?

(A) 300
(B) 320
(C) 340
(D) 360
(E) 400

ATTENDANCE OF READING CLUB MEMBERS

4. The graph above shows attendance of members of a reading club over 7 days. According to the graph, during which of the following two days is the greatest rate of change of attendance of members?

(A) Day 1 and 2
(B) Day 3 and 4
(C) Day 4 and 5
(D) Day 5 and 6
(E) Day 6 and 7

7, 2, 12, a, 6, 4

5. When a number is chosen at random from the six numbers listed above, the probability that this number will be less than 7 is $\frac{2}{3}$. Which of the following could be the value of a ?

(A) 6
(B) 7
(C) 8
(D) 9
(E) 10

6. Let the operations $\uparrow$ and $\downarrow$ be defined as follows.

$$P \uparrow = P + 1$$
$$P \downarrow = P - 1$$

Which of the following is equal to $(5 \uparrow) \times (3 \downarrow)$?

(A) $9 \downarrow$
(B) $10 \uparrow$
(C) $11 \downarrow$
(D) $12 \uparrow$
(E) $13 \downarrow$

GO ON TO THE NEXT PAGE ⇨

7. If set S has the property that x is in S, and $x^2 - 2x + 1$ is also in S, which of the following sets could be S?

(A) $\{1, 2, 3\}$
(B) $\{0, 1, 2\}$
(C) $\{-1, 0, 1\}$
(D) $\{0, 2\}$
(E) $\{1, 2\}$

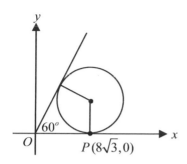

Note: Figure not drawn to scale.

8. The figure above shows the graphs of a line and a circle. If the circle is tangent to the line and the x-axis, what is the area of the circle?

(A) 36π
(B) 48π
(C) 54π
(D) 64π
(E) 81π

GO ON TO THE NEXT PAGE

5 ☐☐☐ 5 ☐☐☐ 5 ☐☐☐ 5

Directions: For Students-Produced Response questions 9-18, use the grid at the bottom of the answer sheet page on which you have answered questions 1-8.

Each of the remaining 10 questions requires you to solve the problem and enter your answer by making the circles in the special grid, as shown in the examples below. You may use any available space for scratchwork.

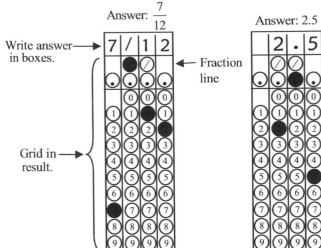

Answer: $\frac{7}{12}$

Write answer in boxes.

Fraction line

Grid in result.

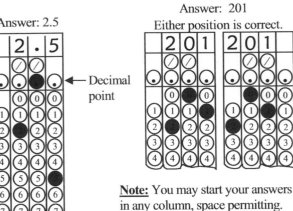

Answer: 2.5

Decimal point

Answer: 201
Either position is correct.

Note: You may start your answers in any column, space permitting. Columns not needed should be left blank.

- Mark no more than one circle in any column.

- Because the answer sheet will be machine-scored, **you will receive credit only if the circles are filled in correctly.**

- Although not required, it is suggested that you write your answer in the boxes at the top of the columns to help you fill in the circles accurately.

- Some problems may have more than one correct answer. In such cases, grid only one answer.

- No question has a negative answer.

- **Mixed numbers** such as $3\frac{1}{2}$ must be gridded as 3.5 or 7/2. (If [3][1][/][2] is gridded, it will be interpreted as $\frac{31}{2}$, not $3\frac{1}{2}$.)

- **Decimal Answers:** If you obtain a decimal answer with more digits than the grid can accommodate, it may be either rounded or truncated, but it must fill the entire grid. For example, if you obtain an answer such as 0.6666…, you should record your result as .666 or .667. **A less accurate value such as .66 or .67 will be scored as incorrect.**

Acceptable ways to grid $\frac{2}{3}$ are:

9. If $\dfrac{a}{b} = 3$ and $\dfrac{b}{c} = \dfrac{1}{5}$, what is the value of $\dfrac{5a}{c}$?

10. If the average (arithmetic mean) of two numbers is 30 and the smaller number is one-third of the larger number, what is the larger number?

GO ON TO THE NEXT PAGE

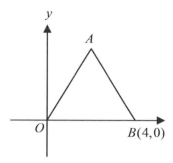

Note: Figure not drawn to scale.

11. In the figure above, $OA = AB$, and the area of $\triangle OAB$ is 10. What is the slope of $\overline{OA}$?

12. If m varies inversely with n^2 and if $m = 0.2$ when $n = 5$, what is the value of m when $n = 0.2$?

13. If x and y are positive integers, and if $x^2 - y^2 = 36$, what is the value of xy ?

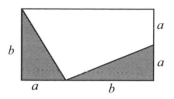

14. In the rectangle above, the sum of the areas of the shaded regions is 4. What is the area of the unshaded region?

15. Mr. Lee drove from A to B at an average speed of 60mph, and drove back from B to A at an average speed of 40mph. What was the average speed for the round trip? (Disregard the mph sign when gridding your answer.)

GO ON TO THE NEXT PAGE

16. The average (arithmetic mean) of the test scores of m girl students is 80 , and the average of the test scores of n boy students is 90. When both scores are combined, the average score is 82.

What is the value of $\dfrac{m}{n}$?

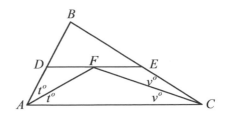

Note: Figure not drawn to scale.

18. In the figure above, $\overline{AF}$ and $\overline{CF}$ bisect $\angle DAC$ and $\angle ECA$ respectively, and $\overline{DE} \parallel \overline{AC}$. If $AB = 18$, $AC = 30$, and $BC = 24$, then what is the perimeter of $\triangle BDE$?

17. If p and q are numbers such that $(p-10)(q-10) = 0$. What is the smallest possible value of $p^2 + q^2$?

STOP

If you finish before time is called, you may check your work on this section only.
Do not turn to any other section in the test.

SECTION 7
Time- 20 minutes
16 Questions

Turn to Section 7 (Page 2) of your answer sheet to answer the questions in this section.

Directions: For this section, solve each problem and decide which is the best of the choices given. Fill in the corresponding circle on the answer sheet. You may use any available space for scratchwork.

1. If $8m^3 = 27$, what is the value of m?

   (A) 8
   (B) 4
   (C) 2
   (D) 1.5
   (E) 0.75

2. If $2(ax + bx) = 48$, what is the value of x when $a + b = 8$?

   (A) 3
   (B) 6
   (C) 12
   (D) 18
   (E) 24

GO ON TO THE NEXT PAGE

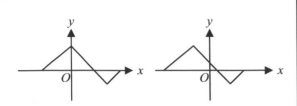

Figure I   Figure 2

3. If the figure I above shows the graph of $y = f(x)$, which of the following functions could describe the graph in figure 2?

(A) $y = f(x-2)$
(B) $y = f(x-1)+1$
(C) $y = f(x+1)$
(D) $y = f(x+1)+1$
(E) $y = f(x+1)-1$

4. If $2t + k$ is an odd integer, where t and k are positive integers, which of the following expressions must be an even integer?

(A) t
(B) $t-1$
(C) $t+1$
(D) k
(E) $k+1$

5. If k is at least 20 more than $\frac{1}{3}$ of p, which of the following describes this relationship?

(A) $k \le \dfrac{p+60}{3}$

(B) $k \le \dfrac{p}{3}+60$

(C) $k \le p+\dfrac{20}{3}$

(D) $k \ge \dfrac{p+60}{3}$

(E) $k \ge \dfrac{p}{3}+60$

6. If $a+b=c$ and $b+c=d$, what is d in terms of a and b?

(A) $a+b$
(B) $2a+b$
(C) $a+2b$
(D) $2a-b$
(E) $a-2b$

GO ON TO THE NEXT PAGE

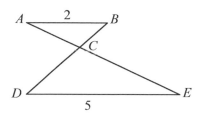

Note: Figure not drawn to scale.

7. In the figure above, $\overline{AB} \parallel \overline{DE}$, $AB = 2$, and $DE = 5$. If the perimeter of $\triangle ABC$ is 7, what is the perimeter of $\triangle CDE$?

    (A) 14
    (B) 15.5
    (C) 16
    (D) 17.5
    (E) 21

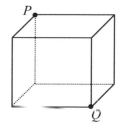

8. In the figure above, the cube has a volume of 64. What is the length of a diagonal (not shown) which cuts through the center of the cube?

    (A) 4
    (B) $4\sqrt{3}$
    (C) 6
    (D) $6\sqrt{2}$
    (E) $8\sqrt{3}$

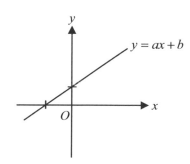

9. The figure above shows the graph of $y = ax + b$, where a and b are constants. Which of the following best represents the graph of the line $y = -2ax - b$?

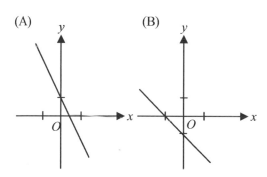

(A) (B)

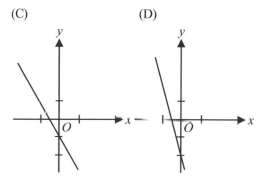

(C) (D)

(E)

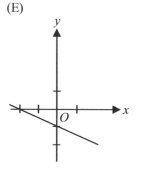

GO ON TO THE NEXT PAGE

10. If $F = k\dfrac{v^2}{r}$, where k is a constant, what happened to the value of F when v is tripled and r is halved?

    (A) F is doubled.
    (B) F is tripled.
    (C) F is multiplied by 8
    (D) F is multiplied by 12
    (E) F is multiplied by 18

11. If $x^3 < x < \sqrt{x}$, which of the following could be the value if x ?

    (A) 4
    (B) 3
    (C) 2
    (D) 1
    (E) 0.25

$$|3-k| = 10$$
$$|m+3| = 6$$

12. In the equations above, what is the greatest value of $k - m$?

    (A) 10
    (B) 14
    (C) 16
    (D) 22
    (E) 24

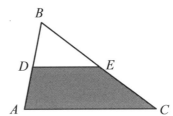

13. In $\triangle ABC$ above, points D and E are the midpoints of $\overline{AB}$ and $\overline{BC}$ respectively. If the area of the shaded region is $\dfrac{2}{3}$, what is the area of $\triangle BDE$?

    (A) $\dfrac{1}{9}$

    (B) $\dfrac{2}{9}$

    (C) $\dfrac{1}{3}$

    (D) $\dfrac{4}{9}$

    (E) $\dfrac{5}{9}$

GO ON TO THE NEXT PAGE ⟶

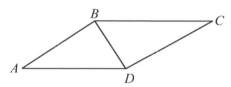

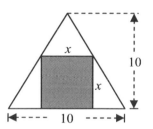

Note: Figure not drawn to scale.

14. In the figure above, $AB = BC = CD = DA = 13$ and $BD = 10$. What is the area of the quadrilateral $ABCD$?

(A) 40
(B) 60
(C) 70
(D) 80
(E) 120

16. In an isosceles triangle with a height 10 and a base 10, a square is inscribed with side x along the base of the triangle as shown above. What is the value of x?

(A) 4
(B) 5
(C) 5.25
(D) 5.5
(E) 6

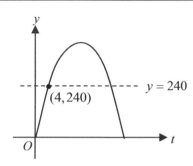

Note: Figure not drawn to scale.

15. The figure above shows the graph of a projectile fired from level ground. After t seconds, its height y, in feet, is given by $y = 80t - 5t^2$.
During what period is the projectile higher than 240 feet?

(A) $4 < t < 6$
(B) $4 < t < 7$
(C) $4 < t < 8$
(D) $4 < t < 10$
(E) $4 < t < 12$

STOP

If you finish before time is called, you may check your work on this section only.
Do not turn to any other section in the test.

TEST 13   ANSWER KEY

#	SECTION 3	SECTION 5	SECTION 7
1	A	D	D
2	D	B	A
3	D	C	C
4	E	A	E
5	E	A	D
6	E	E	C
7	D	B	D
8	C	D	B
9	C	3	C
10	C	45	E
11	B	2.5 or $\frac{5}{2}$	E
12	D	125	D
13	D	80	B
14	B	8	E
15	E	48	E
16	C	4	B
17	A	100	
18	C	42	
19	B		
20	B		

TEST 13   SECTION 3

1. (A)

Since $a + 2b = \frac{7}{3}$, $9(a + 2b) = 9\left(\frac{7}{3}\right) = 21$.

2. (D)

If $k = 81$, $\left|5 - \sqrt{81}\right| = 4$.

3. (D)

$b = 2a + 6$ and $b = c - 3$

$\rightarrow c = b + 3 = (2a + 6) + 3 = 2a + 9$.

4. (E)

E: $f(-2) = -(-2)^3 + 2 = 10$ and

$f(2) = -(2)^3 + 2 = -6$. Therefore,

$f(-2) > f(2)$.

5. (E)

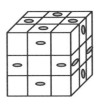

In the figure above, the number of cubes with only two red faces is 12. (Top 4, bottom 4, and middle 4)

6. (E)

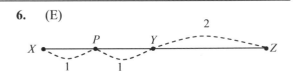

From the figure above, I, II, and III are all correct.

7. (D)

Since $\dfrac{p}{q} = \dfrac{3^a}{3^b} = 3^{a-b} = 3^3 = 27$, $p = 27q$.

8. (C)

Let x be the number of Jane's marbles.

Jane x , $x+n$

Marty $x+m$, $x+m-n$

Therefore, Jane-Marty =
$(x+n)-(x+m-n)=2n-m.$

---

9. (C)

If $x=-0.3$, $x^2=0.09$ and $\dfrac{1}{x^2}=11.1..$

Therefore, $x^2 < \dfrac{1}{x^2}$.

---

10. (C)

The ratio of the lengths = 3:5.

The ratio of the areas $= 9:25 = 9k:25k$.

The area of the shaded region $= 16k = 64\pi$
$\rightarrow k = 4\pi$.

Therefore, the area of the smaller circle =
$9k = 9(4\pi) = 36\pi$. $\pi r^2 = 36\pi$. Therefore,
the radius of the smaller circle is 6.

---

11. (B)

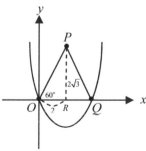

The x-intercepts are 0 and 4. Since $\overline{OQ}=4$
and $\triangle OPR$ is a special triangle, the area of

$\triangle OPQ$ is $\dfrac{4\times2\sqrt{3}}{2}=4\sqrt{3}.$

---

12. (D)

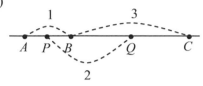

Use a convenient number(ratio). If $PQ=2k$,
then

$BC=3k$. Therefore, $2k=12, k=6.$
$3k = 3(6) = 18.$

---

13. (D)

$2x+w=85$ and $w+60=95.$ Since
$w=35, x=25.$

---

14. (B)

If x is the number of people who attended
the meeting, $x+6$ is the number of people at

the beginning. Therefore, $\dfrac{80}{x}-\dfrac{80}{x+6}=0.8.$

---

15. (E)

Let the two numbers be a and b. Product is
$k=ab$. New product is
$(a+2)(b+2)=ab+2a+2b+4$
$=k+2a+2b+4$
That is, $(a+2)(b+2)=2(a+b)+k+4$.

Therefore, The new product is $k+4$ greater
than twice the sum of two original numbers.

Or

You can use numbers. Let two numbers be
1 and 1. The product is 1 (that is, $k=1$).
New product is $(1+2)(1+2)=9$ and twice of
the sum of two original numbers is
$2(1+1)=4.$ Therefore, $9-4=5.$ When
$k=1,$ only (E) has a number 5.

---

16. (C)

If $x=1 \rightarrow y^2<6, y=1,2$. $\rightarrow(1,1) (1,2)$

If $x=2 \rightarrow y^2<3, y=1$. $\rightarrow(2,1)$

If $x=3 \rightarrow y^2<-2$. No solution.

Therefore, there are three pairs.

---

17. (A)

Since the slope of line ℓ is $\dfrac{3t}{t}=3$, the slope

of line m is $-\dfrac{1}{3}$. Therefore, $\dfrac{3t-10}{t-0}=\dfrac{-1}{3}$.

$\rightarrow 9t-30=-t \rightarrow 10t=30 \rightarrow t=3.$

---

18. (C)

Salt =

$m\%$ of m gallons $= \dfrac{m}{100}\times m = \dfrac{m^2}{100} = 0.01m^2$.

If x gallons of water is added, then
$\dfrac{0.01m^2}{m+x} = \dfrac{10}{100}$. Therefore,
$10(m+x) = 0.01m^2 \times 100 \rightarrow 10m + 10x = m^2$.
$x = \dfrac{m^2 - 10m}{10}$.

19. (B)

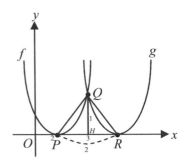

In the figure above, the coordinates of point P and H is (2,0) and (3,0) respectively.
Since $PR = 2$ and $QH = 1$, the area of $\triangle PQR$
is $\dfrac{2 \times 1}{2} = 1$.

20. (B)
$-10t(t-10) = 0$.
x-intercepts are $t = 0$ and $t = 10$.
The midpoint is 5.

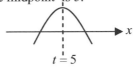

The maximum height occurs at $t = 5$.
Therefore, the maximum height is
$h(5) = -10(5)^2 + 100(5) = 250$.

TEST 13   SECTION 5

1. (D)
Proportion: Since 2 minutes is equal to 120
seconds, $\dfrac{w}{s} = \dfrac{x}{120}$. Therefore, $x = \dfrac{120w}{s}$.

2. (B)
(B) $k = -1$, $|1 - 2(-1)| = -1 + 4 \rightarrow 3 = 3$.

3. (C)

If $b = 20$, then $a = 160$, $c = 20$, and
$d = 160$.
Therefore, $a + c + d = 340$.

4. (A)
Day 1 and 2 $\rightarrow \dfrac{20-10}{10} = 100\%$

Day 3 and 4 $\rightarrow \dfrac{40-25}{25} = 60\%$

Day 4 and 5 $\rightarrow \dfrac{60-40}{40} = 50\%$

Day 5 and 6 $\rightarrow \dfrac{30-60}{60} = 50\%$ decreased

Day 6 and 7 $\rightarrow \dfrac{20-30}{30} = 33.3\%$ decreased

5. (A)
Since four of six numbers must be less than 7,
a is less than 7.

6. (E)
Since $(5\uparrow) = 6$ and $(3\downarrow) = 2$,
$(5\uparrow)(3\downarrow) = 12$.
Therefore, $13\downarrow = 12$.

7. (B)
$\{1,2,3\} \rightarrow 1^2 - 2(1) + 1 = -2$ (False)
$\{0,1,2\} \rightarrow 0^2 - 2(0) + 1 = 1,\ 1^2 - 2(1) + 1 = 0,$
            $2^2 - 2(2) + 1 = 1$ (True)
$\{-1,0,1\} \rightarrow (-1)^2 - 2(-1) + 1 = 3$ (False)
$\{0,2\} \rightarrow 0^2 - 2(0) + 1 = 1$ (False)

8. (D)

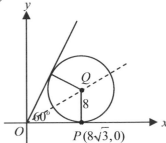

In the figure above, $\triangle OPQ$ is a special
triangle. If $OP = 8\sqrt{3}$, $PQ = 8$. Since 8 is the
radius, the area of the circle is $\pi r^2 = 64\pi$.

9. 3

Since $a = 3b$ and $c = 5b$, $\dfrac{5a}{c} = \dfrac{5(3b)}{5b} = 3$.

---

10. 45

Let smaller number $= a$ and larger number $= 3a$.

$\dfrac{a + 3a}{2} = 2a = 30 \rightarrow a = 15$. Therefore,

$3a = 45$.

---

11. 2.5 or $\dfrac{5}{2}$

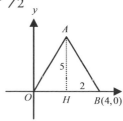

Since the area is 10 and $OB = 4$, height

$\overline{AH} = 5$. Therefore, the slope of $\overline{OA} = \dfrac{5}{2}$.

---

12. 125

Inverse proportion: $(m)(n^2) = K \,(\text{constant})$

Thus, $(0.2)(5^2) = m(0.2^2)$. Therefore,

$m = \dfrac{0.2 \times 25}{0.2^2} = 125$.

---

13. 80

$(x + y)(x - y) = 36$.

$x + y =$	36	18	9	6
$x - y =$	1	2	4	6
$2x =$	~~37~~	20	~~13~~	~~12~~

$x = 10$ and $y = 8$.

---

14. 8

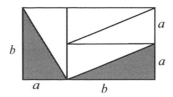

In the figure above, each triangle has an area of 2. Therefore, the unshaded region has an area of 8.

---

15. 48

$\text{Time}_{A \to B} = \dfrac{D}{60}$ and $\text{Time}_{B \to A} = \dfrac{D}{40}$.

The entire time $= \dfrac{D}{60} + \dfrac{D}{40} = \dfrac{5D}{120}$. Therefore,

the average speed $= \dfrac{2D}{\,5D\big/120\,} = 48\text{mph}$.

Or,

use a convenient number. Let distance from A to $B = 120$, because 120 is a multiple of 60 and 40. Now

$\text{Time}_{A \leftrightarrow B} = \dfrac{120}{60} + \dfrac{120}{40} = 5 \text{ hours}$. Therefore,

Average speed $= \dfrac{240}{5} = 48 \text{ mph}$.

---

16. 4

$\text{Sum}_{m \,\text{students}} = 80m$ and $\text{Sum}_{n \,\text{students}} = 90n$.

Therefore, $\text{Sum}_{m + n \,\text{students}} = \dfrac{80m + 90n}{m + n} = 82$.

Thus, $m = 4n$.

$\dfrac{m}{n} = \dfrac{4n}{n} = 4$.

---

17. 100

From $(p - 10)(q - 10) = 0$,

If $p - 10$, q can be any real number. Or

If $q = 10$, p can be any real number.

Therefore, the smallest value of

$p^2 + q^2 = 10^2 + 0 \;or\; 0 + 10^2$.

---

18. 42

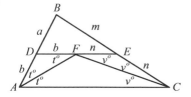

In the figure above, $\overline{DE} \parallel \overline{AC}$,

$a + b = 18$, $m + n = 24$, and $\angle DFA = \angle FAC$

(alternate angles are congruent). $\triangle ADF$ and

$\triangle VEF$ are isosceles triangles. Therefore, the

perimeter of $\triangle BDE$ is $a + b + m + n = 42$.

---

1.   (D)

Since $m^3 = \dfrac{27}{8} = \left(\dfrac{3}{2}\right)^3$, $m = \dfrac{3}{2}$ or 1.5.

2.   (A)

Since $ax + bx = (a+b)x = 24$ and

$a + b = 8$, then $x = 3$.

3.   (C)

Translate: Move to the left, but not up or down.

4.   (E)

t can be either even or odd, but k should be odd. Therefore, $k+1$ must be even.

5.   (D)

$k \geq \dfrac{1}{3}p + 20 = \dfrac{p+60}{3}$.

6.   (C)

$d = b + c = b + (a+b) = a + 2b$.

7.   (D)

$\triangle ABC \sim \triangle ECD$. Since the ratio of the lengths is 2:5, the ratio of the perimeters is also $2:5 = 2k:5k$. Thus, $2k = 7$ or $k = 3.5$.

Therefore, $5k = 17.5$.

8.   (B)

Since the length of a side is 4, the length of diagonal $\overline{PQ}$ is $\sqrt{4^2 + 4^2 + 4^2} = 4\sqrt{3}$.

9.   (C)

From the graph, let $a = 1$ and $b = 1$. the graph of $y = -2x - 1$ will be (C).

10.   (E)

Use convenient numbers. Let $K = 1, r = 1, and\ v = 1$.

Thus, $F = (1)\dfrac{1^2}{1} = 1$. When $v = 3$ and $r = 0.5$,

$F' = (1)\dfrac{3^2}{0.5} = 18$. Therefore, $F' = 18F$.

Or, $F' = k\dfrac{(3v)^2}{\left(\dfrac{r}{2}\right)} = 18k\dfrac{v^2}{r} = 18F$

11.   (E)

If $x = \dfrac{1}{4}$, $x^3 = \dfrac{1}{64}$, and $\sqrt{\dfrac{1}{4}} = \dfrac{1}{2}$. Therefore,

$x^3 < x < \sqrt{x}$.

12.   (D)

$|3 - k| = |k - 3| = 10 \rightarrow k - 3 = 10, -10$

$\rightarrow k = 13, -7$.

$|m + 3| = 6 \rightarrow m + 3 = 6, -6 \rightarrow m = 3, -9$

Therefore, the greatest value of $k - m$ is

$13 - (-9) = 22$.

13.   (B)

The ratio of the lengths of $\triangle BDE$ to $\triangle BAC$ is 1:2.

The ratio of the areas $= 1:4 = k:4k$. Thus, the area of the shaded region is $4k - k = 3k$.

Since $3k = \dfrac{2}{3}$, $k = \dfrac{2}{9}$. Therefore, the area of

$\triangle BDE$ is $k = \dfrac{2}{9}$.

14.   (E)

$ABCD$ is a rhombus.

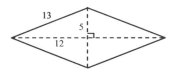

The diagonals of rhombus perpendicularly bisect each other.

In the figure above, the area of the rhombus

is $4 \times \left(\dfrac{5 \times 12}{2}\right) = 120$.

Or,

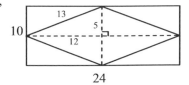

The area of rhombus is half of the area of the rectangle. Therefore, $\dfrac{10 \times 24}{2} = 120$

---

15. (E)

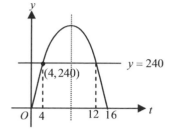

Find x-intercepts. $80t - 5t^2 = -5t(t-16) = 0$
$\to t = 0, 16$. The parabola is exactly symmetric with respect to the axis of symmetry. In the figure above, $y > 240$ for $4 < x < 12$.

---

16. (B)

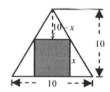

In figure above, two triangles are similar. Therefore,

$\dfrac{x}{10} = \dfrac{10-x}{10} \to 10x = 100 - 10x \to x = 5.$

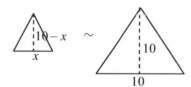

Corresponding sides are in proportion.

$\boxed{END}$

NO MATERIAL ON THIS PAGE

Dr. John Chung's SAT Math

TEST
14

ANSWER SHEET   TEST #:

SECTION 3

1	Ⓐ Ⓑ Ⓒ Ⓓ Ⓔ	11	Ⓐ Ⓑ Ⓒ Ⓓ Ⓔ	21	Ⓐ Ⓑ Ⓒ Ⓓ Ⓔ	31	Ⓐ Ⓑ Ⓒ Ⓓ Ⓔ
2	Ⓐ Ⓑ Ⓒ Ⓓ Ⓔ	12	Ⓐ Ⓑ Ⓒ Ⓓ Ⓔ	22	Ⓐ Ⓑ Ⓒ Ⓓ Ⓔ	32	Ⓐ Ⓑ Ⓒ Ⓓ Ⓔ
3	Ⓐ Ⓑ Ⓒ Ⓓ Ⓔ	13	Ⓐ Ⓑ Ⓒ Ⓓ Ⓔ	23	Ⓐ Ⓑ Ⓒ Ⓓ Ⓔ	33	Ⓐ Ⓑ Ⓒ Ⓓ Ⓔ
4	Ⓐ Ⓑ Ⓒ Ⓓ Ⓔ	14	Ⓐ Ⓑ Ⓒ Ⓓ Ⓔ	24	Ⓐ Ⓑ Ⓒ Ⓓ Ⓔ	34	Ⓐ Ⓑ Ⓒ Ⓓ Ⓔ
5	Ⓐ Ⓑ Ⓒ Ⓓ Ⓔ	15	Ⓐ Ⓑ Ⓒ Ⓓ Ⓔ	25	Ⓐ Ⓑ Ⓒ Ⓓ Ⓔ	35	Ⓐ Ⓑ Ⓒ Ⓓ Ⓔ
6	Ⓐ Ⓑ Ⓒ Ⓓ Ⓔ	16	Ⓐ Ⓑ Ⓒ Ⓓ Ⓔ	26	Ⓐ Ⓑ Ⓒ Ⓓ Ⓔ	36	Ⓐ Ⓑ Ⓒ Ⓓ Ⓔ
7	Ⓐ Ⓑ Ⓒ Ⓓ Ⓔ	17	Ⓐ Ⓑ Ⓒ Ⓓ Ⓔ	27	Ⓐ Ⓑ Ⓒ Ⓓ Ⓔ	37	Ⓐ Ⓑ Ⓒ Ⓓ Ⓔ
8	Ⓐ Ⓑ Ⓒ Ⓓ Ⓔ	18	Ⓐ Ⓑ Ⓒ Ⓓ Ⓔ	28	Ⓐ Ⓑ Ⓒ Ⓓ Ⓔ	38	Ⓐ Ⓑ Ⓒ Ⓓ Ⓔ
9	Ⓐ Ⓑ Ⓒ Ⓓ Ⓔ	19	Ⓐ Ⓑ Ⓒ Ⓓ Ⓔ	29	Ⓐ Ⓑ Ⓒ Ⓓ Ⓔ	39	Ⓐ Ⓑ Ⓒ Ⓓ Ⓔ
10	Ⓐ Ⓑ Ⓒ Ⓓ Ⓔ	20	Ⓐ Ⓑ Ⓒ Ⓓ Ⓔ	30	Ⓐ Ⓑ Ⓒ Ⓓ Ⓔ	40	Ⓐ Ⓑ Ⓒ Ⓓ Ⓔ

SECTION 5

1	Ⓐ Ⓑ Ⓒ Ⓓ Ⓔ	11	Ⓐ Ⓑ Ⓒ Ⓓ Ⓔ	21	Ⓐ Ⓑ Ⓒ Ⓓ Ⓔ	31	Ⓐ Ⓑ Ⓒ Ⓓ Ⓔ
2	Ⓐ Ⓑ Ⓒ Ⓓ Ⓔ	12	Ⓐ Ⓑ Ⓒ Ⓓ Ⓔ	22	Ⓐ Ⓑ Ⓒ Ⓓ Ⓔ	32	Ⓐ Ⓑ Ⓒ Ⓓ Ⓔ
3	Ⓐ Ⓑ Ⓒ Ⓓ Ⓔ	13	Ⓐ Ⓑ Ⓒ Ⓓ Ⓔ	23	Ⓐ Ⓑ Ⓒ Ⓓ Ⓔ	33	Ⓐ Ⓑ Ⓒ Ⓓ Ⓔ
4	Ⓐ Ⓑ Ⓒ Ⓓ Ⓔ	14	Ⓐ Ⓑ Ⓒ Ⓓ Ⓔ	24	Ⓐ Ⓑ Ⓒ Ⓓ Ⓔ	34	Ⓐ Ⓑ Ⓒ Ⓓ Ⓔ
5	Ⓐ Ⓑ Ⓒ Ⓓ Ⓔ	15	Ⓐ Ⓑ Ⓒ Ⓓ Ⓔ	25	Ⓐ Ⓑ Ⓒ Ⓓ Ⓔ	35	Ⓐ Ⓑ Ⓒ Ⓓ Ⓔ
6	Ⓐ Ⓑ Ⓒ Ⓓ Ⓔ	16	Ⓐ Ⓑ Ⓒ Ⓓ Ⓔ	26	Ⓐ Ⓑ Ⓒ Ⓓ Ⓔ	36	Ⓐ Ⓑ Ⓒ Ⓓ Ⓔ
7	Ⓐ Ⓑ Ⓒ Ⓓ Ⓔ	17	Ⓐ Ⓑ Ⓒ Ⓓ Ⓔ	27	Ⓐ Ⓑ Ⓒ Ⓓ Ⓔ	37	Ⓐ Ⓑ Ⓒ Ⓓ Ⓔ
8	Ⓐ Ⓑ Ⓒ Ⓓ Ⓔ	18	Ⓐ Ⓑ Ⓒ Ⓓ Ⓔ	28	Ⓐ Ⓑ Ⓒ Ⓓ Ⓔ	38	Ⓐ Ⓑ Ⓒ Ⓓ Ⓔ
9	Ⓐ Ⓑ Ⓒ Ⓓ Ⓔ	19	Ⓐ Ⓑ Ⓒ Ⓓ Ⓔ	29	Ⓐ Ⓑ Ⓒ Ⓓ Ⓔ	39	Ⓐ Ⓑ Ⓒ Ⓓ Ⓔ
10	Ⓐ Ⓑ Ⓒ Ⓓ Ⓔ	20	Ⓐ Ⓑ Ⓒ Ⓓ Ⓔ	30	Ⓐ Ⓑ Ⓒ Ⓓ Ⓔ	40	Ⓐ Ⓑ Ⓒ Ⓓ Ⓔ

9 10 11 12 13

14 15 16 17 18

Math Scoring Worksheet

A. Section 3

‾‾‾‾‾‾‾‾‾‾   ‾‾‾‾‾‾‾‾‾‾
numer of correct   *number of incorrect*

+   +

B. Section 5 (1-8)

‾‾‾‾‾‾‾‾‾‾   ‾‾‾‾‾‾‾‾‾‾
numer of correct   *number of incorrect*

+

C. Section 5 (9-18)

‾‾‾‾‾‾‾‾‾‾
numer of correct

+   +

D. Section 7

‾‾‾‾‾‾‾‾‾‾   ‾‾‾‾‾‾‾‾‾‾
numer of correct   *number of incorrect*

=   =

E. Total Unrounded Raw Score

‾‾‾‾‾‾‾‾‾‾ − ‾‾‾‾‾‾‾‾‾‾ ÷4 = ‾‾‾‾‾‾‾‾‾‾
numer of correct   *number of incorrect*

F. Total Rounded Raw Score   ‾‾‾‾‾‾‾‾ (See table)

Math Score Range = | ‾‾‾‾‾‾ |

Math Conversion Table

Raw Score	Scaled Score	Raw Score	Scaled Score
54	800	23	490-550
53	780-800	22	480-540
52	760-800	21	470-530
51	740-800	20	460-520
50	720-780	19	450-510
49	700-760	18	450-510
48	690-750	17	440-500
47	680-740	16	430-490
46	670-730	15	420-480
45	660-720	14	420-480
44	650-710	13	410-470
43	650-710	12	400-460
42	640-700	11	390-450
41	630-690	10	380-440
40	620-680	9	390-430
39	610-670	8	380-420
38	610-670	7	370-410
37	600-660	6	360-400
36	590-650	5	340-380
35	580-640	4	320-370
34	570-630	3	310-360
33	560-620	2	300-350
32	560-620	1	270-320
31	550-610	0	240-300
30	540-600	-1	200-290
29	530-590	-2	200-270
28	530-590	-3	200-260
27	520-580	-4	200-240
26	510-570	-5	200-220
25	500-560	-6 and below	200
24	500-560		

SECTION 3
Time- 25 minutes
20 Questions

Turn to Section 3 (Page 1) of your answer sheet to answer the questions in this section.

Directions: For this section, solve each problem and decide which is the best of the choices given. Fill in the corresponding circle on the answer sheet. You may use any available space for scratchwork.

Notes

1. The use of a calculator is permitted.
2. All numbers used are real numbers.
3. Figures that accompany problems in this test are intended to provide information useful in solving the problems. They are drawn as accurately as possible EXCEPT when it is stated in a specific problem that the figure is not drawn to scale. All figure lie in a plane unless other indicated.
4. Unless otherwise specified, the domain of any function f is assumed to be set of all real numbers x for which $f(x)$ is a real number.

Reference Information

$A = \pi r^2$ $\quad A = \ell w \quad$ $A = \dfrac{1}{2}bh$ $\quad V = \ell wh \quad$ $V = \pi r^2 h \quad$ $c^2 = a^2 + b^2 \quad$ Special Right Triangles
$C = 2\pi r$

The numbers of degrees of arc in a circle is $360°$.

The sum of the measures in degrees of the angles is $180°$.

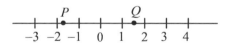

1. In the figure above, if the coordinates of points P and Q are multiplied together, the result will be the coordinate of a point between which two consecutive integers?

    (A) -3 and -2
    (B) -2 and -1
    (C) -1 and 0
    (D) 1 and 2
    (E) 2 and 3

2. Kim owned $\$100$. After winning $\$p$ in one game and losing $\$q$ in the next game, she owned $\$80$. What is the value of $p - q$?

    (A) 10
    (B) 20
    (C) 30
    (D) -10
    (E) -20

GO ON TO THE NEXT PAGE

Note: Figure not drawn to scale.

3. In the figure above, line ℓ and m are parallel. If $y = 40$ and $z = 120$, what is the value of x?

(A) 50
(B) 60
(C) 70
(D) 80
(E) 90

4. If $|k| < 5$, which of the following must be true?

    I. $-5 < k < 5$
    II. $k^2 < 5$
    III. $k^3 < 5$

(A) I only
(B) II only
(C) III only
(D) I and II only
(E) II and III only

    x 3   8 $x+2$ 9   4

5. If the median of six numbers listed above is 6, what is the value of x?

(A) 3
(B) 4
(C) 5
(D) 6
(E) 7

6. What is p percent of $\dfrac{k}{2}$?

(A) $0.5pk$
(B) $0.05pk$
(C) $0.005pk$
(D) $0.0005pk$
(E) $50pk$

GO ON TO THE NEXT PAGE

- 446 -

7. When five consecutive integers are added together, the result is -25. What is the greatest product that can be obtained by multiplying two of these five numbers?

(A) 12
(B) 20
(C) 30
(D) 42
(E) 60

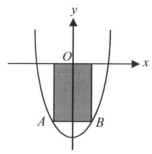

Note: Figure not drawn to scale.

9. The figure above shows the graph of a quadratic function $f(x) = 2x^2 - 18$. If the length of $\overline{AB}$ is 4, what is the area of the shaded region of a rectangle?

(A) 40
(B) 36
(C) 32
(D) 24
(E) 20

8. If $m^{2a} \cdot m^b = m^6$ and $\dfrac{m^a}{m^b} = \dfrac{1}{m^3}$, what is the value of $a + b$?

(A) 2
(B) 4
(C) 5
(D) 7
(E) 9

10. If k is a positive integer, what is the smallest value of k for which $(12k)^{\frac{1}{3}}$ is an integer?

(A) 144
(B) 72
(C) 36
(D) 18
(E) 12

GO ON TO THE NEXT PAGE

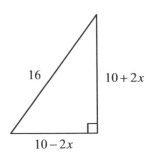

Note: Figure not drawn to scale.

11. In the triangle above, what is the value of $100 - x^2$?

(A) 93
(B) 120
(C) 164
(D) 224
(E) 256

13. If $x > 1$, which of the following increases as x increases?

I. $\dfrac{1}{x}$

II. $\dfrac{x-1}{x}$

III. $\dfrac{1}{x^2 - x}$

(A) I only
(B) II only
(C) III only
(D) II and III only
(E) I, II, and III

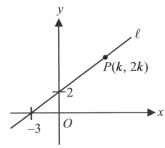

Note: Figure not drawn to scale.

12. In the figure above, line ℓ passes through point P. What is the value of k ?

(A) $\dfrac{1}{2}$

(B) $\dfrac{1}{3}$

(C) $\dfrac{2}{3}$

(D) $\dfrac{3}{2}$

(E) $\dfrac{5}{2}$

14. If k is a constant and $x^2 + kx + (k+1)$ is equivalent to $(x+p)(x+2)$. What is the value of p ?

(A) 3
(B) 2
(C) 0
(D) −1
(E) It cannot be determined from the information given.

GO ON TO THE NEXT PAGE

15. Let the function f be defined by

$f(x) = 2x^2 - 3$. If $\frac{1}{3}f(\sqrt{k}) = 3$, what is the

value of k?

(A) 2
(B) 3
(C) 4
(D) 5
(E) 6

17. At a certain party, an executive committee provided one soda for 8 people, one apple for 4 people, and one cake for 6 people. If the total number of sodas, apples, and cakes was 78, how many people were at the party?

(A) 48
(B) 72
(C) 96
(D) 120
(E) 144

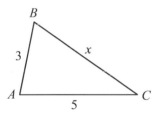

Note: Figure not drawn to scale.

16. In $\triangle ABC$ above, $\angle BAC < 90^o$. Which of the following cannot be the value of x?

(A) 3

(B) $3\frac{1}{2}$

(C) 4

(D) $4\frac{1}{2}$

(E) 6

18. If $a > b$ and $k < 0$, which of the following must be true?

   I. $ak > bk$
   II. $a^2 > b^2$
   III. $\dfrac{a+k}{k} < \dfrac{b+k}{k}$

(A) I only
(B) II only
(C) III only
(D) II and III only
(E) I, II, and III

GO ON TO THE NEXT PAGE

19. If the average of a and b is 5, the average of a, b, and c is 7, and the average of a, b, c, and d is 10, what is the average of c and d?

    (A) 10
    (B) 12
    (C) 15
    (D) 17
    (E) 18

20. For any square, if the area is A square inches and the perimeter is P inches, then A is directly proportional to which of the following?

    (A) $\dfrac{1}{P}$

    (B) $\dfrac{1}{P^2}$

    (C) P

    (D) P^2

    (E) P^3

STOP

If you finish before time is called, you may check your work on this section only.
Do not turn to any other section in the test.

SECTION 5
Time- 25 minutes
18 Questions

Turn to Section 5 (Page 1) of your answer sheet to answer the questions in this section.

Directions: For this section, solve each problem and decide which is the best of the choices given. Fill in the corresponding circle on the answer sheet. You may use any available space for scratchwork.

Notes

1. The use of a calculator is permitted.
2. All numbers used are real numbers.
3. Figures that accompany problems in this test are intended to provide information useful in solving the problems. They are drawn as accurately as possible EXCEPT when it is stated in a specific problem that the figure is not drawn to scale. All figure lie in a plane unless other indicated.
4. Unless otherwise specified, the domain of any function f is assumed to be set of all real numbers x for which $f(x)$ is a real number.

Reference Informatiom

$A = \pi r^2$
$C = 2\pi r$
$A = \ell w$
$A = \frac{1}{2}bh$
$V = \ell wh$
$V = \pi r^2 h$
$c^2 = a^2 + b^2$
Special Right Triangles

The numbers of degrees of arc in a circle is $360°$.

The sum of the measures in degrees of the angles is $180°$.

---

1. If $\dfrac{2k}{2k+13} = \dfrac{14}{27}$, then $k =$

(A) 3
(B) 5
(C) 7
(D) 14
(E) 28

2. If the surface area of a cube is $150x^2$, what is the volume of the cube?

(A) $8x^3$
(B) $27x^3$
(C) $64x^3$
(D) $125x^3$
(E) $150x^3$

GO ON TO THE NEXT PAGE

3. Let the function f be defined by $f(2k) = 2f(k)$ and $f(5) = 10$. Which of the following functions could be the definition of $f(x)$?

(A) $f(x) = x + 5$
(B) $f(x) = x - 5$
(C) $f(x) = 15 - x$
(D) $f(x) = 2x$
(E) $f(x) = x$

4. If a is an odd integer, which of the following is an even integer?

(A) $a + 2$
(B) a^2
(C) a^3
(D) $a^2 - a$
(E) $(a + 2)^2$

5. The average of p and q is 5, and the average of p^2 and q^2 is 35. What is the value of pq?

(A) 10
(B) 15
(C) 30
(D) 45
(E) 70

$$10, \quad 15, \quad 5, \quad 10, \ldots$$

6. In the sequence above, the first term is 10, the second term is 15, and each term after the second is the nonnegative difference between the previous two terms. If a_n is defined as the nth term, what is the value of a_{124}?

(A) 0
(B) 5
(C) 10
(D) 15
(E) 30

GO ON TO THE NEXT PAGE

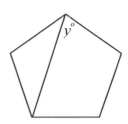

7. The figure above is a pentagon that has sides of equal length. What is the value of y ?

   (A) 36
   (B) 72
   (C) 74
   (D) 75
   (E) 80

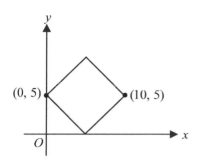

Note: Figure not drawn to scale.

8. In the figure above, what is the perimeter of the square?

   (A) 12
   (B) 20
   (C) 28
   (D) $20\sqrt{2}$
   (E) $40\sqrt{2}$

GO ON TO THE NEXT PAGE

Directions: For Students-Produced Response questions 9-18, use the grid at the bottom of the answer sheet page on which you have answered questions 1-8.

Each of the remaining 10 questions requires you to solve the problem and enter your answer by making the circles in the special grid, as shown in the examples below. You may use any available space for scratchwork.

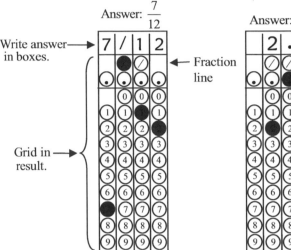

Answer: $\frac{7}{12}$

Write answer in boxes. → Fraction line

Grid in → result.

Answer: 2.5 ← Decimal point

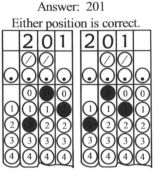

Answer: 201
Either position is correct.

Note: You may start your answers in any column, space permitting. Columns not needed should be left blank.

- Mark no more than one circle in any column.

- Because the answer sheet will be machine-scored, **you will receive credit only if the circles are filled in correctly.**

- Although not required, it is suggested that you write your answer in the boxes at the top of the columns to help you fill in the circles accurately.

- Some problems may have more than one correct answer. In such cases, grid only one answer.

- No question has a negative answer.

- **Mixed numbers** such as $3\frac{1}{2}$ must be gridded as 3.5 or 7/2. (If $\boxed{3\,1\,/\,2}$ is gridded, it will be interpreted as $\frac{31}{2}$, not $3\frac{1}{2}$.)

- **Decimal Answers:** If you obtain a decimal answer with more digits than the grid can accommodate, it may be either rounded or truncated, but it must fill the entire grid. For example, if you obtain an answer such as 0.6666..., you should record your result as .666 or .667. **A less accurate value such as .66 or .67 will be scored as incorrect.**

Acceptable ways to grid $\frac{2}{3}$ are:

9. If $6a - 4b = 7$ and $2a - 8b = 3$, then $a + b =$

10. If $100 \times 10^6 + 0.3 \times 10^k = 1.3 \times 10^8$, what is the value of k?

GO ON TO THE NEXT PAGE ⇨

11. In a community of 85 people, each person owns a dog or a cat or both. If there are 45 dog owners and 53 cat owners, how many people own both a dog and a cat?

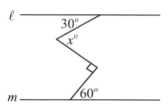

Note: Figure not drawn to scale.

12. In the figure above, line ℓ and m are parallel. What is the value of x ?

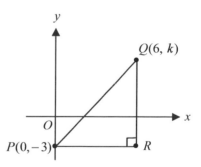

Note: Figure not drawn to sale.

13. In the *xy*-coordinate plane above, $\triangle PQR$ is an isosceles triangle. What is the value of k ?

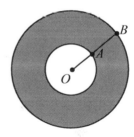

Note: Figure not drawn to scale.

14. In the figure above, two circles have centers at the same point O, and $AB = \dfrac{3}{2} AO$. If the area of the shaded region is 42, what is the area of the unshaded region?

GO ON TO THE NEXT PAGE

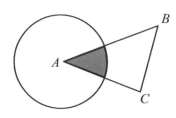

Note: Figure not drawn to scale.

15. In the figure above, the area of the circle is three times the area of the triangle. If the common shaded region is removed, then the area of the rest of the circle would be 36 more than the area of the rest of the triangle. What is the area of $\triangle ABC$?

READING GROUP
ATTENDANCE

Week	Attendance
1	67
2	54
3	35
4	21
5	n

16. The table above shows the attendance at the reading group of the J.C High School. If the median attendance was 35, what is the greatest possible value of n ?

$$A = \{ -10, -5, \ 0, \ p, \ q \}$$

17. If a number x is chosen from set A above, where p and q are different integers, and if the probability that the number x is a member of the solution set of both $2x - 3 > 3$ and $3x - 13 < 5$ is $\dfrac{2}{5}$, then what is the possible value of $p + q$?

18. The relationship between the temperature expressed in Celsius degrees (C) and Fahrenheit degrees (F) is given by $C = \dfrac{5}{9}(F - 32)$. If the Fahrenheit temperature increases by 3 degrees, what is the degree increase in Celsius, in degrees?

STOP

**If you finish before time is called, you may check your work on this section only.
Do not turn to any other section in the test.**

SECTION 7
Time- 20 minutes
16 Questions

Turn to Section 7 (Page 2) of your answer sheet to answer the questions in this section.

Directions: For this section, solve each problem and decide which is the best of the choices given. Fill in the corresponding circle on the answer sheet. You may use any available space for scratchwork.

 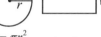
1. If $k = 10$, what is the value of
$$\frac{2k(k+1) - 2k(k-1)}{4}?$$

(A) 10
(B) 20
(C) 50
(D) 100
(E) 200

2. If $4^n = 2^k$, where n and k are positive integers, which of the following could not be the value of k?

(A) 2
(B) 6
(C) 10
(D) 15
(E) 30

GO ON TO THE NEXT PAGE

3. Each time that Sophie counted her marbles by either 3's or 5's she had one left over. But when she counted them by 7's she had none left over. What is the least number of marbles she could have had?

(A) 35
(B) 49
(C) 70
(D) 91
(E) 105

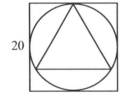

4. In the figure above, a circle is inscribed in a square with sides of length 20, and an equilateral triangle is inscribed in the circle. What is the area of the triangle?

(A) 50
(B) $50\sqrt{3}$
(C) 75
(D) $75\sqrt{3}$
(E) 175

5. The cost of electricity for a month is given by $C = \dfrac{2}{5}K + 25$, where K is the number of kilowatt hours of electricity used, and C is the price, in dollars. If there is a 25 increase in K, which of the following is the increase in C?

(A) 10
(B) 15
(C) 25
(D) 35
(E) 50

6. The ages of three children are consecutive integers. Five years ago, the sum of their ages was n. What is the present age of the oldest child in terms of n?

(A) $\dfrac{n-12}{3}$

(B) $\dfrac{2n-15}{3}$

(C) $\dfrac{n+18}{3}$

(D) $3n-15$

(E) $n+18$

GO ON TO THE NEXT PAGE

7. If $x + y = 5$ and $x - y = 3$, what is the value of $x^2 - y^2 + 2x - 2y$?

(A) 8
(B) 12
(C) 15
(D) 21
(E) 24

9. If $(x+y)^2 - (x-y)^2 = 72$, where x and y are positive integers, which of the following could be a value of $x + y$?

(A) 10
(B) 11
(C) 12
(D) 13
(E) 14

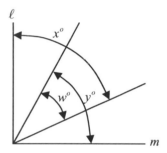

Note: Figure not drawn to scale.

8. In the figure above, line ℓ is perpendicular to line m, and $w = 30$. What is the value of $x + y$?

(A) 60
(B) 90
(C) 120
(D) 140
(E) 150

10. A gas tank with a capacity of g gallons is empty. A pump can deliver h gallons of gas every t seconds. In terms of g, h, and t, how many seconds will it take this pump to fill the tank?

(A) $\dfrac{g}{ht}$

(B) $\dfrac{t}{gh}$

(C) $\dfrac{gh}{t}$

(D) $\dfrac{gt}{h}$

(E) $\dfrac{h}{gt}$

GO ON TO THE NEXT PAGE

11. How many different 4-digit positive integers can be formed using the digits 4, 5, 6, 7, 8, and 9, if the unit digit is 8 or 9 and no digit is repeated within an integer?

(A) 30
(B) 60
(C) 120
(D) 240
(E) 360

13. In the xy-coordinate plane, line m is a reflection of line ℓ about x-axis. If the equation of line ℓ is $y = -\dfrac{2}{5}x + 10$, what is the equation of line m?

(A) $y = -\dfrac{5}{2}x + 10$

(B) $y = \dfrac{5}{2}x - 10$

(C) $y = -\dfrac{2}{5}x + 10$

(D) $y = \dfrac{2}{5}x - 10$

(E) $y = \dfrac{2}{5}x + 10$

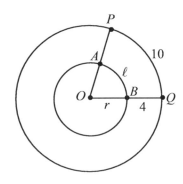

Note: Figure not drawn to scale.

12. The figure above shows circles with radius r and $(r + 4)$ and the same center at point O. If the length of arc PQ is 10 and the length of arc AB is ℓ, what must ℓ equal?

(A) 5

(B) 10

(C) $\dfrac{10r}{r + 4}$

(D) $\dfrac{8r}{10 + r}$

(E) $\pi(r + 4)$

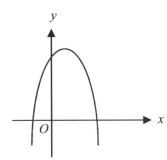

14. The figure above shows the graph of a function f defined by $f(x) = ax^2 + bx + c$, where a, b, and c are constants. Which of the following must be true?

(A) $a > 0$
(B) $b > 0$
(C) $c < 0$
(D) $b < 0$
(E) $bc < 0$

GO ON TO THE NEXT PAGE

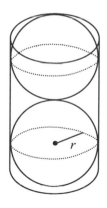

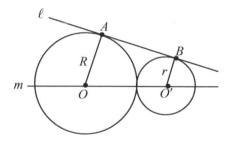

15. In the figure above, two basketballs are packed in the cylindrical box. If each basketball has a radius of r, what is the smallest surface area of the cylindrical box, in terms of r?

(A) $5\pi r^2$

(B) $10\pi r^2$

(C) $12\pi r^2$

(D) $15\pi r^2$

(E) $2\pi r^2 + 10r$

16. In the figure above, two circles are tangent to each other, and line ℓ is tangent to the circles and line m passes through the center of the two circles. If the radius of circle O is R and the radius of circle O' is r, in terms of R and r, what is the length of $\overline{AB}$?

(A) $R + r$

(B) $2\sqrt{Rr}$

(C) $2(R + r)$

(D) $2Rr$

(E) $2(R - r)$

STOP

**If you finish before time is called, you may check your work on this section only.
Do not turn to any other section in the test.**

TEST 14   ANSWER KEY

#	SECTION 3	SECTION 5	SECTION 7
1	A	C	A
2	E	D	D
3	D	D	D
4	A	D	D
5	C	B	A
6	C	A	C
7	D	B	D
8	C	D	C
9	A	1	B
10	D	8	D
11	A	13	C
12	D	60	C
13	B	3	D
14	A	8	B
15	E	18	B
16	E	35	B
17	E	9	
18	C	$\frac{5}{3}$ or 1.66, 1.67	
19	C		
20	D		

TEST 14   SECTION 3

1. (A)
Let $P = -1.8$ and $Q = 1.5$, then
$PQ = -1.8 \times 1.5 = -2.7$.

2. (E)
Since $100 + p - q = 80$, then $p - q = -20$.

3. (D)

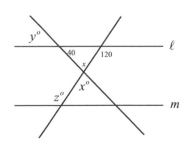

In the figure above, $x + 40 = 120 \rightarrow x = 80$.

4. (A)
$|k| < 5$ is equivalent to $k^2 < 25$ and
$-5 < k < 5$.

5. (C)
If $x = 3$, 3, 3, 4, 5, 8, 9 Median = 4.5
If $x = 4$, 3, 4, 4, 6, 8, 9 Median = 5
If $x = 5$, 3, 4, 5, 7, 8, 9 Median = 6

6. (C)
$$p\% \text{ of } \frac{k}{2} = \frac{p}{100} \times \frac{k}{2} = \frac{pk}{200} = 0.005\,pk.$$

7. (D)
Average (median)$= \dfrac{-25}{5} = -5$. Therefore, the
sequence is as follows.
$-3, -4, -5, -6, -7$.
The greatest product is $(-7)(-6) = 42$.

8. (C)

$m^{2a+b} = m^6 \rightarrow 2a+b = 6$ (1)

$m^{a-b} = m^{-3} \rightarrow a-b = -3$ (2)

(1)+(2) $3a = 3 \rightarrow a = 1$ and

$b = 4$. Therefore, $a+b = 5$.

---

9. (A)

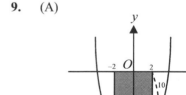

In the figure above, $f(2) = 2(2^?) - 18 - 10$.

Thus the height of the rectangle is 10.
Therefore, the area of the rectangle is
$4 \times 10 = 40$.

---

10. (D)

To be an integer, $12k$ must be a cubed
number.

$12k = 2 \cdot 2 \cdot 3 \cdot k \rightarrow 2 \cdot 2 \cdot 3 \cdot (2 \cdot 3 \cdot 3)$.

Therefore, k should be $2 \cdot 3 \cdot 3 = 18$.

---

11. (A)

Pythagorean theorem:

$(10+2x)^2 + (10-2x)^2 = 16^2$.

$\rightarrow 200 + 8x^2 = 256 \rightarrow x^2 = 7$. Therefore,

$100 - x^2 = 100 - 7 = 93$.

---

12. (D)

Constant slope: $\dfrac{2k-2}{k-0} = \dfrac{2}{3} \rightarrow k = \dfrac{3}{2}$.

---

13. (B)

Use convenient numbers: x increased from
2 to 3.

(I) $\dfrac{1}{2} \rightarrow \dfrac{1}{3}$ (Decreased)

(II) $\dfrac{1}{2} \rightarrow \dfrac{2}{3}$ (Increased)

(III) $\dfrac{1}{2} \rightarrow \dfrac{1}{6}$ (Decreased)

---

14. (A)

$(x+p)(x+2) = x^2 + (p+2)x + 2p$.

Since $k = p+2$ and $k+1 = 2p$,

$p+2+1 = 2p$.

Therefore, $p = 3$.

---

15. (E)

$\dfrac{1}{3}f(\sqrt{k}) = \dfrac{1}{3}(2k-3) \rightarrow 2k-3 = 9 \rightarrow k = 6$.

---

16. (E)

If $\angle A = 90^o$, then $x = 5$. Therefore x must be
less than 5 and greater than 2(triangle
inequality).

---

17. (E)

If the number of people is n, $\dfrac{n}{8} + \dfrac{n}{4} + \dfrac{n}{6} = 78$.

$\dfrac{3n+6n+4n}{24} = 78 \rightarrow \dfrac{13n}{24} = 78$.

$n = \dfrac{24 \times 78}{13} = 144$.

---

18. (C)

I. $ak > bk$ (false)

II. $a^2 > b^2$ (false)   for $a = 1$ and $b = -3$

III. $\dfrac{a+k}{k} < \dfrac{b+k}{k}$ (true)

$\dfrac{a}{k} + 1 < \dfrac{b}{k} + 1 \rightarrow \dfrac{a}{k} < \dfrac{b}{k}$

---

19. (C)

$\dfrac{a+b}{2} = 5 \rightarrow a+b = 10$.

$\dfrac{a+b+c}{3} = 7 \rightarrow a+b+c = 21$.

$\dfrac{a+b+c+d}{4} = 10 \rightarrow a+b+c+d = 40$.

Therefore, $c = 11$ and $d = 19$.

$\dfrac{c+d}{2} = \dfrac{11+19}{2} = 15$.

Or,

$a+b = 10$ and $c+d = 30$

$\dfrac{c+d}{2} = \dfrac{30}{2} = 15$

---

20. (D)

Let the length of a square be x.

$A = x^2$ and $P = 4x$

Express A in terms of P.

Since $x = \dfrac{P}{4}$, $A = \left(\dfrac{P}{4}\right)^2 = \dfrac{1}{16}P^2$.

Therefore, A is directly proportional to P^2.

TEST 14   SECTION 5

1. (C)

$\dfrac{2k}{2k+13} = \dfrac{14}{27} \rightarrow 2k = 14 \rightarrow k = 7$.

$2k + 13 = 2(7) + 13 = 27$.

---

2. (D)

$\dfrac{150x^2}{6} = 25x^2$ (area of a face). The length of
an edge is $5x$. Therefore, the volume is
$(5x)^3 = 125x^3$.

---

3. (D)

$f(x) = 2x,\ 2f(x) = 4x.\ f(2k) = 2(2k) = 4k.$

$f(5) = 2(5) = 10.$

---

4. (D)

Use a good number: Let $a = 1$ or 3.

(A) $a + 2 = 3, 5$ (odd)   (B) $a^2 = 1, 9$ (odd)

(C) $a^3 = 1, 27$ (odd)   (D) $a^2 - a = 0, 6$ (even)

(E) $(a+2)^2 = 9, 25$ (odd)

---

5. (B)

$\dfrac{p+q}{2} = 5 \rightarrow p + q = 10.$

$\dfrac{p^2 + q^2}{2} = 35 \rightarrow p^2 + q^2 = 70.$

$(p+q)^2 = p^2 + q^2 + 2pq = 100$

Therefore, $2pq = 30 \rightarrow pq = 15.$

---

6. (A)

$10,\ 15,\ 5,\ 10,\ \underline{5,\ 5,\ 0,\ 5,\ 5,\ 0,\ 5,}\ \dots\ a_{124}$

After 4th term $5,5,0$ $5,5,0$ $5,5,0$ ….
124-4=120th term after 4th term. Since 120 is
multiple of 3, 120th term is 0. Actually
$a_{124} = 0.$

---

7. (B)

In the figure above, each interior angle is
$\dfrac{540}{5} = 108$. Therefore, $y = 108 - 36 = 72.$

---

8. (D)

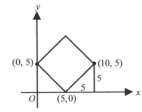

In the figure above, the length of an edge is
$5\sqrt{2}$. Therefore, the perimeter of the square
is $4 \times 5\sqrt{2} = 20\sqrt{2}.$

---

9. 1

$\begin{array}{r} 6a - 4b = 7 \\ - \underline{\,2a - 8b = 3\,} \\ 4a + 4b = 4 \end{array}$ Therefore, $a + b = 1.$

---

10. 8

$100 \times 10^6 + 0.3 \times 10^k = 10^8 + 0.3 \times 10^k$

$= 1.3 \times 10^8 = 1 \times 10^8 + 0.3 \times 10^8.$

Therefore, $0.3 \times 10^k = 0.3 \times 10^8$ or $k = 8.$

---

11. 13

$(45 + 53) - 85 = 13.$

12. 60

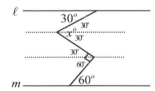

In the figure above, $x = 30 + 30 = 60$.

---

13. 3

Since $PQ = 6$ and $QR = 6$, $k = -3 + 6 = 3$.

---

14. 8

Because the ratio of the lengths of the two circles is 2:5, the ratio of the areas is $4 : 25 = 4k : 25k$. The area of the shaded region is $21k = 42. \rightarrow k = 2$.

Therefore, the unshaded area is $4k = 8$.

---

15. 18

If the area of the triangle is x, the area of the circle is $3x$. Let the common area is k. Then $3x - k = (x - k) + 36 \rightarrow 2x = 36 \rightarrow x = 18$.

---

16. 35

n	21	<u>35</u>	54	67
		Median		

Or

21	n	<u>35</u>	54	67
		Median		

Therefore, the maximum value of n is 35.

---

17. 9

When you simplify inequalities, $3 < x < 6$.

Since probability is $\dfrac{2}{5}$, two numbers p and q are greater by 3 and less by 6. Therefore, $p = 4$ and $q = 5$, or $q = 4$ and $p = 5$.

$p + q = 9$.

---

18. $\dfrac{5}{3}$ or 1.66, 1.67

Slope $= \dfrac{\triangle C}{\triangle F} = \dfrac{\triangle C}{3} = \dfrac{5}{9}$. Therefore,

$\triangle C = \dfrac{15}{9} = \dfrac{5}{3}$.

---

1. (A)

$\dfrac{2k(k+1) - 2k(k-1)}{4} = \dfrac{4k}{4} = k$. Therefore, $k = 10$.

---

2. (D)

Since $4^n = 2^{2n} = 2^k$, $k = 2n$. Therefore, k is an even number.

---

3. (D)

The number must be 1 more than a multiple of 15.

16,31,46,61,76,91….. This number is also a multiple of 7. The least multiple of 7 is 91.

---

4. (D)

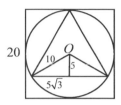

The radius of the circle is 10 and the triangle is a special triangle. Therefore, the area of the triangle is $3\left(\dfrac{10\sqrt{3} \times 5}{2}\right) = 75\sqrt{3}$.

---

5. (A)

Slope $\dfrac{2}{5} = \dfrac{\triangle C}{\triangle K} = \dfrac{\triangle C}{25}$. Therefore, $\triangle C = 10$.

---

6. (C)

Let the ages be $k-2, k-1$, and k. Five years ago the ages were $k-7, k-6,$ and $k-5$. Because $k-7+k-6+k-5 = 3k-18 = n$, $k = \dfrac{n+18}{3}$.

---

7. (D)

$x^2 - y^2 = (x+y)(x-y) = 15$ and $2x - 2y = 2(x - y) = 6$.

Therefore, 15+6 = 21.

---

8. (C)

$x + y = 90 + w \rightarrow 90 + 30 = 120.$

---

9. (B)

$(x+y)^2 - (x-y)^2 = 4xy$

$\rightarrow 4xy = 72 \rightarrow xy = 18.$

Therefore, $(x, y) \rightarrow (1, 18)\ (2, 9)\ (3, 6)$.

$x + y = 19, 11, \text{ or } 9$.

---

10. (D)

Proportion: $\dfrac{g}{x} = \dfrac{h}{t}$. Therefore, $x = \dfrac{gt}{h}$.

---

11. (C)

□□□8 or □□□9

$5 \times 4 \times 3 = 60 \qquad 5 \times 3 \times 2 = 60$

Therefore, $60 + 60 = 120$.

---

12. (C)

The ratio of the lengths is $r : (r+4)$.

Therefore, $\dfrac{r}{r+4} = \dfrac{\ell}{10}$. $\ell = \dfrac{10r}{r+4}$.

---

13. (D)

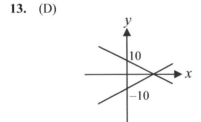

In the figure above, $y = \dfrac{2}{5}x - 10.$

---

14. (B)

Axis of symmetry $= -\dfrac{b}{2a} > 0$ and y-intercept

$= c > 0$.

The graph opens downward, $a < 0$. Therefore, $b > 0$.

---

15. (B)

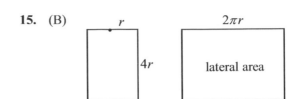

In the figure above, the areas of top and bottom are $2\left(\pi r^2\right)$, and the lateral area is

$2\pi r \times 4r = 8\pi r^2$.

Therefore, total surface area is $10\pi r^2$.

---

16. (B)

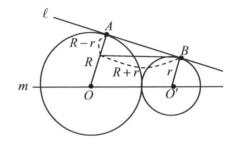

In the figure above,

$AB = \sqrt{(R+r)^2 - (R-r)^2} = \sqrt{4Rr} = 2\sqrt{Rr}$.

$\boxed{END}$

Dr. John Chung's SAT Math

TEST
15

ANSWER SHEET　　　TEST #:

SECTION 3

1 Ⓐ Ⓑ Ⓒ Ⓓ Ⓔ	11 Ⓐ Ⓑ Ⓒ Ⓓ Ⓔ	21 Ⓐ Ⓑ Ⓒ Ⓓ Ⓔ	31 Ⓐ Ⓑ Ⓒ Ⓓ Ⓔ
2 Ⓐ Ⓑ Ⓒ Ⓓ Ⓔ	12 Ⓐ Ⓑ Ⓒ Ⓓ Ⓔ	22 Ⓐ Ⓑ Ⓒ Ⓓ Ⓔ	32 Ⓐ Ⓑ Ⓒ Ⓓ Ⓔ
3 Ⓐ Ⓑ Ⓒ Ⓓ Ⓔ	13 Ⓐ Ⓑ Ⓒ Ⓓ Ⓔ	23 Ⓐ Ⓑ Ⓒ Ⓓ Ⓔ	33 Ⓐ Ⓑ Ⓒ Ⓓ Ⓔ
4 Ⓐ Ⓑ Ⓒ Ⓓ Ⓔ	14 Ⓐ Ⓑ Ⓒ Ⓓ Ⓔ	24 Ⓐ Ⓑ Ⓒ Ⓓ Ⓔ	34 Ⓐ Ⓑ Ⓒ Ⓓ Ⓔ
5 Ⓐ Ⓑ Ⓒ Ⓓ Ⓔ	15 Ⓐ Ⓑ Ⓒ Ⓓ Ⓔ	25 Ⓐ Ⓑ Ⓒ Ⓓ Ⓔ	35 Ⓐ Ⓑ Ⓒ Ⓓ Ⓔ
6 Ⓐ Ⓑ Ⓒ Ⓓ Ⓔ	16 Ⓐ Ⓑ Ⓒ Ⓓ Ⓔ	26 Ⓐ Ⓑ Ⓒ Ⓓ Ⓔ	36 Ⓐ Ⓑ Ⓒ Ⓓ Ⓔ
7 Ⓐ Ⓑ Ⓒ Ⓓ Ⓔ	17 Ⓐ Ⓑ Ⓒ Ⓓ Ⓔ	27 Ⓐ Ⓑ Ⓒ Ⓓ Ⓔ	37 Ⓐ Ⓑ Ⓒ Ⓓ Ⓔ
8 Ⓐ Ⓑ Ⓒ Ⓓ Ⓔ	18 Ⓐ Ⓑ Ⓒ Ⓓ Ⓔ	28 Ⓐ Ⓑ Ⓒ Ⓓ Ⓔ	38 Ⓐ Ⓑ Ⓒ Ⓓ Ⓔ
9 Ⓐ Ⓑ Ⓒ Ⓓ Ⓔ	19 Ⓐ Ⓑ Ⓒ Ⓓ Ⓔ	29 Ⓐ Ⓑ Ⓒ Ⓓ Ⓔ	39 Ⓐ Ⓑ Ⓒ Ⓓ Ⓔ
10 Ⓐ Ⓑ Ⓒ Ⓓ Ⓔ	20 Ⓐ Ⓑ Ⓒ Ⓓ Ⓔ	30 Ⓐ Ⓑ Ⓒ Ⓓ Ⓔ	40 Ⓐ Ⓑ Ⓒ Ⓓ Ⓔ

SECTION 5

1 Ⓐ Ⓑ Ⓒ Ⓓ Ⓔ	11 Ⓐ Ⓑ Ⓒ Ⓓ Ⓔ	21 Ⓐ Ⓑ Ⓒ Ⓓ Ⓔ	31 Ⓐ Ⓑ Ⓒ Ⓓ Ⓔ
2 Ⓐ Ⓑ Ⓒ Ⓓ Ⓔ	12 Ⓐ Ⓑ Ⓒ Ⓓ Ⓔ	22 Ⓐ Ⓑ Ⓒ Ⓓ Ⓔ	32 Ⓐ Ⓑ Ⓒ Ⓓ Ⓔ
3 Ⓐ Ⓑ Ⓒ Ⓓ Ⓔ	13 Ⓐ Ⓑ Ⓒ Ⓓ Ⓔ	23 Ⓐ Ⓑ Ⓒ Ⓓ Ⓔ	33 Ⓐ Ⓑ Ⓒ Ⓓ Ⓔ
4 Ⓐ Ⓑ Ⓒ Ⓓ Ⓔ	14 Ⓐ Ⓑ Ⓒ Ⓓ Ⓔ	24 Ⓐ Ⓑ Ⓒ Ⓓ Ⓔ	34 Ⓐ Ⓑ Ⓒ Ⓓ Ⓔ
5 Ⓐ Ⓑ Ⓒ Ⓓ Ⓔ	15 Ⓐ Ⓑ Ⓒ Ⓓ Ⓔ	25 Ⓐ Ⓑ Ⓒ Ⓓ Ⓔ	35 Ⓐ Ⓑ Ⓒ Ⓓ Ⓔ
6 Ⓐ Ⓑ Ⓒ Ⓓ Ⓔ	16 Ⓐ Ⓑ Ⓒ Ⓓ Ⓔ	26 Ⓐ Ⓑ Ⓒ Ⓓ Ⓔ	36 Ⓐ Ⓑ Ⓒ Ⓓ Ⓔ
7 Ⓐ Ⓑ Ⓒ Ⓓ Ⓔ	17 Ⓐ Ⓑ Ⓒ Ⓓ Ⓔ	27 Ⓐ Ⓑ Ⓒ Ⓓ Ⓔ	37 Ⓐ Ⓑ Ⓒ Ⓓ Ⓔ
8 Ⓐ Ⓑ Ⓒ Ⓓ Ⓔ	18 Ⓐ Ⓑ Ⓒ Ⓓ Ⓔ	28 Ⓐ Ⓑ Ⓒ Ⓓ Ⓔ	38 Ⓐ Ⓑ Ⓒ Ⓓ Ⓔ
9 Ⓐ Ⓑ Ⓒ Ⓓ Ⓔ	19 Ⓐ Ⓑ Ⓒ Ⓓ Ⓔ	29 Ⓐ Ⓑ Ⓒ Ⓓ Ⓔ	39 Ⓐ Ⓑ Ⓒ Ⓓ Ⓔ
10 Ⓐ Ⓑ Ⓒ Ⓓ Ⓔ	20 Ⓐ Ⓑ Ⓒ Ⓓ Ⓔ	30 Ⓐ Ⓑ Ⓒ Ⓓ Ⓔ	40 Ⓐ Ⓑ Ⓒ Ⓓ Ⓔ

Grid-in response boxes numbered 9, 10, 11, 12, 13, 14, 15, 16, 17, 18 (each with digit bubbles 0–9 and fraction/decimal markers).

Math Scoring Worksheet

A. Section 3

_____   _____
numer of correct   *number of incorrect*

+   +

B. Section 5 (1-8)

_____   _____
numer of correct   *number of incorrect*

+

C. Section 5 (9-18)

_____
numer of correct

+   +

D. Section 7

_____   _____
numer of correct   *number of incorrect*

=   =

E. Total Unrounded Raw Score

_____ − _____ ÷4 = _____
numer of correct   *number of incorrect*

F. Total Rounded Raw Score   _____ (See table)

Math Score Range = | _____ |

Math Conversion Table

Raw Score	Scaled Score	Raw Score	Scaled Score
54	800	23	490-550
53	780-800	22	480-540
52	760-800	21	470-530
51	740-800	20	460-520
50	720-780	19	450-510
49	700-760	18	450-510
48	690-750	17	440-500
47	680-740	16	430-490
46	670-730	15	420-480
45	660-720	14	420-480
44	650-710	13	410-470
43	650-710	12	400-460
42	640-700	11	390-450
41	630-690	10	380-440
40	620-680	9	390-430
39	610-670	8	380-420
38	610-670	7	370-410
37	600-660	6	360-400
36	590-650	5	340-380
35	580-640	4	320-370
34	570-630	3	310-360
33	560-620	2	300-350
32	560-620	1	270-320
31	550-610	0	240-300
30	540-600	-1	200-290
29	530-590	-2	200-270
28	530-590	-3	200-260
27	520-580	-4	200-240
26	510-570	-5	200-220
25	500-560	-6 and below	200
24	500-560		

SECTION 3
Time- 25 minutes
20 Questions

Turn to Section 3 (Page 1) of your answer sheet to answer the questions in this section.

Directions: For this section, solve each problem and decide which is the best of the choices given. Fill in the corresponding circle on the answer sheet. You may use any available space for scratchwork.

Notes

1. The use of a calculator is permitted.
2. All numbers used are real numbers.
3. Figures that accompany problems in this test are intended to provide information useful in solving the problems. They are drawn as accurately as possible EXCEPT when it is stated in a specific problem that the figure is not drawn to scale. All figure lie in a plane unless other indicated.
4. Unless otherwise specified, the domain of any function f is assumed to be set of all real numbers x for which $f(x)$ is a real number.

Reference Informatiom

$A = \pi r^2$
$C = 2\pi r$
$A = \ell w$
$A = \frac{1}{2} bh$
$V = \ell wh$
$V = \pi r^2 h$
$c^2 = a^2 + b^2$
Special Right Triangles

The numbers of degrees of arc in a circle is $360°$.

The sum of the measures in degrees of the angles is $180°$.

---

1. If $|x - 5| = 2x - 1$, which of the following is the value of x ?

(A) -4
(B) -2
(C) 0
(D) 2
(E) 4

2. Let the function f be defined by $f(x) = \sqrt{5x}$. If $f(2k) = 10$, what is the value of k ?

(A) 10
(B) 20
(C) 30
(D) 40
(E) 50

GO ON TO THE NEXT PAGE

Dr. John Chung's SAT Math Test 15

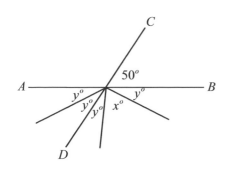

Note: Figure not drawn to scale.

3. In the figure above, if AB and CD are lines, what is the value of x?

(A) 50
(B) 60
(C) 70
(D) 75
(E) 80

4. For how many values of n, where n is a positive integer less than 10, is $\dfrac{(n+1)^2}{2}$ an integer?

(A) One
(B) Two
(C) Three
(D) Four
(E) Five

5. If the ratio of a to b is $7:8$ and the ratio of c to d is 7:5, what is the value of $\dfrac{ad}{bc}$?

(A) $\dfrac{5}{8}$

(B) $\dfrac{49}{40}$

(C) $\dfrac{8}{5}$

(D) $\dfrac{40}{49}$

(E) $\dfrac{7}{8}$

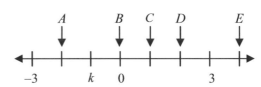

6. On the number line above, which of the following corresponding letters indicates the value of m so that $\dfrac{k+m}{k-m} > 0$?

(A) A
(B) B
(C) C
(D) D
(D) E

GO ON TO THE NEXT PAGE

7. If the legs of right triangle PQR have lengths 5 and 12, what is the perimeter of a right triangle with each side twice the length of its corresponding side in $\triangle PQR$?

(A) 30
(B) 45
(C) 56
(D) 60
(E) 64

9. If $k = m^4$, where m is a positive integer, which of the following could be a value of k ?

(A) 1.6×10^{16}
(B) 1.6×10^{17}
(C) 1.6×10^{18}
(D) 1.6×10^{19}
(E) 1.6×10^{20}

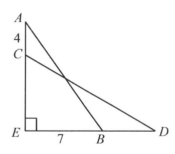

Note: Figure not drawn to scale.

8. In the figure above, $AB = CD = 25$, $EB = 7$, and $AC = 4$. What is the length of $\overline{BD}$?

(A) 4
(B) 5
(C) 6
(D) 7
(E) 8

10. The first term of a sequence is a. Every term after the first is equal to d plus the preceding term. If the seventh term is 20 and the fourth term is 11, what is the value of $a + d$?

(A) 5
(B) 6
(C) 7
(D) 8
(E) 9

GO ON TO THE NEXT PAGE

11. Which of the following sets of numbers does <u>not</u> have the property that the product of any two numbers in the set is also a number in the set?

(A) The set of even integers
(B) The set of odd integers
(C) The set of positive numbers
(D) The set of positive even integers
(E) The set of prime numbers

12. If $4^{x+1} = 8 \times 2^k$, what is the value of k in terms of x?

(A) $x + 2$
(B) $2x - 1$
(C) $2x + 1$
(D) $2x + 4$
(E) $2x + 6$

13. For a positive integer n, let $\odot n = (n+1)^2$ when n is an even integer, and let $\odot n = (n-1)^2$ when n is an odd integer.
If $\odot(k) - \odot(k+1) = 9$, where k is a positive even integer, what is the value of k?

(A) 4
(B) 5
(C) 6
(D) 7
(E) 8

14. For the positive integers a and b, $ab < 40$ and $a + b > 20$. Which of the following could be a value of a?

(A) 10
(B) 15
(C) 17
(D) 19
(E) 40

GO ON TO THE NEXT PAGE

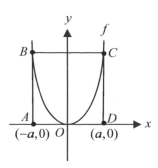

Note: Figure not drawn to scale.

15. The figure above shows the graph of f defined by $f(x) = x^2$. If $ABCD$ is a square, what is the value of a?

(A) 2
(B) 2.5
(C) 3
(D) 3.5
(E) 4

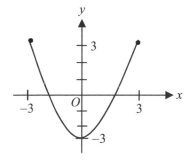

16. The graph of the function f in the xy-plane is shown above. For how many values of k between -3 and 3 does $|f(k)| = 2$?

(A) Four
(B) Three
(C) Two
(D) One
(E) None

17. A chemist has a solution consisting of a ounces of alcohol and w ounces of water. He wants to change the solution to x percent alcohol by adding b ounces of alcohol. Which of the following equations could be the equation to determine the value of x?

(A) $\dfrac{a+b}{w} = \dfrac{x}{100}$

(B) $\dfrac{a+b}{w+a} = \dfrac{x}{100}$

(C) $\dfrac{a+b}{w+b} = \dfrac{x}{100}$

(D) $\dfrac{a+b}{w+a+b} = \dfrac{x}{100}$

(E) $\dfrac{a}{w+a+b} = \dfrac{x}{100}$

18. If $ax^2 + bx + 36 = (3x + n)^2$ for all values of x, where $n < 0$, what is the value of $a + b$?

(A) -36
(B) -27
(C) 9
(D) 27
(E) 36

GO ON TO THE NEXT PAGE

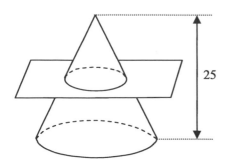

Note: Figure not drawn to scale.

19. In the figure above, the height of the circular cone is 25 and the area of its base is 25π. A cut parallel to the circular base is made completely through the cone as shown above. If the area of the base of the small cone is 9π, what is the height of the small cone?

(A) 4
(B) 6
(C) 8
(D) 12
(E) 15

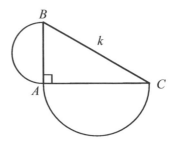

20. In the figure above, AB and BC are the diameters of the two semicircles. If the length of $\overline{BC}$ is k, what is the sum of the areas of the semicircles in terms of k?

(A) πk^2

(B) $\dfrac{\pi k^2}{2}$

(C) $\dfrac{\pi k^2}{4}$

(D) $\dfrac{\pi k^2}{8}$

(E) $\dfrac{\pi k^2}{12}$

STOP

**If you finish before time is called, you may check your work on this section only.
Do not turn to any other section in the test.**

SECTION 5
Time- 25 minutes
18 Questions

Turn to Section 5 (Page 1) of your answer sheet to answer the questions in this section.

Directions: For this section, solve each problem and decide which is the best of the choices given. Fill in the corresponding circle on the answer sheet. You may use any available space for scratchwork.

Notes

1. The use of a calculator is permitted.
2. All numbers used are real numbers.
3. Figures that accompany problems in this test are intended to provide information useful in solving the problems. They are drawn as accurately as possible EXCEPT when it is stated in a specific problem that the figure is not drawn to scale. All figure lie in a plane unless other indicated.
4. Unless otherwise specified, the domain of any function f is assumed to be set of all real numbers x for which $f(x)$ is a real number.

Reference Informatiom

$A = \pi r^2$
$C = 2\pi r$
$A = \ell w$
$A = \frac{1}{2}bh$
$V = \ell w h$
$V = \pi r^2 h$
$c^2 = a^2 + b^2$
Special Right Triangles

The numbers of degrees of arc in a circle is $360°$.

The sum of the measures in degrees of the angles is $180°$.

---

1. Kevin and John are removing marbles from a container. Kevin removes $\frac{1}{4}$ of the marbles from the container, and John removes $\frac{1}{3}$ of the remaining marbles. After that, 20 marbles are left in the container. How many marbles were originally in the container?

(A) 30
(B) 40
(C) 60
(D) 64
(E) 80

2. If $c < b < a < 0$, which of the following products is the greatest?

(A) a^2
(B) ab
(C) b^2
(D) bc
(E) c^2

GO ON TO THE NEXT PAGE

3. 40 percent of the teachers at a meeting are male. Of the male teachers at the meeting, 20 percent teach mathematics. Which of the following could be the possible number of teachers at the meeting?

(A) 60
(B) 64
(C) 72
(D) 75
(E) 80

4. If k is an integer and $\dfrac{k+5}{30} < \dfrac{7}{20} < \dfrac{k}{10}$, then

$k =$

(A) 5
(B) 6
(C) 7
(D) 8
(E) 9

5. If $m^x \cdot m^5 = n^x \cdot n^5$ and $m = \dfrac{1}{2}n$, what is the value of x?

(A) 10
(B) 5
(C) 0
(D) -5
(E) -10

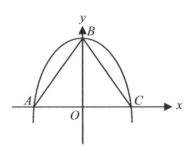

Note: Figure not drawn to scale.

6. The figure above shows the graph of f defined by $f(x) = 9 - x^2$. What is the perimeter of $\triangle ABC$?

(A) 18
(B) 27
(C) $6 + 6\sqrt{10}$
(D) $6 + 12\sqrt{10}$
(E) $12 + 12\sqrt{2}$

GO ON TO THE NEXT PAGE

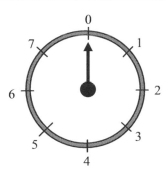

7. On the dial above, the numbers are equally spaced. The arrow rotates clockwise at a constant rate of 160° every 10 seconds. If the arrow starts at 0, how many times will the arrow pass the number 3 on the dial in 150 seconds?

(A) Five
(B) Six
(C) Seven
(D) Eight
(E) Nine

8. In a sequence, the first term is 27 and each term after the first is 3 times the previous term. Which of the following expresses the nth term of the sequence, where n is a positive integer?

(A) 3^n
(B) 3^{n+1}
(C) 3^{n+2}
(D) 27^n
(E) $27 + 3^{n-1}$

GO ON TO THE NEXT PAGE

5 □□□ 5 □□□ 5 □□□ 5

Directions: For Students-Produced Response questions 9-18, use the grid at the bottom of the answer sheet page on which you have answered questions 1-8.

Each of the remaining 10 questions requires you to solve the problem and enter your answer by making the circles in the special grid, as shown in the examples below. You may use any available space for scratchwork.

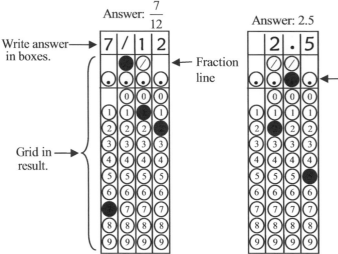

Answer: $\frac{7}{12}$

Write answer in boxes.

Fraction line

Grid in result.

Answer: 2.5

Decimal point

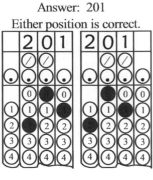

Answer: 201
Either position is correct.

Note: You may start your answers in any column, space permitting. Columns not needed should be left blank.

- Mark no more than one circle in any column.

- Because the answer sheet will be machine-scored, **you will receive credit only if the circles are filled in correctly.**

- Although not required, it is suggested that you write your answer in the boxes at the top of the columns to help you fill in the circles accurately.

- Some problems may have more than one correct answer. In such cases, grid only one answer.

- No question has a negative answer.

- **Mixed numbers** such as $3\frac{1}{2}$ must be gridded as 3.5 or 7/2. (If [3 1 / 2] is gridded, it will be interpreted as $\frac{31}{2}$, not $3\frac{1}{2}$.)

- **Decimal Answers:** If you obtain a decimal answer with more digits than the grid can accommodate, it may be either rounded or truncated, but it must fill the entire grid. For example, if you obtain an answer such as 0.6666…, you should record your result as .666 or .667. **A less accurate value such as .66 or .67 will be scored as incorrect.**

Acceptable ways to grid $\frac{2}{3}$ are:

9. If $\dfrac{1}{a-b} = \dfrac{a+b}{5}$, where a and b are positive integers, what is the value of a?

10. If a is increased by 20 percent and b is decreased by 20 percent, the resulting numbers are equal. What is the value of $\dfrac{a}{b}$?

GO ON TO THE NEXT PAGE

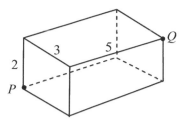

11. In the figure above, the dimensions of the rectangular box are 2, 3, and 5. How many different paths are there of total length 10 from P to Q along the edges?

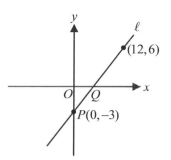

Note: Figure not drawn to scale.

13. In the figure above, line ℓ passes through point P and Q. What is the area of $\triangle OPQ$?

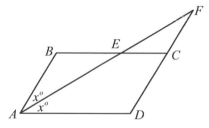

Note: Figure not drawn to scale.

12. In the figure above, $ABCD$ is a parallelogram and $\overline{AB} \parallel \overline{DF}$. If $AB = 4$ and $AD = 7$, what is the length of $\overline{CF}$?

14. For all positive integers n, let $[\![n]\!]$ be defined as the product of all factors of n. What is the value of $\dfrac{[\![48]\!]}{[\![24]\!]}$?

GO ON TO THE NEXT PAGE

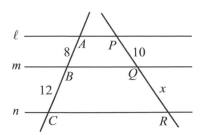

Note: Figure not drawn to scale.

15. In the figure above, line ℓ, m, and n are parallel. If $AB = 8$, $BC = 12$, $PQ = 10$, and $QR = x$, what is the value of x?

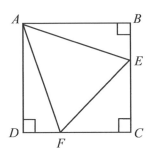

Note: Figure not drawn to scale.

16. In the figure above, $\triangle AEF$ is an equilateral triangle and $AF = 10$. What is the area of $\triangle CEF$?

17. A car travels 40 miles per hour from city A to city B, 60 miles per hour from city B to city A, and takes 3 hours for the entire trip. What is its average speed, in miles per hour, for the trip?

18. In the figure above, how many different arrangements are possible that begins and ends with a ●   ?

STOP

If you finish before time is called, you may check your work on this section only.
Do not turn to any other section in the test.

SECTION 7
Time- 20 minutes
16 Questions

Turn to Section 7 (Page 2) of your answer sheet to answer the questions in this section.

Directions: For this section, solve each problem and decide which is the best of the choices given. Fill in the corresponding circle on the answer sheet. You may use any available space for scratchwork.

Notes

1. The use of a calculator is permitted.
2. All numbers used are real numbers.
3. Figures that accompany problems in this test are intended to provide information useful in solving the problems. They are drawn as accurately as possible EXCEPT when it is stated in a specific problem that the figure is not drawn to scale. All figure lie in a plane unless other indicated.
4. Unless otherwise specified, the domain of any function f is assumed to be set of all real numbers x for which $f(x)$ is a real number.

Reference Informatiom

$A = \pi r^2$
$C = 2\pi r$
$A = \ell w$
$A = \dfrac{1}{2}bh$
$V = \ell w h$
$V = \pi r^2 h$
$c^2 = a^2 + b^2$
Special Right Triangles

The numbers of degrees of arc in a circle is $360°$.

The sum of the measures in degrees of the angles is $180°$.

---

1. If $e = 1 - f$, then what is the average (arithmetic mean) of e and f ?

(A) 3

(B) 2

(C) 1

(D) $\dfrac{1}{2}$

(E) $\dfrac{1}{4}$

2. If $pqr = 5$, which of the following CANNOT be a value for p ?

(A) 5
(B) $\sqrt{5}$
(C) 0
(D) -1
(E) -5

GO ON TO THE NEXT PAGE

3. If $\dfrac{100-n}{25}$ is a positive integer, for how many possible positive values of n is the statement true?

(A) 1
(B) 2
(C) 3
(D) 4
(E) 5

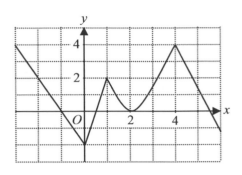

5. The figure above shows the graph of the function g in the xy-plane. If $|g(k)| = g(1)$, which of the following could be the value of k ?

(A) 5
(B) 4
(C) 2
(D) 0
(E) -3

x	$f(x)$
0	20
1	21
−1	21

4. Each of the ordered pairs in the table above is on the graph of the function f in the xy-plane. Which of the following could be true for all x ?

(A) $f(x) = 20x$
(B) $f(x) = 20 + x$
(C) $f(x) = 20 - x$
(D) $f(x) = 20 + x^2$
(E) $f(x) = 20 - x^2$

6. Peter bikes on a 100 mile path at an average of 10 miles per hour for the first k hours. In terms of k, how many miles remain to be traveled?

(A) $10k$
(B) $40k$
(C) $10(10-k)$
(D) $100(10-k)$
(E) $10(k-100)$

GO ON TO THE NEXT PAGE

7. Kevin was 3 years younger than Lee two years ago. If n represents Lee's age now, what was Kevin's age 10 years ago, in terms of n?

(A) $n-13$
(B) $n-10$
(C) $n-7$
(D) $n+10$
(E) $n+13$

8. If $\left((k^2 \div 2) \div 3\right) \div 4$ is an integer, which of the following could be a value of k?

(A) 6
(B) 8
(C) 10
(D) 12
(E) 14

X = set of multiples of 5.
Y = set of multiples of 4.
Z = set of multiples of 3.

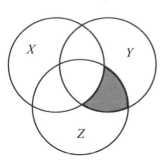

9. X, Y, and Z represent three sets of numbers, as defined above. Which of the following is the number that belongs to the shaded region, in the figure above?

(A) 120
(B) 70
(C) 60
(D) 56
(E) 48

10. If $x^{\frac{2}{5}} = \left(x^2\right)^{k-1}$ for all positive values of x, what is the value of k?

(A) $\dfrac{6}{5}$

(B) $\dfrac{3}{5}$

(C) $\dfrac{2}{3}$

(D) $\dfrac{4}{3}$

(E) $\dfrac{5}{3}$

GO ON TO THE NEXT PAGE

11. The total daily profit p, in dollars, for selling n units of a certain toy is given by the function $p(n) = \dfrac{255n - 500}{n} + k$, where k is a constant. If 100 units were sold for the total profit of \$500, what is the value of k ?

(A) 300
(B) 250
(C) 100
(D) –250
(E) –300

$$g(x) = kf(x) - 10$$

12. The function g above is defined in terms of another function $f(x) = x^2 - 10$, where k is a constant. If $g(5) = 10$, what is the value of k ?

(A) $\dfrac{4}{3}$

(B) $\dfrac{2}{3}$

(C) $\dfrac{3}{5}$

(D) 3

(E) 5

$$x^2 + (k+1)x + 4 = (x+h)^2$$

13. In the equation above, k and h are positive constants. If the equation is true for all values of x, what is the value of $k + h$?

(A) 3
(B) 4
(C) 5
(D) 7
(E) 9

14. If $4 \le x \le 12$ and $-1 \le y \le 1$, which of the following gives the set of all possible values of xy ?

(A) $-4 \le xy \le 12$
(B) $\;\;\,0 \le xy \le 12$
(C) $-12 \le xy \le -4$
(D) $-12 \le xy \le 4$
(E) $-12 \le xy \le 12$

GO ON TO THE NEXT PAGE

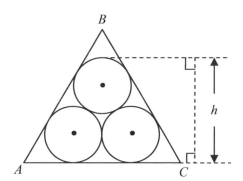

15. In the figure above, $\triangle ABC$ is an equilateral triangle and each circle is tangent to the other two circles. If each circle has diameter 10, what is the distance h?

(A) $10\sqrt{3}$

(B) $15\sqrt{3}$

(C) $15 + 5\sqrt{3}$

(D) $10 + 10\sqrt{3}$

(E) $10 + 5\sqrt{3}$

16. If p and q are positive numbers and $\dfrac{p}{q} < pq$, which of the following must be true?

    I. $p > 1$

    II. $q > 1$

    III. $p > q$

(A) I only

(B) II only

(C) III only

(D) I and II only

(E) I, II, and III

STOP

If you finish before time is called, you may check your work on this section only.
Do not turn to any other section in the test.

TEST 15

#	SECTION 3	SECTION 5	SECTION 7
1	D	B	D
2	A	E	C
3	E	D	C
4	E	A	D
5	A	D	D
6	B	C	C
7	D	C	A
8	E	C	D
9	B	3	E
10	A	$\frac{2}{3}$ or .666, .667	A
11	E	6	B
12	B	3	A
13	A	6	C
14	D	768	E
15	A	15	E
16	A	25	B
17	D	48	
18	B	60	
19	E		
20	D		

TEST 15   SECTION 3

1. (D)
If $x = 2$, $|2 - 5| = 2(2) - 1 \rightarrow 3 = 3$.

2. (A)
$$f(2k) = \sqrt{5(2k)} = \sqrt{10k} \rightarrow \sqrt{10k} \Rightarrow k = 10$$

3. (E)
Because $y = 25$ and $x + 2y = 130$, $x = 80$.

4. (E)
In order to be an integer, $(n+1)^2$ should be even.
Therefore, $n = 1,3,5,7,9$.

5. (A)
$\dfrac{a}{b} = \dfrac{7}{8}$ and $\dfrac{c}{d} = \dfrac{7}{5}$.
$$\frac{ad}{bc} = \left(\frac{a}{b}\right)\left(\frac{d}{c}\right) = \left(\frac{7}{8}\right)\left(\frac{5}{7}\right) = \frac{5}{8}.$$

6. (B)
$k = -1$.
If $m = A = -2$, then $\dfrac{-1-2}{-1+2} < 0$.
If $m = B = 0$, then $\dfrac{-1+0}{-1-0} > 0$.

7. (D)
The perimeter of $\triangle PQR$ is $5 + 12 + 13 = 30$.
Therefore, the perimeter of the similar triangle is $2 \times 30 = 60$.

8.   (E)

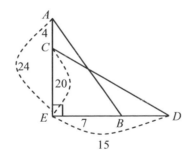

In the figure above, by Pythagorean Theorem,
$BD = 15 - 7 = 8$.

---

9.   (B)

$$1.6 \times 10^{17} = 16 \times 10^{16} = \left(2 \times 10^{4}\right)^{4}.$$

---

10.   (A)

$a_1 = a$

$a_2 = a + d$

$a_3 = a + 2d$

Thus, $a_7 = a + 6d = 20$ and

$a_4 = a + 3d = 11$.

$a_7 - a_4 \rightarrow 3d = 9 \ d = 3$.

$a + 3(3) = 11 \ \rightarrow a = 2$.

Therefore, $a + d = 5$.

---

11.   (E)

(A) even×even = even   (B) odd×odd=odd

(C) positive×positive = positive

(D) p.even×p.even = p.even

(E) 3(prime)×7(prime) = 21(composite)

---

12.   (B)

$$2^{2(x+1)} = 2^3 \cdot 2^k \ \rightarrow 2x + 2 = 3 + k \ \rightarrow k = 2x - 1.$$

---

13.   (A)

Because k is an even positive integer,

$\odot k = (k+1)^2$ and $\odot (k+1) = (k+1-1)^2$.

$\odot k - \odot(k+1) = (k+1)^2 - k^2 = 2k + 1$.

Therefore, $2k + 1 = 9 \ \rightarrow k = 4$.

---

14.   (D)

If $a = 10$, $10b < 40 \rightarrow b < 4$ and

$10 + b > 20$ (false).

If $a = 15$, $15b < 40 \rightarrow b < 2.6$ and

$15 + b > 20$ (false).

If $a = 17$, $17b < 40 \rightarrow b < 2.3$ and

$17 + b > 20$ (false).

If $a = 19$, $19b < 40 \rightarrow b < 2.1$ and

$19 + b > 20$ (true).

If $a = 40$, $40b < 40 \rightarrow b < 1$ and

$40 + b > 20$ (false,

b is not a positive integer).

---

15.   (A)

$CD = f(a) = a^2$ must be equal to $2a$.

Therefore,

$a^2 - 2a = 0 \ \rightarrow a(a - 2) = 0 \ \rightarrow a = \cancel{0}$ or 2.

---

16.   (A)

$|f(k)| = 2 \ \rightarrow f(k) = 2 \ \rightarrow \ f(k) = -2$.

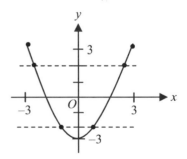

In the graph above, there are four
intersections.

---

17.   (D)

$\dfrac{\text{solute}}{\text{solute+water}} = \dfrac{x}{100} \ \rightarrow$ % of solution.

$\dfrac{a}{w + a} \ \rightarrow \dfrac{(a) + b}{(w + a) + b} = \dfrac{x}{100}$.

---

18.   (B)

$n^2 = 36 \ \rightarrow n = -6$.

$(3x - 6)^2 = 9x^2 - 36x + 36$. Therefore,

$a = 9$ and $b = -36$. $a + b = -27$.

---

19. (E)

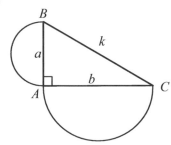

Ratio of the areas = 9:25
Ratio of the lengths =3:5

Therefore, $\dfrac{h}{25} = \dfrac{3}{5} \rightarrow h = 15$.

20. (D)

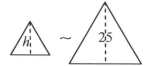

In the figure above, $a^2 + b^2 = k^2$. The sum of the areas =

$$\frac{1}{2}\left[\pi\left(\frac{a}{2}\right)^2 + \pi\left(\frac{b}{2}\right)^2\right] = \frac{\pi\left(a^2+b^2\right)}{8} = \frac{\pi k^2}{8}.$$

TEST 15   SECTION 5

1. (B)
Let x = number of marbles.

$$1 - \left(\frac{1}{4} + \frac{1}{3} \times \frac{3}{4}\right) = \frac{1}{2}. \qquad \frac{1}{2}x = 20 \rightarrow x = 40.$$

2. (E)
$-3 < -2 < -1 < 0$, $c^2 = 9$ is the greatest.

3. (D)
Let the number of teachers be x. The number of math teachers =
$\dfrac{20}{100}\left(\dfrac{40}{100}\right)x = \dfrac{2}{25}x$. Since the number must

be an integer, a possible number is a multiple of 25.

4. (A)
Backsolving: If $k = 5$, then
$$\frac{10}{30} < \frac{7}{20} < \frac{5}{10} \rightarrow 0.33 < 0.35 < 0.5 \text{ (true)}$$

Or, algebraically
$$\frac{2k+10}{60} < \frac{21}{60} < \frac{6k}{60} \rightarrow 3.5 < k < 5.5.$$

5. (D)
Since $n = 2m$ and $m^{x+5} = (2m)^{x+5}$,
$m^{x+5} = 2^{x+5} m^{x+5}$.
Therefore, $2^{x+5} = 1 \rightarrow x = -5$.

6. (C)
$9 - x^2 = 0 \rightarrow x = 3, -3$. $f(0) = 9$.

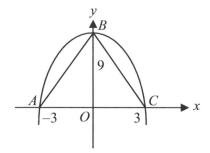

In the figure above, $OB = 9$ and
$BC = \sqrt{9^2 + 3^2} = 3\sqrt{10}$. Therefore, the
perimeter of $\triangle ABC$ is
$3\sqrt{10} + 3\sqrt{10} + 6 = 6 + 6\sqrt{10}$.

7. (C)
Proportion: $\dfrac{160^o}{10} = \dfrac{x}{150} \rightarrow x = 2400^o$.

$\dfrac{2400^o}{360^o} = 6\dfrac{2}{3}$. Therefore, 6 and $\dfrac{2}{3}$ rotations.

Therefore, there are seven passes.

8. (C)
$a_1 = 10$, $a_2 = 10 \times 3$, $a_3 = 10 \times 3^2 \ldots \ldots$
Therefore, $a_n = 10 \times 3^{n-1}$.

9. 3
Cross multiplication.
$(a+b)(a-b) = 5$. Since a and b are integers,
$a+b = 5$, $a-b = 1$. Therefore, $a = 3$ and
$b = 2$. (using addition)

10. $\frac{2}{3}$ or .666, .667

Since $1.2a = 0.8b$, $a = \frac{2}{3}b$. Therefore,

$\frac{a}{b} = \frac{2}{3}$.

---

11. 6

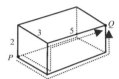

In the figure above, there are three paths at point P. Each path has two different paths to point Q.

Therefore, $3 \times 2 = 6$.

---

12. 3

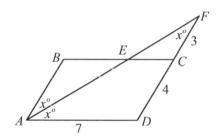

In the figure above,
$\angle EFC \cong \angle EAD$ (alternate angles are congruent, because $AB \parallel DF$). That is,
$\triangle AFD$ is an isosceles triangle.
Therefore, $FC = 3$.

---

13. 6

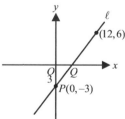

In the figure above, the slope of line ℓ is

$\frac{6-(-3)}{12-0} = \frac{3}{4}$ and the equation is $y = \frac{3}{4}x - 3$.

x-intercept is 4. Since $OQ = 4$, the area of

$\triangle OPQ$ is $\frac{3 \times 4}{2} = 6$.

Or, use similarity as follows.

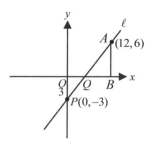

Since $\triangle OPQ \sim \triangle ABQ$, $\frac{OP}{AB} = \frac{3}{6} = \frac{1}{2}$.

$\frac{OQ}{BQ} = \frac{1}{2}$ and $OB = 12$. $OQ = 12\left(\frac{1}{3}\right) = 4$.

Therefore, the area is 6.

---

14. 768

$$\frac{[\![48]\!]}{[\![24]\!]} = \frac{(1 \times 2 \times 3 \times 4 \times 6 \times 8 \times 12 \times 16 \times 24) \times 48}{(1 \times 2 \times 3 \times 4 \times 6 \times 8 \times 12 \times 24)}$$

$\frac{16 \times 48}{1} = 768$.

---

15. 15

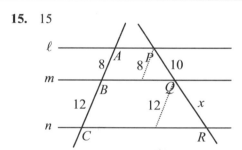

In the figure the two triangles are similar.

Therefore, $\frac{8}{12} = \frac{10}{x} \rightarrow x = 15$.

---

16. 25

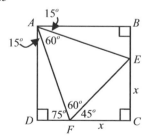

In the figure above, $x^2 + x^2 = 10^2 \rightarrow x^2 = 50$. Therefore, the area of $\triangle CEF$ is

$$\frac{x^2}{2} = \frac{50}{2} = 25.$$

Or, use the ratio of special triangle. Isosceles has the ratio $s, s, s\sqrt{2}$.

$s\sqrt{2} = 10 \rightarrow s = \dfrac{10}{\sqrt{2}}$. Therefore the area is

$$\frac{1}{2}\left(\frac{10}{\sqrt{2}}\right)\left(\frac{10}{\sqrt{2}}\right) = 25.$$

17. 48

Let the distance be D, then

$$\frac{D}{40} + \frac{D}{60} = 3 \rightarrow \frac{5D}{120} = 3 \rightarrow D = 72.$$

Average Speed $= \dfrac{2 \times 72(\text{mile})}{3(\text{hours})} = 48\text{mph}.$

18. 60

$5\times\ 4\times\ 3\times\ 2\times\ 1 = 120$

But two circles are identical, $\dfrac{120}{2!} = 60$.

TEST 15   SECTION 7

1. (D)

Since $e + f = 1$, $\dfrac{e+f}{2} = \dfrac{1}{2}$.

2. (C)

Since the product is 5, p, q, and r cannot be zero.

3. (C)

Since $\dfrac{100-n}{25}$ is an integer, $100-n$ must be a multiple of 25. $100-n = 25, 50, 75, 100\ldots$.
Therefore, $n = 75, 50, 25$.

4. (D)

Substitute the values of x, and check.

5. (D)

Since $g(1) = 2$,

$|g(k)| = 2 \rightarrow g(k) = 2$ or -2. Therefore, $k = 1$ or 0 for integer value.

6. (C)

$100 - (10k) = 10(10 - k)$.

7. (A)

	$\leftarrow 2$	Now
Kevin	$n-2-3$	
Lee	$n-2$	n

Therefore, 10 years ago, Kevin's age is $n - 5 - 8 = n - 13$.

8. (D)

Because $\dfrac{k^2}{2 \cdot 3 \cdot 4} = \text{integer}$, $k^2 = $ multiple of 24.

$12^2 = 144$ is a multiple of 24.

9. (E)

The number in the shaded region is a multiple of 12, but not a multiple of 60.

10. (A)

$x^{\frac{2}{5}} = x^{2k-2} \rightarrow \dfrac{2}{5} = 2k - 2 \rightarrow k = \dfrac{6}{5}$.

11. (B)

$500 = \dfrac{255 \times 100 - 500}{100} + k \rightarrow 500 = 250 + k$

Therefore, $k = 250$.

12. (A)

$g(5) = kf(5) - 10$ and $f(5) = 5^2 - 10 = 15$.

Therefore, $k \times 15 - 10 = 10 \rightarrow k = \dfrac{4}{3}$.

13. (C)

Since $x^2 + (k+1)x + 4 = x^2 + 2hx + h^2$, where $h > 0$,

$h = 2$ and $k + 1 = 2h = 4$. Therefore, $k = 3$, and $k + h = 2 + 3 = 5$.

14. (E)

The lowest one $< xy <$ The largest one

Therefore, $(12)(-1) < xy < (12)(1)$.

15. (E)

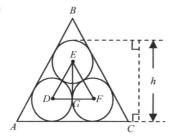

Special triangle: $2s,\ s,\ s\sqrt{3}$

Because $DE = 20$, $s = DG = 5$ and

$s\sqrt{3} = EG = 5\sqrt{3}$. Therefore, $h = 10 + 5\sqrt{3}$.

---

16. (B)

Because p and q are positive,

$$\frac{p}{q} < pq \quad \rightarrow \quad p < pq^2 \rightarrow 1 < q^2 \rightarrow 1 < q.$$

$\boxed{END}$

NO MATERIAL ON THIS PAGE

Dr. John Chung's SAT Math

TEST
16

ANSWER SHEET   TEST #:

SECTION 3

1 Ⓐ Ⓑ Ⓒ Ⓓ Ⓔ	11 Ⓐ Ⓑ Ⓒ Ⓓ Ⓔ	21 Ⓐ Ⓑ Ⓒ Ⓓ Ⓔ	31 Ⓐ Ⓑ Ⓒ Ⓓ Ⓔ
2 Ⓐ Ⓑ Ⓒ Ⓓ Ⓔ	12 Ⓐ Ⓑ Ⓒ Ⓓ Ⓔ	22 Ⓐ Ⓑ Ⓒ Ⓓ Ⓔ	32 Ⓐ Ⓑ Ⓒ Ⓓ Ⓔ
3 Ⓐ Ⓑ Ⓒ Ⓓ Ⓔ	13 Ⓐ Ⓑ Ⓒ Ⓓ Ⓔ	23 Ⓐ Ⓑ Ⓒ Ⓓ Ⓔ	33 Ⓐ Ⓑ Ⓒ Ⓓ Ⓔ
4 Ⓐ Ⓑ Ⓒ Ⓓ Ⓔ	14 Ⓐ Ⓑ Ⓒ Ⓓ Ⓔ	24 Ⓐ Ⓑ Ⓒ Ⓓ Ⓔ	34 Ⓐ Ⓑ Ⓒ Ⓓ Ⓔ
5 Ⓐ Ⓑ Ⓒ Ⓓ Ⓔ	15 Ⓐ Ⓑ Ⓒ Ⓓ Ⓔ	25 Ⓐ Ⓑ Ⓒ Ⓓ Ⓔ	35 Ⓐ Ⓑ Ⓒ Ⓓ Ⓔ
6 Ⓐ Ⓑ Ⓒ Ⓓ Ⓔ	16 Ⓐ Ⓑ Ⓒ Ⓓ Ⓔ	26 Ⓐ Ⓑ Ⓒ Ⓓ Ⓔ	36 Ⓐ Ⓑ Ⓒ Ⓓ Ⓔ
7 Ⓐ Ⓑ Ⓒ Ⓓ Ⓔ	17 Ⓐ Ⓑ Ⓒ Ⓓ Ⓔ	27 Ⓐ Ⓑ Ⓒ Ⓓ Ⓔ	37 Ⓐ Ⓑ Ⓒ Ⓓ Ⓔ
8 Ⓐ Ⓑ Ⓒ Ⓓ Ⓔ	18 Ⓐ Ⓑ Ⓒ Ⓓ Ⓔ	28 Ⓐ Ⓑ Ⓒ Ⓓ Ⓔ	38 Ⓐ Ⓑ Ⓒ Ⓓ Ⓔ
9 Ⓐ Ⓑ Ⓒ Ⓓ Ⓔ	19 Ⓐ Ⓑ Ⓒ Ⓓ Ⓔ	29 Ⓐ Ⓑ Ⓒ Ⓓ Ⓔ	39 Ⓐ Ⓑ Ⓒ Ⓓ Ⓔ
10 Ⓐ Ⓑ Ⓒ Ⓓ Ⓔ	20 Ⓐ Ⓑ Ⓒ Ⓓ Ⓔ	30 Ⓐ Ⓑ Ⓒ Ⓓ Ⓔ	40 Ⓐ Ⓑ Ⓒ Ⓓ Ⓔ

SECTION 5

1 Ⓐ Ⓑ Ⓒ Ⓓ Ⓔ	11 Ⓐ Ⓑ Ⓒ Ⓓ Ⓔ	21 Ⓐ Ⓑ Ⓒ Ⓓ Ⓔ	31 Ⓐ Ⓑ Ⓒ Ⓓ Ⓔ
2 Ⓐ Ⓑ Ⓒ Ⓓ Ⓔ	12 Ⓐ Ⓑ Ⓒ Ⓓ Ⓔ	22 Ⓐ Ⓑ Ⓒ Ⓓ Ⓔ	32 Ⓐ Ⓑ Ⓒ Ⓓ Ⓔ
3 Ⓐ Ⓑ Ⓒ Ⓓ Ⓔ	13 Ⓐ Ⓑ Ⓒ Ⓓ Ⓔ	23 Ⓐ Ⓑ Ⓒ Ⓓ Ⓔ	33 Ⓐ Ⓑ Ⓒ Ⓓ Ⓔ
4 Ⓐ Ⓑ Ⓒ Ⓓ Ⓔ	14 Ⓐ Ⓑ Ⓒ Ⓓ Ⓔ	24 Ⓐ Ⓑ Ⓒ Ⓓ Ⓔ	34 Ⓐ Ⓑ Ⓒ Ⓓ Ⓔ
5 Ⓐ Ⓑ Ⓒ Ⓓ Ⓔ	15 Ⓐ Ⓑ Ⓒ Ⓓ Ⓔ	25 Ⓐ Ⓑ Ⓒ Ⓓ Ⓔ	35 Ⓐ Ⓑ Ⓒ Ⓓ Ⓔ
6 Ⓐ Ⓑ Ⓒ Ⓓ Ⓔ	16 Ⓐ Ⓑ Ⓒ Ⓓ Ⓔ	26 Ⓐ Ⓑ Ⓒ Ⓓ Ⓔ	36 Ⓐ Ⓑ Ⓒ Ⓓ Ⓔ
7 Ⓐ Ⓑ Ⓒ Ⓓ Ⓔ	17 Ⓐ Ⓑ Ⓒ Ⓓ Ⓔ	27 Ⓐ Ⓑ Ⓒ Ⓓ Ⓔ	37 Ⓐ Ⓑ Ⓒ Ⓓ Ⓔ
8 Ⓐ Ⓑ Ⓒ Ⓓ Ⓔ	18 Ⓐ Ⓑ Ⓒ Ⓓ Ⓔ	28 Ⓐ Ⓑ Ⓒ Ⓓ Ⓔ	38 Ⓐ Ⓑ Ⓒ Ⓓ Ⓔ
9 Ⓐ Ⓑ Ⓒ Ⓓ Ⓔ	19 Ⓐ Ⓑ Ⓒ Ⓓ Ⓔ	29 Ⓐ Ⓑ Ⓒ Ⓓ Ⓔ	39 Ⓐ Ⓑ Ⓒ Ⓓ Ⓔ
10 Ⓐ Ⓑ Ⓒ Ⓓ Ⓔ	20 Ⓐ Ⓑ Ⓒ Ⓓ Ⓔ	30 Ⓐ Ⓑ Ⓒ Ⓓ Ⓔ	40 Ⓐ Ⓑ Ⓒ Ⓓ Ⓔ

Grid-in answer boxes numbered: 9, 10, 11, 12, 13, 14, 15, 16, 17, 18
(each with columns for digits 0–9 and fraction/decimal markers)

1	Ⓐ Ⓑ Ⓒ Ⓓ Ⓔ	11	Ⓐ Ⓑ Ⓒ Ⓓ Ⓔ	21	Ⓐ Ⓑ Ⓒ Ⓓ Ⓔ	31	Ⓐ Ⓑ Ⓒ Ⓓ Ⓔ
2	Ⓐ Ⓑ Ⓒ Ⓓ Ⓔ	12	Ⓐ Ⓑ Ⓒ Ⓓ Ⓔ	22	Ⓐ Ⓑ Ⓒ Ⓓ Ⓔ	32	Ⓐ Ⓑ Ⓒ Ⓓ Ⓔ
3	Ⓐ Ⓑ Ⓒ Ⓓ Ⓔ	13	Ⓐ Ⓑ Ⓒ Ⓓ Ⓔ	23	Ⓐ Ⓑ Ⓒ Ⓓ Ⓔ	33	Ⓐ Ⓑ Ⓒ Ⓓ Ⓔ
4	Ⓐ Ⓑ Ⓒ Ⓓ Ⓔ	14	Ⓐ Ⓑ Ⓒ Ⓓ Ⓕ	24	Ⓐ Ⓑ Ⓒ Ⓓ Ⓔ	34	Ⓐ Ⓑ Ⓒ Ⓓ Ⓔ
5	Ⓐ Ⓑ Ⓒ Ⓓ Ⓔ	15	Ⓐ Ⓑ Ⓒ Ⓓ Ⓔ	25	Ⓐ Ⓑ Ⓒ Ⓓ Ⓔ	35	Ⓐ Ⓑ Ⓒ Ⓓ Ⓔ
6	Ⓐ Ⓑ Ⓒ Ⓓ Ⓔ	16	Ⓐ Ⓑ Ⓒ Ⓓ Ⓔ	26	Ⓐ Ⓑ Ⓒ Ⓓ Ⓔ	36	Ⓐ Ⓑ Ⓒ Ⓓ Ⓔ
7	Ⓐ Ⓑ Ⓒ Ⓓ Ⓔ	17	Ⓐ Ⓑ Ⓒ Ⓓ Ⓔ	27	Ⓐ Ⓑ Ⓒ Ⓓ Ⓔ	37	Ⓐ Ⓑ Ⓒ Ⓓ Ⓔ
8	Ⓐ Ⓑ Ⓒ Ⓓ Ⓔ	18	Ⓐ Ⓑ Ⓒ Ⓓ Ⓔ	28	Ⓐ Ⓑ Ⓒ Ⓓ Ⓔ	38	Ⓐ Ⓑ Ⓒ Ⓓ Ⓔ
9	Ⓐ Ⓑ Ⓒ Ⓓ Ⓔ	19	Ⓐ Ⓑ Ⓒ Ⓓ Ⓔ	29	Ⓐ Ⓑ Ⓒ Ⓓ Ⓔ	39	Ⓐ Ⓑ Ⓒ Ⓓ Ⓔ
10	Ⓐ Ⓑ Ⓒ Ⓓ Ⓔ	20	Ⓐ Ⓑ Ⓒ Ⓓ Ⓔ	30	Ⓐ Ⓑ Ⓒ Ⓓ Ⓔ	40	Ⓐ Ⓑ Ⓒ Ⓓ Ⓔ

Math Scoring Worksheet

A. Section 3

_____   _____
numer of correct   number of incorrect

+   +

B. Section 5 (1-8)

_____   _____
numer of correct   number of incorrect

+

C. Section 5 (9-18)

_____
numer of correct

+   +

D. Section 7

_____   _____
numer of correct   number of incorrect

=   =

E. Total Unrounded Raw Score

_____ − _____ ÷4 = _____
numer of correct   number of incorrect

F. Total Rounded Raw Score   _____ (See table)

Math Score Range =   [_____]

Math Conversion Table

Raw Score	Scaled Score	Raw Score	Scaled Score
54	800	23	490-550
53	780-800	22	480-540
52	760-800	21	470-530
51	740-800	20	460-520
50	720-780	19	450-510
49	700-760	18	450-510
48	690-750	17	440-500
47	680-740	16	430-490
46	670-730	15	420-480
45	660-720	14	420-480
44	650-710	13	410-470
43	650-710	12	400-460
42	640-700	11	390-450
41	630-690	10	380-440
40	620-680	9	390-430
39	610-670	8	380-420
38	610-670	7	370-410
37	600-660	6	360-400
36	590-650	5	340-380
35	580-640	4	320-370
34	570-630	3	310-360
33	560-620	2	300-350
32	560-620	1	270-320
31	550-610	0	240-300
30	540-600	-1	200-290
29	530-590	-2	200-270
28	530-590	-3	200-260
27	520-580	-4	200-240
26	510-570	-5	200-220
25	500-560	-6 and below	200
24	500-560		

SECTION 3
Time- 25 minutes
20 Questions

Turn to Section 3 (Page 1) of your answer sheet to answer the questions in this section.

Directions: For this section, solve each problem and decide which is the best of the choices given. Fill in the corresponding circle on the answer sheet. You may use any available space for scratchwork.

Notes

1. The use of a calculator is permitted.
2. All numbers used are real numbers.
3. Figures that accompany problems in this test are intended to provide information useful in solving the problems. They are drawn as accurately as possible EXCEPT when it is stated in a specific problem that the figure is not drawn to scale. All figure lie in a plane unless other indicated.
4. Unless otherwise specified, the domain of any function f is assumed to be set of all real numbers x for which $f(x)$ is a real number.

Reference Information

 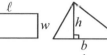

$A = \pi r^2$ $A = \ell w$ $A = \frac{1}{2}bh$ $V = \ell wh$ $V = \pi r^2 h$ $c^2 = a^2 + b^2$ Special Right Triangles
$C = 2\pi r$

The numbers of degrees of arc in a circle is $360°$.

The sum of the measures in degrees of the angles is $180°$.

1. If $a+b = 5$ and $a-b = 3$, What is the value of $(a^2 + 5) - b^2$?

(A) 10
(B) 15
(C) 20
(D) 22
(E) 25

2. If it takes a people 10 hours to finish a certain task, how many hours would it take for a^2 people to finish the same task?

(A) 5

(B) 10

(C) $\frac{5}{a}$

(D) $\frac{10}{a}$

(E) $\frac{a}{10}$

GO ON TO THE NEXT PAGE

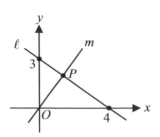

Note: Figure not drawn to scale.

3. In the figure above, line ℓ is perpendicular to line m. If the x- and y-intercept of line ℓ is 4 and 3 respectively, what is the length of $\overline{OP}$?

(A) 1.2
(B) 1.6
(C) 1.8
(D) 2.0
(E) 2.4

4. If the average (arithmetic mean) of t, u, and v is 14, and the average of $2u$ and $2v$ is 30, what is the value of t?

(A) 12
(B) 14
(C) 16
(D) 20
(E) 24

5. What is the largest three-digit number that is the square of a prime number?

(A) 997
(B) 961
(C) 900
(D) 899
(E) 841

$A(-3, 2)$
$B(1, 2)$
$C(6, -1)$

6. The coordinates of points A, B, and C in the xy-plane are given above. What is the area of $\triangle ABC$?

(A) 6
(B) 8
(C) 10
(D) 12
(E) 18

GO ON TO THE NEXT PAGE

7. The terms in a sequence are given by $a_n = a + (n-1)d$, where n is the number of terms, a is the first term, and each term after the first is found by adding 8 to the term immediately preceding it. If the first term is 10, which term in the sequence is equal to 210 ?

   (A) The 15th
   (B) The 18th
   (C) The 25th
   (D) The 26th
   (E) The 27th

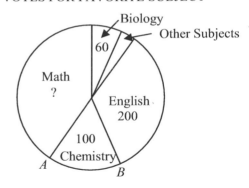

VOTES FOR FAVORITE SUBJECT

8. In the figure above, line ℓ is perpendicular to line m. Which of the following pairs of angle measures is not sufficient for determining all six angle measures?

   (A) x and y
   (B) x and b
   (C) y and a
   (D) y and c
   (E) z and a

9. The circle graph above shows the results of a vote for favorite subjects at J.C High School. Chemistry came in third with 100 votes. If 600 students voted for the survey, how many students could have possibly voted for Math?

   (A) 200
   (B) 220
   (C) 240
   (D) 270
   (E) 290

10. If point $M(a, b)$ is the midpoint of the line segment connecting point $A(5, 10)$ and point $B(15, 4)$, what is the value of $a + b$?

    (A) 17
    (B) 18
    (C) 19
    (D) 24
    (E) 28

GO ON TO THE NEXT PAGE

11. How many days are there in w weeks and h hours?

(A) $\dfrac{h}{7} + 24w$

(B) $\dfrac{w}{7} + 24h$

(C) $7w + 24h$

(D) $7w + \dfrac{h}{24}$

(E) $\dfrac{w}{24} + \dfrac{h}{7}$

---

12. A convenience store changes the price of candy bars from 5 for 2 dollars to 8 for 4 dollars. What can be said about the price change?

(A) The price remains unchanged.
(B) It reflects a decrease of 10% in the price of one candy bar.
(C) It reflects an increase of 25% in the price of one candy bar.
(D) It reflects an increase of 20% in the price of one candy bar.
(E) It reflects a decrease of 25% in the price of one candy bar.

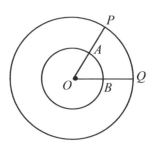

Note: Figure not drawn to scale.

13. In the figure above, two circles have centers at point O. If the ratio of the lengths of $\overline{OA}$ to $\overline{AP}$ is $2:3$ and the length of $\overparen{AB}$ is 10, what is the length of $\overparen{PQ}$?

(A) 15
(B) 20
(C) 25
(D) 30
(E) 35

---

14. If $x^{12} = 5000$ and $\dfrac{x^{11}}{y} = 10$, what is the value of xy?

(A) 500
(B) 100
(C) 50
(D) 10
(E) 5

GO ON TO THE NEXT PAGE

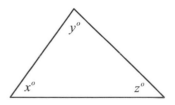

Note: Figure not drawn to scale.

15. In the figure above, $y > 60$ and $z = x + 10$. If x is an integer, what is the greatest possible value of x?

(A) 54
(B) 55
(C) 56
(D) 61
(E) 70

16. For all positive integers p and q, let Φ be defined by $p\Phi q = (p+q)^2 - (p-q)^2$.

If a and b are positive integers, which of the following CANNOT be the value of $a\Phi b$?

(A) 4
(B) 8
(C) 12
(D) 18
(E) 20

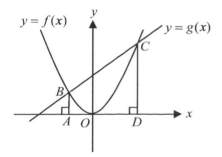

$$f(x) = x^2$$
$$g(x) = x + 2$$

17. The figure above shows the graphs of the functions f and g, and the functions f and g are defined above. What is the area of quadrilateral $ABCD$?

(A) 5
(B) 7.5
(C) 9
(D) 12.5
(E) 15

$$x + 3y = 9$$
$$3x + 9y = k$$

18. For the system of equations above, if $k = 27$, how many solutions do the equations have?

(A) None
(B) One
(C) Two
(D) Three
(E) More than three

GO ON TO THE NEXT PAGE

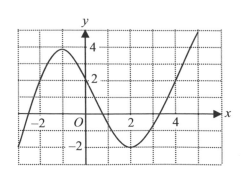

19. The function $y = f(x)$, defined for $-3 \le x \le 5$, is graphed above. Which of the following gives all values of x for which $-2 \le f(x) \le 2$?

(A) $-3 \le x \le 4$
(B) $0 \le x \le 5$
(C) $0 \le x \le 2$
(D) $-2 \le x \le 0$ and $4 \le x \le 5$
(E) $-3 \le x \le -2$ and $0 \le x \le 4$

20. If $\dfrac{a}{b}$ is negative, which of the following must be positive?

    I. $(a+b)(a-b)$
    II. $1-ab$
    III. $\dfrac{b-a}{a}$

(A) I only
(B) II only
(C) III only
(D) I and II only
(E) I, II, and III

STOP

**If you finish before time is called, you may check your work on this section only.
Do not turn to any other section in the test.**

SECTION 5
Time- 25 minutes
18 Questions

Turn to Section 5 (Page 1) of your answer sheet to answer the questions in this section.

Directions: For this section, solve each problem and decide which is the best of the choices given. Fill in the corresponding circle on the answer sheet. You may use any available space for scratchwork.

Notes

1. The use of a calculator is permitted.
2. All numbers used are real numbers.
3. Figures that accompany problems in this test are intended to provide information useful in solving the problems. They are drawn as accurately as possible EXCEPT when it is stated in a specific problem that the figure is not drawn to scale. All figure lie in a plane unless other indicated.
4. Unless otherwise specified, the domain of any function f is assumed to be set of all real numbers x for which $f(x)$ is a real number.

Reference Informatiom

$A = \pi r^2$
$C = 2\pi r$
$A = \ell w$
$A - \frac{1}{2}bh$
$V = \ell wh$
$V = \pi r^2 h$
$c^2 = a^2 + b^2$
Special Right Triangles

The numbers of degrees of arc in a circle is 360°.

The sum of the measures in degrees of the angles is 180°.

40, 35, 30, …

1. In the sequence above, the first term is 40 and each term after the first is 5 less than the one immediately preceding it. What is the sum of the first 8 terms in the sequence?

(A) 150
(B) 180
(C) 210
(D) 240
(E) 250

2. Points A, B, C, and D lie on a line in that order. If $AB = 2BC = 3CD = k$ and $CD = 6$, what is the distance from A to C?

(A) 27
(B) 33
(C) 38
(D) 42
(E) 54

GO ON TO THE NEXT PAGE

3. A father is t inches taller than his son. If their total height is p, in terms of p and t, what is the father's height, in inches?

(A) $p - t$

(B) $p + t$

(C) $\dfrac{p - t}{2}$

(D) $\dfrac{p + t}{2}$

(E) $2p - t$

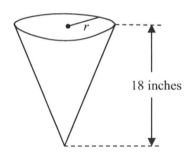

4. In the figure above, a container in the shape of a right circular cone is 18 inches high and has a capacity of 24 quarts. What is the number of quarts of liquid in the container when it is filled to a height of 9 inches?

(A) 3
(B) 6
(C) 9
(D) 12
(E) 16

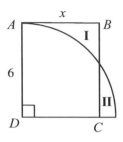

Note: Figure not drawn to scale.

5. The figure above shows a rectangle $ABCD$ and a quarter of a circle with a radius of 6. If the area of region **I** is equal to the area of region **II**, what is the value of x, the length of $\overline{AB}$?

(A) 3
(B) π
(C) 1.5π
(D) 2π
(E) 4.5π

6. If a, b, and c are positive and $a^{-2}b^2c^2 > a^{-3}b^3c^2$, which of the following must be true?

    I. $a > b$
    II. $a > c$
    III. $b > c$

(A) I only
(B) II only
(C) III only
(D) I and II only
(E) I, II, and III

GO ON TO THE NEXT PAGE

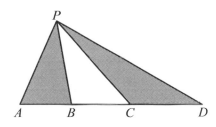

Note: Figure not drawn to scale.

7. In the triangle above, the sum of the areas of the shaded regions is 10. If $AB : BC : CD = 2 : 2 : 3$, what is the area of $\triangle PBC$?

    (A) 2
    (B) 3
    (C) 4
    (D) 5
    (E) 6

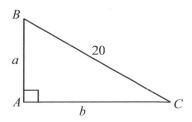

Note: Figure not drawn to scale.

8. In $\triangle ABC$ above, which of the following inequalities is true?

    (A) $0 < (a+b)^2 \le 40$
    (B) $40 < (a+b)^2 \le 80$
    (C) $80 < (a+b)^2 \le 200$
    (D) $200 < (a+b)^2 \le 400$
    (F) $400 < (a+b)^2$

GO ON TO THE NEXT PAGE

5 ☐☐☐ 5 ☐☐☐ 5 ☐☐☐ 5

Directions: For Students-Produced Response questions 9-18, use the grid at the bottom of the answer sheet page on which you have answered questions 1-8.

Each of the remaining 10 questions requires you to solve the problem and enter your answer by making the circles in the special grid, as shown in the examples below. You may use any available space for scratchwork.

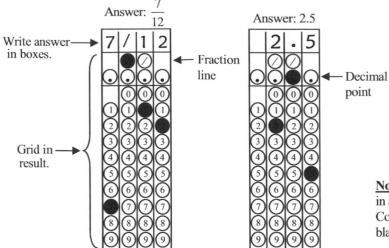

Answer: $\frac{7}{12}$

Write answer in boxes.

Fraction line

Grid in result.

Answer: 2.5

Decimal point

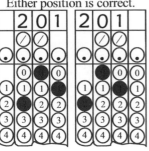

Answer: 201
Either position is correct.

Note: You may start your answers in any column, space permitting. Columns not needed should be left blank.

- Mark no more than one circle in any column.

- Because the answer sheet will be machine-scored, **you will receive credit only if the circles are filled in correctly.**

- Although not required, it is suggested that you write your answer in the boxes at the top of the columns to help you fill in the circles accurately.

- Some problems may have more than one correct answer. In such cases, grid only one answer.

- No question has a negative answer.

- **Mixed numbers** such as $3\frac{1}{2}$ must be gridded as 3.5 or 7/2. (If ☐3☐1☐/☐2 is gridded, it will be interpreted as $\frac{31}{2}$, not $3\frac{1}{2}$.)

- **Decimal Answers:** If you obtain a decimal answer with more digits than the grid can accommodate, it may be either rounded or truncated, but it must fill the entire grid. For example, if you obtain an answer such as 0.6666..., you should record your result as .666 or .667. **A less accurate value such as .66 or .67 will be scored as incorrect.**

Acceptable ways to grid $\frac{2}{3}$ are:

9. What is the smallest number greater than 1000 that is a square of an integer?

10. If a and b are positive even integers and $\sqrt{a} + \sqrt{b} = 8$, what is one possible value of $a+b$?

GO ON TO THE NEXT PAGE ⟩

11. Let the function g be defined by $g(x) = 2f(x) - 3$, where $f(x)$ is a linear function. If $g(1) = 3$ and $g(3) = 9$, what is the slope of the linear function f?

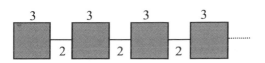

13. The figure above shows some squares, each measuring 3 inches, placed at one end of a 128-inch line segment. If there are 2-inch spaces between consecutive squares, what is the maximum number of such squares that can be placed on the 128-inch segment?

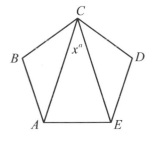

12. In the figure above, regular pentagon $ABCDE$ is divided into three nonoverlapping triangles. If $\angle ACE = x^o$, what is the value of x?

14. If a is a positive even integer and $5 < \sqrt{5a} < 10$, for how many values of a is the inequality true?

GO ON TO THE NEXT PAGE

15. The average (arithmetic mean) of ten different positive integers is 30. If m is the greatest of these integers, what is the greatest possible value of m?

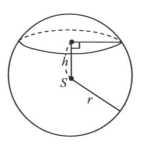

17. In the figure above, a plane passed through a sphere S creating a circular region. If the area of the circular region is 144π and $h = 5$, what is the value of r, the radius of the sphere?

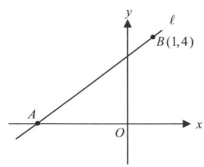

Note: Figure not drawn to scale.

16. In the xy-coordinate plane above, points A and B lie on line ℓ. If $AB = 5$, what is the y-intercept of line ℓ?

Category	Number of students
Have a sister	20
Have a brother	17
Have both a sister and brother	5
Do not have a sister or brother	8

18. The table above gives the number of students who have a sister, brother, both, or none from Mr. Lee's class. What is the probability that a randomly selected student from the class has both a sister and brother?

STOP
If you finish before time is called, you may check your work on this section only.
Do not turn to any other section in the test.

SECTION 7
Time- 20 minutes
16 Questions

Turn to Section 7 (Page 2) of your answer sheet to answer the questions in this section.

Directions: For this section, solve each problem and decide which is the best of the choices given. Fill in the corresponding circle on the answer sheet. You may use any available space for scratchwork.

Notes

1. The use of a calculator is permitted.
2. All numbers used are real numbers.
3. Figures that accompany problems in this test are intended to provide information useful in solving the problems. They are drawn as accurately as possible EXCEPT when it is stated in a specific problem that the figure is not drawn to scale. All figure lie in a plane unless other indicated.
4. Unless otherwise specified, the domain of any function f is assumed to be set of all real numbers x for which $f(x)$ is a real number.

Reference Informatiom

$A = \pi r^2$
$C = 2\pi r$
$A = \ell w$
$A = \dfrac{1}{2}bh$
$V = \ell wh$
$V = \pi r^2 h$
$c^2 = a^2 + b^2$
Special Right Triangles

The numbers of degrees of arc in a circle is $360°$.

The sum of the measures in degrees of the angles is $180°$.

---

1. If $ax + bx = 48$, what is the value of $2a + 2b$ when $x = 4$?

(A) 12
(B) 18
(C) 24
(D) 48
(E) 196

2. If $|x - 10| = -5$, for how many values of x is the equation true?

(A) None
(B) One
(C) Two
(D) Four
(E) More than Four

GO ON TO THE NEXT PAGE

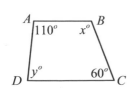

Note: Figure not drawn to scale.

3. In the figure above, $\overline{AB} \parallel \overline{CD}$. What is the value of $x - y$?

(A) 50
(B) 55
(C) 60
(D) 65
(E) 80

4. What is the length of the diagonal of a cube with surface area $54x^2$?

(A) $3x$
(B) $6x$
(C) $3x\sqrt{2}$
(D) $3x\sqrt{3}$
(E) $27x$

5. For the function f , $y = f(x)$ is inversely proportional to x . If $f(5) = 24$, what is the value of $f(10)$?

(A) 6
(B) 12
(C) 24
(D) 36
(E) 48

$$|x| + |y| = 5$$

6. Which of the following points is NOT on the graph above?

(A) $(0,5)$
(B) $(-2,3)$
(C) $(-3,-2)$
(D) $(3,-4)$
(E) $(-1,-4)$

GO ON TO THE NEXT PAGE

7. If $\sqrt{ab} = 6$, where a and b are positive integers, which of the following could NOT be a value of $a - b$?

    (A) 16
    (B) 9
    (C) 5
    (D) 4
    (E) 0

8. In the xy-coordinate plane, line ℓ is parallel to the x-axis and passes through the point $(-4, -8)$. Which of the following is an equation of line ℓ?

    (A) $x = -4$
    (B) $y = -8$
    (C) $y = -4$
    (D) $x = -8$
    (E) $y = x - 4$

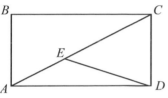

Note: Figure not drawn to scale.

9. In the rectangle above, the length of $\overline{AE}$ is equal to $\frac{2}{5}$ of the length of $\overline{AC}$. If the area of $\triangle AED$ is 18, what is the area of $\triangle CED$?

    (A) 24
    (B) 27
    (C) 36.5
    (D) 40.5
    (E) 45

10. An arrow is shot upward on the moon with an initial velocity of 60 meters per second and returns to the surface after 60 seconds. If the height is given by the formula $h = t(60 - t)$, what is the maximum height that the arrow reaches?

    (A) 300 meters
    (B) 600 meters
    (C) 900 meters
    (D) 1200 meters
    (E) 1500 meters

GO ON TO THE NEXT PAGE

COST VS. WEIGHT
FOR 12 BOXES

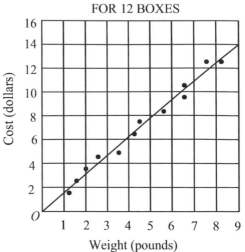

Weight (pounds)

11. For 12 boxes of text books of various weights, the cost and weight of each box from mailing company X are displayed in the scatter plot above and the line of best fit for the data is shown. Which of the following is closest to the average (arithmetic mean) cost per pound for the 12 boxes?

(A) $0.50
(B) $1.00
(C) $1.50
(D) $1.75
(E) $1.90

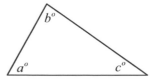

Note: Figure not drawn to scale.

12. In the figure above, $b = a + 10$ and $c < 20$. Which of the following could be a value of a?

(A) 45
(B) 55
(C) 75
(D) 80
(E) 90

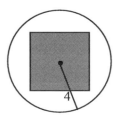

Note: Figure not drawn to scale.

13. In the figure above, a square lies in a circle. When a point is chosen at random from the circle, the probability that the chosen point is in the shaded region is $\dfrac{1}{2}$. If the radius of the circle is 4, what is the length of a side of the square?

(A) 4
(B) 6
(C) $2\sqrt{\pi}$
(D) $\sqrt{6\pi}$
(E) $2\sqrt{2\pi}$

GO ON TO THE NEXT PAGE

14. If p gallons of 10 percent antifreeze solution is added to q gallons of 20 percent antifreeze solution, what is the percent antifreeze of the resulting solution in terms of p and q ?

(A) $\dfrac{p+q}{2}$

(B) $\dfrac{10p+20q}{30}$

(C) $\dfrac{10p+20q}{p+q}$

(D) $\dfrac{10p+20q}{100}$

(E) $\dfrac{100p+200q}{p+q}$

15. If $\sqrt{a^2+b^2}=a-b$, where a and b are non-negative numbers, which of the following could be true?

    I. $a=0$
    II. $b=0$
    III. $a \geq b$

(A) I only
(B) II only
(C) III only
(D) II and III only
(E) I, II, and III

Note: Figure not drawn to scale.

16. In the figure above, $ABCD$ and $AEFG$ are squares with integer-length sides of b and a respectively. If the area of the shaded region is 28, what is the area of square $ABCD$?

(A) 64
(B) 81
(C) 100
(D) 144
(E) 196

STOP

If you finish before time is called, you may check your work on this section only.
Do not turn to any other section in the test.

#	SECTION 3	SECTION 5	SECTION 7
1	C	B	C
2	D	A	A
3	E	D	A
4	A	A	D
5	B	C	B
6	A	A	D
7	D	C	D
8	E	E	B
9	B	1024	B
10	A	32 or 40	C
11	D	$3/2$ or 1.5	C
12	C	36	D
13	C	26	E
14	A	7	C
15	A	255	D
16	D	$8/3$ or 2.66, 2.67	A
17	B	13	
18	E	$1/8$ or .125	
19	E		
20	B		

TEST 16   SECTION 3

1. (C)

$a + b = 5$

$\dfrac{a - b = 3}{2a = 8}$ → $a = 4$ and $b = 1$. Therefore,

$(4^2 + 5) - 1^2 = 21 - 1 = 20$.

2. (D)

Inverse proportion:

$a \times 10 = a^2 \times h$ → $h = \dfrac{10}{a}$.

3. (E)

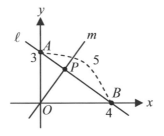

In the figure above, the area of $\triangle AOB$ is

$\dfrac{3 \times 4}{2} = \dfrac{5 \times OP}{2}$ → $OP = \dfrac{12}{5} = 2.4$.

4. (A)

$\dfrac{t + u + v}{3} = 14$ and $\dfrac{2u + 2v}{2} = 30$. $u + v = 30$.

Therefore, $\dfrac{t + 30}{3} = 14$ → $t = 12$.

5. (B)

Prime number--- 2,3,5,.......29, 31, 37

$31^2 = 961$, but $37^2 = 1369$ (4-digits number).

---

6. (A)

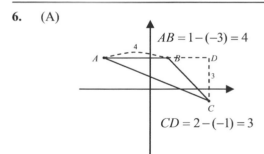

$AB = 1 - (-3) = 4$

$CD = 2 - (-1) = 3$

In the figure above, the area of $\triangle ABC$ is

$\dfrac{4 \times 3}{2} = 6$.

---

7. (D)

$10 + (n-1)8 = 210 \rightarrow n = 26$.

---

8. (E)

For (A), $x = c$, $y = b$, then $z = 90 - (x+y)$.

Therefore all of the angles can be found.

But for (E), only a and z can be cleared, not the other angles.

---

9. (B)

It must be greater than 200, but less than 240.

---

10. (A)

$M(a,b) = \left(\dfrac{5+15}{2}, \dfrac{10+4}{2} \right) = (10,7)$. So,

$a + b = 17$.

---

11. (D)

w weeks $= 7 \times w$ days and h hours $= \dfrac{h}{24}$ days.

---

12. (C)

$\dfrac{2^\$}{5} = 40$ cents and $\dfrac{4^\$}{8} = 50$ cents.

% increase $= \dfrac{50 - 40}{40} \times 100 = 25\%$

---

13. (C)

The ratio of the lengths of two circles is $2:5 = 2k:5k$. The sectors of the circles, sector OAB and OPQ are similar. So,

$2k = 10 \rightarrow k = 5$, then $5k = 5(5) = 25$.

---

14. (A)

$x \cdot x^{11} = 5000 \rightarrow x^{11} = \dfrac{5000}{x}$.

So, $\dfrac{x^{11}}{y} = \dfrac{5000/x}{y} = \dfrac{5000}{xy} = 10$. Therefore

$xy = 500$.

---

15. (A)

Because $y > 60$, $x + z < 120$.

$x + z = x + (x+10) = 2x + 10 < 120$.

$2x < 110 \rightarrow x < 55$.

Therefore, the greatest integer value of x is 54.

---

16. (D)

$a\Phi b = (a+b)^2 - (a-b)^2 = 4ab$. Therefore, the value of $a\Phi b$ is a multiple of 4. 18 is not a multiple of 4.

---

17. (B)

First find the x-intersections.

$x^2 = x + 2 \rightarrow x^2 - x - 2 = (x-2)(x+1) = 0$.

Therefore, there are two intersections at $x = -1$ and $x = 2$.

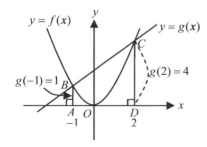

In the figure above, $AD = 3$,

$AB = g(-1) = 1$, and $CD = g(2) = 4$.

The area of $ABCD$ is $\dfrac{(4+1)3}{2} = 7.5$.

---

18. (E)

$\begin{aligned} x + 3y &= 9 \\ 3x + 9y &= 27 \end{aligned} \rightarrow \dfrac{1}{3} = \dfrac{3}{9} = \dfrac{9}{27}$.

Therefore, the systems of equations have infinite solutions.

---

19. (E)

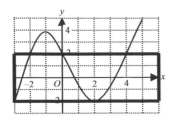

In the graph above, for $-2 \le y \le 2$, the graph is defined as following intervals, $-3 \le x \le -2$ and $0 \le x \le 4$.

20. (B)

$\dfrac{a}{b} < 0$ is equivalent to $(a > 0, b < 0)$ or $(a < 0, b > 0)$. Use numbers.

Check when $a = 2$ and $b = -1$, or $a = -1$ and $b = 2$.

  I. $(2-1)(2+1) = 3 > 0$,

    $(-1+2)(-1-2) = -3 < 0$

  II. $1 - (2)(-1) = 3 > 0$, $1 - (-1)(2) = 3 > 0$

  III. $\dfrac{-1-2}{2} = -1.5 < 0$, $\dfrac{2-(-1)}{-1} = -3 < 0$

Or, algebraically,

  I. $(a+b)(a-b) = a^2 - b^2$.

    If $b = -3$ and $a = 2$

    It is negative .

  II. $1 - ab > 0$, because $ab < 0$.

  III. $\dfrac{b-a}{a} = \dfrac{b}{a} - 1 < 0$. Because $\dfrac{a}{b} < 0$.

TEST 16   SECTION 5

1. (B)

$40 + 35 + \ldots = 5(8+7+6+5+4+3+2+1) =$
$5 \times 36 = 180$.

2. (A)

$AB = 3CD = 3(6) = 18$. $BC = \dfrac{AB}{2} = \dfrac{18}{2} = 9$.

Therefore, $AC = AB + BC = 18 + 9 = 27$.

3. (D)

Let father's height be x, then son's height is $x - t$.

So, $x + (x - t) = p$ $\rightarrow 2x = p + t$ $\rightarrow x = \dfrac{p+t}{2}$.

---

4. (A)

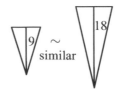

In the figures above, the ratio of lengths is $9:18 = 1:2$.

Then the ratio of the volumes is
$1^3 : 2^3 = 1:8 = k:8k$.
If $8k = 24$, $k = 3$.

---

5. (C)

The area of a quarter circle is equal to the area of the rectangle. Therefore,

$\dfrac{\pi \cdot 6^2}{4} = 6x$ $\rightarrow x = \dfrac{3\pi}{2} = 1.5\pi$.

---

6. (A)

Because a, b, and c are all positive,

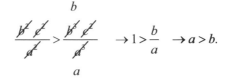

$\rightarrow 1 > \dfrac{b}{a}$ $\rightarrow a > b$.

---

7. (C)

The areas of the triangles is $2k : 2k : 3k$.
Because $2k + 3k = 5k = 10$ $\rightarrow k = 2$.
Therefore, the area of $\triangle ABC$ is $2k = 4$.

---

8. (E)

Because $a^2 + b^2 = 400$,
$(a+b)^2 = a^2 + b^2 + 2ab > a^2 + b^2 = 400$

---

9. 1024

Because $\sqrt{1000} \approx 31.6$, the least square is $32^2 = 1024$.

---

10. 32 or 40

$\sqrt{a} + \sqrt{b} = 8$

$2 + 6 = 8$ $\rightarrow a = 4$, $b = 36$ $\rightarrow a + b = 40$.

$4 + 4 = 8$ $\rightarrow a = 16$, $b = 16$ $\rightarrow a + b = 32$.

---

11. $\frac{3}{2}$ or 1.5

$g(1) = 2f(1) - 3 = 3 \rightarrow f(1) = 3 \rightarrow (1,3)$

$g(3) = 2f(3) - 3 = 9 \rightarrow f(3) = 6 \rightarrow (3,6)$

Therefore, the slope is $\frac{6-3}{3-1} = \frac{3}{2}$.

12. 36

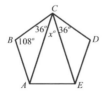

In the figure above, one of the interior angles is 108° and $\triangle ABC$ is an isosceles triangle. Therefore, $x = 108 - (36 + 36) = 36$.

13. 26

Let the number of a square be n, then the number of 2-inch space is $n-1$. Therefore, the length is
$3n + 2(n-1) = 128 \rightarrow 5n = 130 \rightarrow n = 26$.

14. 7

$5 < \sqrt{5a} < 10 \rightarrow 25 < 5a < 100 \rightarrow 5 < a < 20$.
Therefore, the positive integers will be
$6, 8, 10, 12, 14, 16, 18. \rightarrow 7$ numbers.

15. 255

Sum of the numbers $= 30 \times 10 = 300$. If 9 of the numbers are different least numbers like 1,2,3,4,5,6,7,8,9, then the last number will be the greatest one. $1 + 2 + 3 + ..9 = 45$.
$300 - 45 = 255$.

16. $\frac{8}{3}$ or 2.66, 2.67

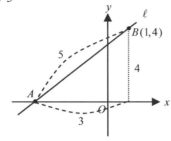

In the figure above, the equation of line ℓ is $y = \frac{4}{3}x + b$, where b is an y-intercept. Point

B lies on this line. Therefore,
$4 = \frac{4}{3}(1) + b \rightarrow b = \frac{8}{3}$.

17. 13

The radius R of the circle is,
$\pi R^2 = 144\pi \rightarrow R = 12$.

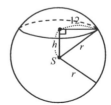

In the figure above, the radius of the sphere is $r = \sqrt{12^2 + 5^2} = 13$.

18. $\frac{1}{8}$ or .125

The total number of students=
$(20 + 17 - 5) + 8 = 40$.

Therefore, $P = \frac{5}{40} = \frac{1}{8}$.

TEST 16   SECTION 7

1. (C)

$(a+b)x = 48 \rightarrow (a+b)(4) = 48$

$\rightarrow (a+b) = 12$

Therefore, $2(a+b) = 24$.

2. (A)

Distance cannot have a negative value.
No solution.

3. (A)

Because sum of the interior angles is 180, $x = 120$ and $y = 70$. Therefore, $x - y = 50$.

4. (D)

The area of one face is $\frac{54x^2}{6} = 9x^2$. So, the length of an edge is $\sqrt{9x^2} = 3x$. Therefore, the length of the diagonal is
$\sqrt{(3x)^2 + (3x) +^2 (3x)^2} = 3x\sqrt{3}$.

5. (B)

When $x = 5$, y is 24. When $x = 10$, what is y?

$5 \times 24 = 10 \times y \ \rightarrow \ y = 12$.

---

6. (D)

For $(3, -4)$, $|3| + |-4| \neq 5$.

---

7. (D)

Because $ab = 36$, the pairs of $\{a, b\}$ are

$\{1, 36\}, \{2, 18\}, \{3, 12\}, \{4, 9\}, \{6, 6\}$.

Therefore, $a - b = 35, 16, 9, 5$, or 0.

---

8. (B)

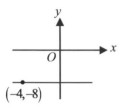

In the plane above, the equation of the line is $y = -8$.

---

9. (B)

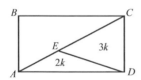

In the figure above, the ratio of $AE : EC$ is $2 : 3$. With the same height, the ratio of the areas is also $2 : 3 = 2k : 3k$. So, $2k = 18 \ \rightarrow \ k = 9$. Therefore, the area of $\triangle CED$ is $3k = 27$.

---

10. (C)

The maximum height occurs in the middle of two x-intercepts. $x = \dfrac{0 + 60}{2} = 30$.

Therefore, the maximum height is $h(30) = 30(60 - 30) = 900$.

---

11. (C)

The average cost per pound = the slope of the line.

$\dfrac{14 - 0}{9 - 0} = \dfrac{14}{9} \simeq 1.55$. The closest number is 1.5.

---

12. (D)

Since $0 < b < 20$, $160 < a + b < 180$.

Therefore,

$160 < a + (a + 10) < 180 \ \rightarrow \ 150 < 2a < 170$

Therefore, $75 < a < 85$.

Or, backsolving

If $a = 45$, $b = 55$, then $c = 80$ (false), and also

If $a = 90$, $b = 100$, then $a + b > 180$. (false)

If $a = 80$, $b = 90$, then $c = 10 < 20$. (true)

---

13. (E)

$P = \dfrac{\text{area of } \square}{\text{area of } \bigcirc} = \dfrac{1}{2}$. Therefore, $\dfrac{x^2}{16\pi} = \dfrac{1}{2}$.

$x^2 = 8\pi \ \rightarrow \ x = 2\sqrt{2\pi}$.

---

14. (C)

1--Antifreeze = 10% of $\times p = 0.1p$.

2--Antifreeze = 20% of $q = 0.2q$.

When mixed up, the % of the solution is

$\left(\dfrac{0.1p + 0.2q}{p + q} \right) \times 100 = \left(\dfrac{10p + 20q}{P + q} \right)\%$.

---

15. (D)

$\sqrt{a^2 + b^2} = a - b$

$\rightarrow \ a^2 + b^2 = a^2 + b^2 - 2ab$

Therefore, $ab = 0$ and $a - b \geq 0 \ \rightarrow \ a \geq b$.

If $a = 0$, then $\sqrt{b^2} = b \neq -b$. Not true.

If $b = 0$, then $\sqrt{a^2} = a$. True. (II) is correct.

Also, $a \geq b(= 0)$. (III) is correct.

16. (A)

Since the difference between the areas is

$b^2 - a^2 = 28 \rightarrow (b+a)(b-a) = 28.$

Therefore,

$\begin{bmatrix} b+a = \\ b-a = \end{bmatrix} \begin{cases} 28 \\ 1 \end{cases}, \begin{cases} 14 \\ 2 \end{cases}, \begin{cases} 7 \\ 4 \end{cases}$

Because a and b are positive integers,
Solve the system of equations.

$b + a = 14$

$\dfrac{b - a = 2}{2b \quad = 16} \quad b = 8 \text{ and } a = 6$

Therefore, the area of $ABCD$ is 64.

Or,

Trial and error, find the number as follows.

$b^2 = 1, 4, 9, 16, 25, 36, 49, 64, 81, 100...$

$a^2 = 1, 4, 9, 16, 25, 36, 49, 64, 81, 100..$

Difference between 64 and 36 is 28.

$\boxed{END}$

NO MATERIAL ON THIS PAGE

Dr. John Chung's SAT Math

TEST
17

ANSWER SHEET   TEST #:

SECTION 3

1	Ⓐ Ⓑ Ⓒ Ⓓ Ⓔ	11	Ⓐ Ⓑ Ⓒ Ⓓ Ⓔ	21	Ⓐ Ⓑ Ⓒ Ⓓ Ⓔ	31	Ⓐ Ⓑ Ⓒ Ⓓ Ⓔ
2	Ⓐ Ⓑ Ⓒ Ⓓ Ⓔ	12	Ⓐ Ⓑ Ⓒ Ⓓ Ⓔ	22	Ⓐ Ⓑ Ⓒ Ⓓ Ⓔ	32	Ⓐ Ⓑ Ⓒ Ⓓ Ⓔ
3	Ⓐ Ⓑ Ⓒ Ⓓ Ⓔ	13	Ⓐ Ⓑ Ⓒ Ⓓ Ⓔ	23	Ⓐ Ⓑ Ⓒ Ⓓ Ⓔ	33	Ⓐ Ⓑ Ⓒ Ⓓ Ⓔ
4	Ⓐ Ⓑ Ⓒ Ⓓ Ⓔ	14	Ⓐ Ⓑ Ⓒ Ⓓ Ⓔ	24	Ⓐ Ⓑ Ⓒ Ⓓ Ⓔ	34	Ⓐ Ⓑ Ⓒ Ⓓ Ⓔ
5	Ⓐ Ⓑ Ⓒ Ⓓ Ⓔ	15	Ⓐ Ⓑ Ⓒ Ⓓ Ⓔ	25	Ⓐ Ⓑ Ⓒ Ⓓ Ⓔ	35	Ⓐ Ⓑ Ⓒ Ⓓ Ⓔ
6	Ⓐ Ⓑ Ⓒ Ⓓ Ⓔ	16	Ⓐ Ⓑ Ⓒ Ⓓ Ⓔ	26	Ⓐ Ⓑ Ⓒ Ⓓ Ⓔ	36	Ⓐ Ⓑ Ⓒ Ⓓ Ⓔ
7	Ⓐ Ⓑ Ⓒ Ⓓ Ⓔ	17	Ⓐ Ⓑ Ⓒ Ⓓ Ⓔ	27	Ⓐ Ⓑ Ⓒ Ⓓ Ⓔ	37	Ⓐ Ⓑ Ⓒ Ⓓ Ⓔ
8	Ⓐ Ⓑ Ⓒ Ⓓ Ⓔ	18	Ⓐ Ⓑ Ⓒ Ⓓ Ⓔ	28	Ⓐ Ⓑ Ⓒ Ⓓ Ⓔ	38	Ⓐ Ⓑ Ⓒ Ⓓ Ⓔ
9	Ⓐ Ⓑ Ⓒ Ⓓ Ⓔ	19	Ⓐ Ⓑ Ⓒ Ⓓ Ⓔ	29	Ⓐ Ⓑ Ⓒ Ⓓ Ⓔ	39	Ⓐ Ⓑ Ⓒ Ⓓ Ⓔ
10	Ⓐ Ⓑ Ⓒ Ⓓ Ⓔ	20	Ⓐ Ⓑ Ⓒ Ⓓ Ⓔ	30	Ⓐ Ⓑ Ⓒ Ⓓ Ⓔ	40	Ⓐ Ⓑ Ⓒ Ⓓ Ⓔ

SECTION 5

1	Ⓐ Ⓑ Ⓒ Ⓓ Ⓔ	11	Ⓐ Ⓑ Ⓒ Ⓓ Ⓔ	21	Ⓐ Ⓑ Ⓒ Ⓓ Ⓔ	31	Ⓐ Ⓑ Ⓒ Ⓓ Ⓔ
2	Ⓐ Ⓑ Ⓒ Ⓓ Ⓔ	12	Ⓐ Ⓑ Ⓒ Ⓓ Ⓔ	22	Ⓐ Ⓑ Ⓒ Ⓓ Ⓔ	32	Ⓐ Ⓑ Ⓒ Ⓓ Ⓔ
3	Ⓐ Ⓑ Ⓒ Ⓓ Ⓔ	13	Ⓐ Ⓑ Ⓒ Ⓓ Ⓔ	23	Ⓐ Ⓑ Ⓒ Ⓓ Ⓔ	33	Ⓐ Ⓑ Ⓒ Ⓓ Ⓔ
4	Ⓐ Ⓑ Ⓒ Ⓓ Ⓔ	14	Ⓐ Ⓑ Ⓒ Ⓓ Ⓔ	24	Ⓐ Ⓑ Ⓒ Ⓓ Ⓔ	34	Ⓐ Ⓑ Ⓒ Ⓓ Ⓔ
5	Ⓐ Ⓑ Ⓒ Ⓓ Ⓔ	15	Ⓐ Ⓑ Ⓒ Ⓓ Ⓔ	25	Ⓐ Ⓑ Ⓒ Ⓓ Ⓔ	35	Ⓐ Ⓑ Ⓒ Ⓓ Ⓔ
6	Ⓐ Ⓑ Ⓒ Ⓓ Ⓔ	16	Ⓐ Ⓑ Ⓒ Ⓓ Ⓔ	26	Ⓐ Ⓑ Ⓒ Ⓓ Ⓔ	36	Ⓐ Ⓑ Ⓒ Ⓓ Ⓔ
7	Ⓐ Ⓑ Ⓒ Ⓓ Ⓔ	17	Ⓐ Ⓑ Ⓒ Ⓓ Ⓔ	27	Ⓐ Ⓑ Ⓒ Ⓓ Ⓔ	37	Ⓐ Ⓑ Ⓒ Ⓓ Ⓔ
8	Ⓐ Ⓑ Ⓒ Ⓓ Ⓔ	18	Ⓐ Ⓑ Ⓒ Ⓓ Ⓔ	28	Ⓐ Ⓑ Ⓒ Ⓓ Ⓔ	38	Ⓐ Ⓑ Ⓒ Ⓓ Ⓔ
9	Ⓐ Ⓑ Ⓒ Ⓓ Ⓔ	19	Ⓐ Ⓑ Ⓒ Ⓓ Ⓔ	29	Ⓐ Ⓑ Ⓒ Ⓓ Ⓔ	39	Ⓐ Ⓑ Ⓒ Ⓓ Ⓔ
10	Ⓐ Ⓑ Ⓒ Ⓓ Ⓔ	20	Ⓐ Ⓑ Ⓒ Ⓓ Ⓔ	30	Ⓐ Ⓑ Ⓒ Ⓓ Ⓔ	40	Ⓐ Ⓑ Ⓒ Ⓓ Ⓔ

Grid-in response boxes numbered 9, 10, 11, 12, 13, 14, 15, 16, 17, 18 — each with four columns of bubbles (0–9) with fraction slash and decimal point options.

SECTION

7

1	Ⓐ Ⓑ Ⓒ Ⓓ Ⓔ	11	Ⓐ Ⓑ Ⓒ Ⓓ Ⓔ	21	Ⓐ Ⓑ Ⓒ Ⓓ Ⓔ	31	Ⓐ Ⓑ Ⓒ Ⓓ Ⓔ
2	Ⓐ Ⓑ Ⓒ Ⓓ Ⓔ	12	Ⓐ Ⓑ Ⓒ Ⓓ Ⓔ	22	Ⓐ Ⓑ Ⓒ Ⓓ Ⓔ	32	Ⓐ Ⓑ Ⓒ Ⓓ Ⓔ
3	Ⓐ Ⓑ Ⓒ Ⓓ Ⓔ	13	Ⓐ Ⓑ Ⓒ Ⓓ Ⓔ	23	Ⓐ Ⓑ Ⓒ Ⓓ Ⓔ	33	Ⓐ Ⓑ Ⓒ Ⓓ Ⓔ
4	Ⓐ Ⓑ Ⓒ Ⓓ Ⓔ	14	Ⓐ Ⓑ Ⓒ Ⓓ Ⓔ	24	Ⓐ Ⓑ Ⓒ Ⓓ Ⓔ	34	Ⓐ Ⓑ Ⓒ Ⓓ Ⓔ
5	Ⓐ Ⓑ Ⓒ Ⓓ Ⓔ	15	Ⓐ Ⓑ Ⓒ Ⓓ Ⓔ	25	Ⓐ Ⓑ Ⓒ Ⓓ Ⓔ	35	Ⓐ Ⓑ Ⓒ Ⓓ Ⓔ
6	Ⓐ Ⓑ Ⓒ Ⓓ Ⓔ	16	Ⓐ Ⓑ Ⓒ Ⓓ Ⓔ	26	Ⓐ Ⓑ Ⓒ Ⓓ Ⓔ	36	Ⓐ Ⓑ Ⓒ Ⓓ Ⓔ
7	Ⓐ Ⓑ Ⓒ Ⓓ Ⓔ	17	Ⓐ Ⓑ Ⓒ Ⓓ Ⓔ	27	Ⓐ Ⓑ Ⓒ Ⓓ Ⓔ	37	Ⓐ Ⓑ Ⓒ Ⓓ Ⓔ
8	Ⓐ Ⓑ Ⓒ Ⓓ Ⓔ	18	Ⓐ Ⓑ Ⓒ Ⓓ Ⓔ	28	Ⓐ Ⓑ Ⓒ Ⓓ Ⓔ	38	Ⓐ Ⓑ Ⓒ Ⓓ Ⓔ
9	Ⓐ Ⓑ Ⓒ Ⓓ Ⓔ	19	Ⓐ Ⓑ Ⓒ Ⓓ Ⓔ	29	Ⓐ Ⓑ Ⓒ Ⓓ Ⓔ	39	Ⓐ Ⓑ Ⓒ Ⓓ Ⓔ
10	Ⓐ Ⓑ Ⓒ Ⓓ Ⓔ	20	Ⓐ Ⓑ Ⓒ Ⓓ Ⓔ	30	Ⓐ Ⓑ Ⓒ Ⓓ Ⓔ	40	Ⓐ Ⓑ Ⓒ Ⓓ Ⓔ

Math Scoring Worksheet

A. Section 3

_____   _____
numer of correct   number of incorrect

+   +

B. Section 5 (1-8)

_____   _____
numer of correct   number of incorrect

+

C. Section 5 (9-18)

_____
numer of correct

+   +

D. Section 7

_____   _____
numer of correct   number of incorrect

=   =

E. Total Unrounded Raw Score

_____   −   _____   ÷4 = _____
numer of correct   number of incorrect

F. Total Rounded Raw Score   _____   (See table)

Math Score Range =   [_____ — _____]

Math Conversion Table

Raw Score	Scaled Score	Raw Score	Scaled Score
54	800	23	490-550
53	780-800	22	480-540
52	760-800	21	470-530
51	740-800	20	460-520
50	720-780	19	450-510
49	700-760	18	450-510
48	690-750	17	440-500
47	680-740	16	430-490
46	670-730	15	420-480
45	660-720	14	420-480
44	650-710	13	410-470
43	650-710	12	400-460
42	640-700	11	390-450
41	630-690	10	380-440
40	620-680	9	390-430
39	610-670	8	380-420
38	610-670	7	370-410
37	600-660	6	360-400
36	590-650	5	340-380
35	580-640	4	320-370
34	570-630	3	310-360
33	560-620	2	300-350
32	560-620	1	270-320
31	550-610	0	240-300
30	540-600	-1	200-290
29	530-590	-2	200-270
28	530-590	-3	200-260
27	520-580	-4	200-240
26	510-570	-5	200-220
25	500-560	-6 and below	200
24	500-560		

SECTION 3
Time- 25 minutes
20 Questions

Turn to Section 3 (Page 1) of your answer sheet to answer the questions in this section.

Directions: For this section, solve each problem and decide which is the best of the choices given. Fill in the corresponding circle on the answer sheet. You may use any available space for scratchwork.

Notes

1. The use of a calculator is permitted.
2. All numbers used are real numbers.
3. Figures that accompany problems in this test are intended to provide information useful in solving the problems. They are drawn as accurately as possible EXCEPT when it is stated in a specific problem that the figure is not drawn to scale. All figure lie in a plane unless other indicated.
4. Unless otherwise specified, the domain of any function f is assumed to be set of all real numbers x for which $f(x)$ is a real number.

Reference Information

$A = \pi r^2$
$C = 2\pi r$ $A = \ell w$ $A = \dfrac{1}{2}bh$ $V = \ell wh$ $V = \pi r^2 h$ $c^2 = a^2 + b^2$ Special Right Triangles

The numbers of degrees of arc in a circle is 360°.

The sum of the measures in degrees of the angles is 180°.

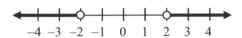

1. In the figure above, which of the following represents the shaded portion of the figure shown?

(A) $-2 < x < 2$
(B) $|x| < 2$
(C) $|x| \geq 2$
(D) $|x| > 2$
(E) $x^2 \leq 4$

2. If the average of x, $x+1$, and 10 is 7, what is the value of $x+1$?

(A) 5
(B) 6
(C) 7
(D) 8
(E) 9

GO ON TO THE NEXT PAGE

3. If k represents an odd integer, which of the following also represents an odd integer?

(A) $k+3$
(B) $2(k+3)+k$
(C) k^2+k-2
(D) $(k+1)(k-1)$
(E) $(k+1)(k+2)$

SALES OF COMPUTERS
AT COMPANY K

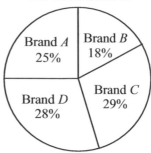

5. The circle graph represents all the computers that were sold at Company K in 2008, according to their brands. If the company sold 100 computers of brand C more than brand A, how many computers of brand B were sold ?

(A) 300
(B) 450
(C) 1000
(D) 1800
(E) 2500

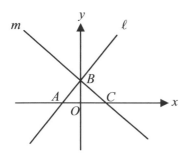

Note: Figure not drawn to scale.

4. In the figure above, line m is perpendicular to line ℓ at point B on the y-axis. If line ℓ is defined by the equation $y = x+2$, what is the area of $\triangle ABC$?

(A) 4
(B) 6
(C) 8
(D) 10
(E) 12

6. If $5^{2-x^2} = \dfrac{1}{25}$, what is a possible value of x ?

(A) -1
(B) 0
(C) 1
(D) 2
(E) 4

GO ON TO THE NEXT PAGE

7. If $\sqrt{x+a+b} = \sqrt{x-a-b}$, which of the following must be true?

    (A) $a = 0$
    (B) $b = 0$
    (C) $ab = 0$
    (D) $a + b = 0$
    (E) $a - b = 0$

8. Harry bought a bag of 10 pound flour for $80, a bag of 25 pound flour for $150, and a bag of 50 pound flour. If the average (arithmetic mean) cost of the flour in all three bags was $6.00 per pound, what was the price of the bag of 50 pound flour?

    (A) $200.00
    (B) $237.50
    (C) $230.00
    (D) $235.50
    (E) $280.00

9. If $a < 0$ and $b > 0$, which of the following could not be true?

    (A) $a + b = 0$

    (B) $a - b = 0$

    (C) $ab < 0$

    (D) $\dfrac{a}{b} < 0$

    (E) $b + a > 0$

10. If a rectangular solid that has a width of 8, a length of 8, and a height of 16 is divided into 16 identical cubes, what is the length of the edge of one of the cubes?

    (A) 2
    (B) 3
    (C) 4
    (D) 6
    (E) 8

GO ON TO THE NEXT PAGE

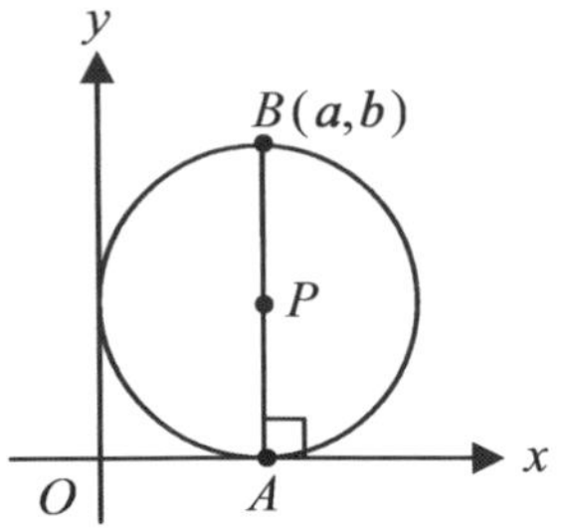

11. In the figure above, the circle has its center at P and $\overline{AB}$ is perpendicular to the x-axis. What is the area of the circle, in terms of b ?

(A) $2\pi b^2$

(B) πb^2

(C) $\dfrac{\pi b^2}{2}$

(D) $\dfrac{\pi b^2}{4}$

(E) $\dfrac{\pi b^2}{8}$

12. For positive numbers a and b, 2 is the remainder when a is divided by 5 and 4 is the remainder when b is divided by 5. If ab is divided by 5, what is the remainder?

(A) 0
(B) 1
(C) 2
(D) 3
(E) 4

2, 8, 26, 80 ...

13. In the sequence above, the first term is 2. Which of the following expressions represents the nth term of the sequence?

(A) $n \times 2^n$
(B) $2n^2$
(C) $3^n - n$
(D) $3^n - 1$
(E) $4n - 2$

14. If the average (arithmetic mean) of four consecutive even integers a, b, c, and d is 17, what is the median of five numbers a, b, c, d, and 18 ?

(A) 16
(B) 17
(C) 18
(D) 19
(E) 20

GO ON TO THE NEXT PAGE

15. Which of the following could be the lengths of the sides of a triangle?

(A) 3, 5, and 8
(B) 3, 6, and 10
(C) 4, 5, and 2
(D) 5, 9, and 15
(E) 10, 20, and 8

17. Let the operation ▲ be defined by $a▲b = a^b - b^a$ for all positive integers a and b, where $a \neq b$. If $2▲x = x▲2$, where x is a positive integer, what is the value of x?

(A) 8
(B) 6
(C) 4
(D) 2
(E) 1

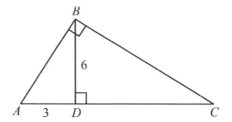

Note: Figure not drawn to scale.

16. In the figure above, $AD = 3$ and $BD = 6$. What is the area of $\triangle ABC$?

(A) 36
(B) $40\sqrt{3}$
(C) 45
(D) $72\sqrt{2}$
(E) 90

18. The areas of the top, the side, and the front of a rectangular solid are x, y, and z square inches, respectively. What is the volume of the rectangular solid, in cubic inches?

(A) $(xyz)^{\frac{1}{3}}$
(B) $(xyz)^{\frac{1}{2}}$
(C) xyz
(D) $(xyz)^2$
(E) $(xyz)^3$

GO ON TO THE NEXT PAGE

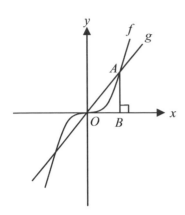

19. The figure above shows the graph of
$f(x) = ax^3$ and $g(x) = x$ for some constant a.

If the area of $\triangle OAB$ is equal to $\dfrac{1}{8}$, what is the

value of a?

(A) 4

(B) 2

(C) $\dfrac{1}{2}$

(D) $\dfrac{1}{4}$

(E) $\dfrac{1}{8}$

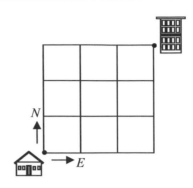

20. In the figure above, Lee is going from home to school along the paths. If he always goes upward (northerly direction) or to the right (easterly direction), how many different paths are there from his house to the school?

(A) Ten
(B) Twenty
(C) Thirty
(D) Forty
(E) Fifty

STOP

**If you finish before time is called, you may check your work on this section only.
Do not turn to any other section in the test.**

SECTION 5
Time- 25 minutes
18 Questions

Turn to Section 5 (Page 1) of your answer sheet to answer the questions in this section.

Directions: For this section, solve each problem and decide which is the best of the choices given. Fill in the corresponding circle on the answer sheet. You may use any available space for scratchwork.

Notes

1. The use of a calculator is permitted.
2. All numbers used are real numbers.
3. Figures that accompany problems in this test are intended to provide information useful in solving the problems. They are drawn as accurately as possible EXCEPT when it is stated in a specific problem that the figure is not drawn to scale. All figure lie in a plane unless other indicated.
4. Unless otherwise specified, the domain of any function f is assumed to be set of all real numbers x for which $f(x)$ is a real number.

Reference Information

$A = \pi r^2$ $A = \ell w$ $A = \frac{1}{2}bh$ $V = \ell wh$ $V = \pi r^2 h$ $c^2 = a^2 + b^2$ Special Right Triangles
$C = 2\pi r$

The numbers of degrees of arc in a circle is $360°$.

The sum of the measures in degrees of the angles is $180°$.

---

1. If $|x| < 2$, then which of the following is equivalent to $|x^2 - 4|$?

    (A) 0
    (B) $x^2 - 4$
    (C) $x^2 + 4$
    (D) $4 - x^2$
    (E) $-4 - x^2$

2. If pencils cost p cents each and notebooks cost n dollars each, which of the following expressions gives the total cost, in dollars, of 10 pencils and 5 notebooks?

    (A) $10p + 500n$
    (B) $10p + 50n$
    (C) $p + 5n$
    (D) $0.1p + 5n$
    (E) $0.01p + 5n$

GO ON TO THE NEXT PAGE

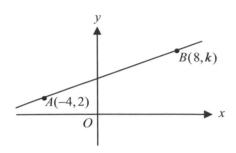

Note: Figure not drawn to scale.

3. In the xy-plane above, the distance between point A and B is equal to 13. What is the value of k, the y-coordinate of point B?

(A) 6
(B) 7
(C) 8
(D) 9
(E) 10

4. For all positive values of a and b, let $a \odot b$ be defined by $a \odot b = \dfrac{ab}{a+b}$.

If $(10 \odot 10)^2 - (8 \odot 8)^2 = k$, where k is a constant, what is the value of k?

(A) 1
(B) 2
(C) 4
(D) 6
(E) 9

5. If 120 days ago was Friday, then 86 days from today falls on what day of the week?

(A) Monday
(B) Tuesday
(C) Thursday
(D) Friday
(E) Saturday

THE NUMBER OF CARS
SOLD BY LEE EACH MONTH

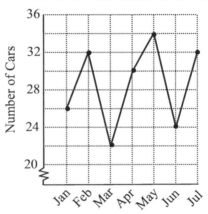

Month

6. According to the graph above, between which two months was there the greatest rate of change in the number of cars sold from the previous month?

(A) Between Jan and Feb
(B) Between Feb and Mar
(C) Between Mar and Apr
(D) Between May and Jun
(E) Between Jun and Jul

GO ON TO THE NEXT PAGE

7. In the xy-plane, the line with equation $x + 2y = 5$ is perpendicular to the line with equation $ax + by = c$, where a, b, and c are constants. Which of the following must be true?

   (A) $a = 0$
   (B) $b = 0$
   (C) $c = 0$
   (D) $a + 2b = 0$
   (E) $a - 2b = 0$

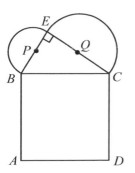

Note: Figure not drawn to scale.

8. In the figure above, the area of semicircle P is 10π and the area of semicircle Q is 20π. What is the area of the square $ABCD$?

   (A) 140
   (B) 180
   (C) 200
   (D) 220
   (E) 240

GO ON TO THE NEXT PAGE

5 □□□ 5 □□□ 5 □□□ 5

Directions: For Students-Produced Response questions 9-18, use the grid at the bottom of the answer sheet page on which you have answered questions 1-8.

Each of the remaining 10 questions requires you to solve the problem and enter your answer by making the circles in the special grid, as shown in the examples below. You may use any available space for scratchwork.

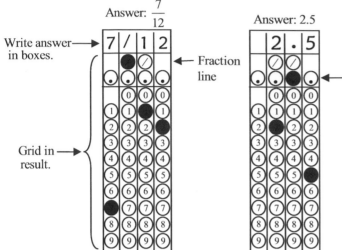

Answer: $\frac{7}{12}$

Write answer in boxes.

Fraction line

Grid in result.

Answer: 2.5

Decimal point

Answer: 201

Either position is correct.

Note: You may start your answers in any column, space permitting. Columns not needed should be left blank.

• Mark no more than one circle in any column.

• Because the answer sheet will be machine-scored, **you will receive credit only if the circles are filled in correctly.**

• Although not required, it is suggested that you write your answer in the boxes at the top of the columns to help you fill in the circles accurately.

• Some problems may have more than one correct answer. In such cases, grid only one answer.

• No question has a negative answer.

• **Mixed numbers** such as $3\frac{1}{2}$ must be gridded as 3.5 or 7/2. (If ⬚3⬚1⬚/⬚2⬚ is gridded, it will be interpreted as $\frac{31}{2}$, not $3\frac{1}{2}$.)

• **Decimal Answers:** If you obtain a decimal answer with more digits than the grid can accommodate, it may be either rounded or truncated, but it must fill the entire grid. For example, if you obtain an answer such as 0.6666…, you should record your result as .666 or .667. **A less accurate value such as .66 or .67 will be scored as incorrect.**

Acceptable ways to grid $\frac{2}{3}$ are:

9. If $ax + 10 = 25$ when $x = 5$, what is the value of $ax + 10$ when $x = 50$?

10. If x, $x + 3$, and 5 are the lengths of sides of a triangle, where x is an integer, what is the smallest value of x?

GO ON TO THE NEXT PAGE ⟹

11. If $a, b, c, d,$ and e are integers such that $a < b < 0 < c < d < e < 10$, what is the greatest possible value of $a + c$?

13. If $32\sqrt{2} = x\sqrt{y}$, where x and y are positive integers and $y > x$, what is the least possible value of $x + y$?

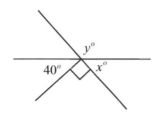

Note: Figure not drawn to scale.

12. In the figure above, what is the value of $y - x$?

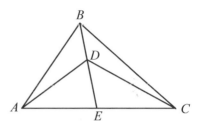

Note: Figure not drawn to scale.

14. In $\triangle ABC$ above, $\dfrac{BD}{DE} = \dfrac{1}{3}$.

What is the value of the fraction

of $\dfrac{\text{area } \triangle ABC}{\text{area } \triangle ADC}$?

GO ON TO THE NEXT PAGE

15. Let the function f be defined by $f(x) = 4 + x^2$. If $f\left(\dfrac{k}{2}\right) = 3k - 1$, what is one possible value of k?

17. If the length of a rectangle is $\dfrac{1}{8}$ of the perimeter of the rectangle, the width is what fraction of the perimeter of the rectangle?

18. How many 4-digit positive integers, between 5000 and 10,000, have only odd integers as digits?

16. If $n = 288k$, where n is a positive integer, and n is the cube of an integer, what is the least possible value of k?

STOP
If you finish before time is called, you may check your work on this section only.
Do not turn to any other section in the test.

SECTION 7
Time- 20 minutes
16 Questions

Turn to Section 7 (Page 2) of your answer sheet to answer the questions in this section.

Directions: For this section, solve each problem and decide which is the best of the choices given. Fill in the corresponding circle on the answer sheet. You may use any available space for scratchwork.

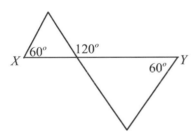

1. In the figure above, the length of $\overline{XY}$ is equal to 10. What is the perimeter of the figure?

(A) 30
(B) 40
(C) 50
(D) 60
(E) 70

2. The cost of k pieces of candy is $1.20. At this rate, what is the cost, in dollars, of p pieces of this candy?

(A) $1.2p$

(B) $1.2kp$

(C) $\dfrac{1.2k}{p}$

(D) $\dfrac{1.2p}{k}$

(E) $\dfrac{120p}{k}$

GO ON TO THE NEXT PAGE

3. If $a > b$ and $a(a-b) = 0$, which of the following must be true?

      I. $a = 0$
      II. $a > 0$
      III. $b < 0$

(A) I only
(B) II only
(C) I and II only
(D) I and III only
(E) I, II, and III

4. If $|10 - x| \le 4$, which of the following is NOT a possible value of x?

(A) 8
(B) 10
(C) 12
(D) 14
(E) 16

3, 9, 27, 81 ....

5. In the sequence above, the first term is 3 and each term after the first is 3 times the preceding term. Which of the following is an expression for the 200th term of the sequence?

(A) $3(199)$
(B) $3(200)$
(C) $3(3^{200})$
(D) 3^{200}
(E) 200^3

6. If $12x + 5 = k + (k+m)x$ for any value of x, where k and m are constants, what is the value of m?

(A) 5
(B) 6
(C) 7
(D) 12
(E) It cannot be determined from the information given.

GO ON TO THE NEXT PAGE

7. In the equation $S = 4\pi r^2 h$, if the value of S is doubled, and the value of r is doubled, then the value of h is multiplied by

(A) $\dfrac{1}{4}$

(B) $\dfrac{1}{2}$

(C) 2

(D) 4

(E) 8

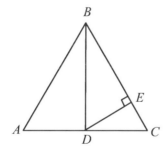

Note: Figure not drawn to scale.

8. In the figure above, if $\triangle ABC$ is an equilateral triangle and $AB = 8$, what is the area of $\triangle CDE$?

(A) $\sqrt{3}$
(B) $2\sqrt{3}$
(C) $4\sqrt{3}$
(D) 6
(E) 8

9. The probability of choosing a black marble from a jar containing only black and red marbles is $\dfrac{1}{4}$. Which of the following could be the number of red marbles?

(A) 20
(B) 22
(C) 24
(D) 32
(E) 35

10. In the figure above, a wheel made 2,000 revolutions and the length of AB, the traveling distance, is $5,000$ feet. What is the radius, in feet, of the wheel?

(A) 3π

(B) $\dfrac{7\pi}{4}$

(C) $\dfrac{3}{2\pi}$

(D) $\dfrac{5}{4\pi}$

(E) 6π

GO ON TO THE NEXT PAGE

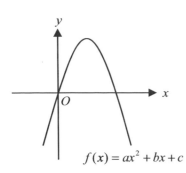

$f(x) = ax^2 + bx + c$

11. The figure above shows the graph of the function f given by $f(x) = ax^2 + bx + c$. Which of the following graphs best represents the graph of $y = ax + b$?

(A)

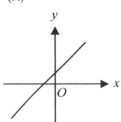

(B)

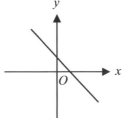

(C)

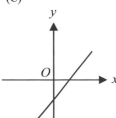

(D)

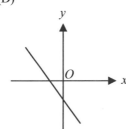

(E)

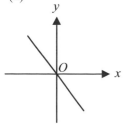

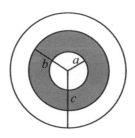

12. The target shown above is composed of 3 concentric circular regions. If the ratio of the radii is $a : b : c$, and the area of the shaded region is 20, what is the area, in terms of a, b, and c, of the largest circle?

(A) 10

(B) $\dfrac{10(c^2 - a^2)}{b^2}$

(C) $\dfrac{10c^2}{b^2 - a^2}$

(D) $\dfrac{20c^2}{b^2 - a^2}$

(E) $\dfrac{10(b^2 - a^2)}{c^2}$

13. From a group of four boys and three girls, a two-person committee will be chosen at random. What is the probability that the committee will consists of two boys or two girls?

(A) $\dfrac{3}{7}$

(B) $\dfrac{2}{5}$

(C) $\dfrac{3}{8}$

(D) $\dfrac{3}{5}$

(E) $\dfrac{1}{2}$

GO ON TO THE NEXT PAGE

14. If the average (arithmetic mean) of 6 different positive integers is 20 and the smallest of the integers is 16, what is the greatest possible value of the integers?

(A) 26
(B) 28
(C) 30
(D) 36
(E) 40

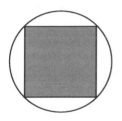

16. In the figure above, a square is inscribed in a circle. If the area of the circle is 16, what is the area of the square?

(A) 18π

(B) 16π

(C) 12π

(D) $\dfrac{32}{\pi}$

(E) $\dfrac{16}{\pi}$

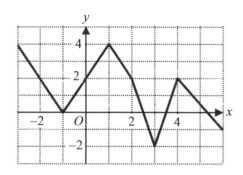

15. The graph of $y = g(x)$ is shown above for $-3 \le x \le 6$. If $g(k) = 2$, how many possible values of k are there?

(A) None
(B) One
(C) Two
(D) Three
(E) Four

STOP

If you finish before time is called, you may check your work on this section only.
Do not turn to any other section in the test.

TEST 17 — ANSWER KEY

#	SECTION 3	SECTION 5	SECTION 7
1	D	D	A
2	B	D	D
3	B	B	D
4	A	E	E
5	B	A	D
6	D	C	C
7	D	D	B
8	E	E	B
9	B	160	C
10	C	2	D
11	D	5	B
12	D	80	D
13	D	40	A
14	C	$\frac{4}{3}$ or 1.33, 1.34	C
15	C	2 or 10	E
16	C	6	D
17	C	$\frac{3}{8}$ or .375	
18	B	375	
19	A		
20	B		

TEST 17   SECTION 3

1. (D)

$x < -2$ or $x > 2$ is equivalent to $|x| > 2$.

---

2. (B)

$\dfrac{x+(x+1)+10}{3} = 7$, $x = 5$, then

$x + 1 = 5 + 1 = 6$

---

3. (B)

Choose $k = 1$

(A) $1 + 3 = 4$ even   (B) $2(1+3) + 1 = 9$ odd

(C) $1 + 1 - 2 = 0$ even

(D) $(1+1)(1-1) = 0$ even

(E) $(1+1)(1+2) = 6$ even

---

4. (A)

From $y = x + 2$,

y – intercept = 2, x - intercept = -2

The equation of line m is $y = -x + 2$,

x-intercept = 2

Therefore, $AC = 4$, $BO = 2$

The area of

$$\triangle ABC = \frac{1}{2}(AC \times BO) = \frac{1}{2}(4 \times 2) = 4$$

---

5. (B)

$29\% - 25\% = 4\%$. Let the total number of computers be x.

Then 4% of $x = 0.04x = 100$, $x = 2500$.

Number of Brand $C = 2500 \times 0.18 = 450$

---

6. (D)

$5^{2-x^2} = \dfrac{1}{25} = 5^{-2}$, $2 - x^2 = -2$, $x^2 = 4$.

Therefore $x = \pm 2$

---

7. (D)

$\cancel{x} + a + b = \cancel{x} - a - b$, $2(a+b) = 0$. That is

$a + b = 0$

Choice (C) is the answer for " could be true".

---

8. (E)

$\text{Average} = \dfrac{\text{Total amount}}{\text{Total weight}} = \dfrac{\$80 + \$150 + x}{85} = 6$

$x = 510 - (80 + 150) = \$280$

---

9. (B)

(A) If $a = -1$ and $b = 1$, $a + b = 0$: could be
true.

(B) $a < 0$, $b > 0$. $a - b < 0$ (C) and (D) are
true.

(E) If $b = 3$ and $a = -1$, then $b + a > 0$: could
be true.

---

10. (C)

$\dfrac{8 \times 8 \times 16}{n \times n \times n} = 2 \times 2 \times 4$. Therefore, $n = 4$

---

11. (D)

The radius of the circle $= \dfrac{b}{2}$.

The area $= \pi \left(\dfrac{b}{2}\right)^2 = \dfrac{\pi b^2}{4}$

---

12. (D)

Select the possible number: $a = 7$ and $b = 9$,
then $ab = 63$. When you divide 63 by 5, the
remainder is 3.

---

13. (D)

When $n = 1$, $3^n - 1 = 3 - 1 = 2$. When $n = 2$,
$3^2 - 1 = 8$.

---

14. (C)

The four numbers are 14, 16, 18, 20. Therefore,
the median of five numbers 12, 16, 18, 18, 20 is
18

---

15. (C)

Triangular inequality is $a \sim b < c < a + b$

$5 - 4 < 2 < 5 + 4$, true

---

16. (C)

$BD^2 = AD \cdot DC$, $36 = 3 \times DC$, $DC = 12$,
$AC = 3 + 12 = 15$.

The area of $\triangle ABC = \dfrac{6 \times 15}{2} = 45$.

---

17. (C)

$2 \blacktriangle x = 2^x$, $x \blacktriangle 2 = x^2$, that is $2^x = x^2$. (C) is
correct.

$2^4 = 4^2 = 16$

---

18. (B)

Let dimensions of the solid be a, b, and c.

$ab = x$, $bc = y$, and $ca = z$. That results

$(ab)(bc)(ca) = xyz$. $(abc)^2 = xyz$. Therefore,

the volume $abc = \sqrt{xyz}$.

---

19. (A)

$OA = AB = x$, $\dfrac{x \cdot x}{2} = \dfrac{1}{8}$, $x = \dfrac{1}{2}$. The

coordinates of point A are $\left(\dfrac{1}{2}, \dfrac{1}{2}\right)$. Point

A lies on the graph of f.

$\dfrac{1}{2} = a \times \left(\dfrac{1}{2}\right)^3 = \dfrac{a}{8}$. Therefore, $a = 4$.

---

20. (B)

Add the number at the intersections.

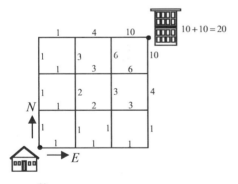

Or $\dfrac{6!}{3!3!} = 20$: arrange $aaabbb$ in order

---

1. (D)

$|x| < 2$ is equivalent to $x^2 < 4$, $x^2 - 4 < 0$.

Therefore, $|x^2 - 4| = -(x^2 - 4) = 4 - x^2$

---

2. (D)

p cents $= \dfrac{p}{100}$ or $0.01p$

Total cost $= \dfrac{p}{100} \times 10 + n(5) = 0.1p + 5n$

---

3. (B)

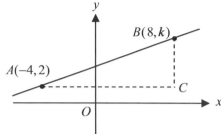

$AC = 8 - (-4) = 12$, $BC = k - 2$, $AB = 13$

$BC = \sqrt{13^2 - 12^2} = 5$, therefore,

$k - 2 = 5$, $k = 7$

Or, use the distance formula

$13 = \sqrt{(8 - (-4))^2 + (k-2)^2}$,

$169 = 144 + (k-2)^2$

$(k-2)^2 = 25$, $k - 2 = 5$, $k = 7$

---

4. (E)

$(10 \odot 10)^2 = \left(\dfrac{10 \cdot 10}{10 + 10}\right)^2 = 25$,

$(8 \odot 8)^2 = \left(\dfrac{8 \cdot 8}{8 + 8}\right)^2 = 16$

Therefore, $25 - 16 = 9 = k$

---

5. (A)

120+86= 206 days. Every 7th day falls on

Friday. $\dfrac{206}{7} = 29 \, R \, 3$. 3 more days after

Friday is Monday

---

6. (C)

(A) $\dfrac{32-26}{26} \simeq 0.23 = 23\%$ increase

(B) $\dfrac{22-32}{32} = -0.3125 = 31.25\%$ decrease

(C) $\dfrac{30-22}{22} \simeq 0.3636 = 36.36\%$ increase

(D) $\dfrac{24-34}{34} = -29.41$

(E) $\dfrac{32-24}{32} = 0.25 = 25\%$

---

7. (D)

For perpendicular lines, slope $m_1 = -\dfrac{1}{m_2}$ or

$m_1 \cdot m_2 = -1$

The slope of $x + 2y = 5$ is $-\dfrac{1}{2}$

The slope of $ax + by = c$ is $-\dfrac{a}{b}$

$-\dfrac{1}{2} = \dfrac{b}{a}$ Therefore, $a = -2b$, $a + 2b = 0$

---

8. (E)

$BC^2 = BE^2 + CE^2$,

$\pi \left(\dfrac{BE}{2}\right)^2 = 2 \cdot 10\pi$, $BE^2 = 80$

$\pi \left(\dfrac{CE}{2}\right)^2 = 2 \cdot 20\pi$, $CE^2 = 160$

Therefore, the area of the square

$BC^2 = 80 + 160 = 240$

---

9. 160

$5a + 10 = 25$, $\Rightarrow a = 3$

$ax + 10 = 3(50) + 10 = 160$

---

10. 2

Triangular inequality

$x < (x+3) + 5$, $x + 3 < x + 5$, and $5 < x + (x+3)$

The solution is $2x > 2$, $x > 1$. The smallest

integer value of x is 2.

---

11. 5

a and c must be the greatest values . That is

$b = -1$, $a = -2$ and c must be 7. $a + c = 5$

---

12. 80

$x = 50$, $y = 180 - x = 130$, $x - y = 80$.

---

13. 40

$32\sqrt{2} = 16\sqrt{8} = 8\sqrt{32} = 4\sqrt{128} = 2\sqrt{512}$

Because $y > x$, $8\sqrt{32}$ has a least value of

$x + y$.

$8 + 32 = 40$

14. $\dfrac{4}{3}$

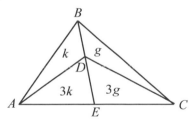

Since $BD:DE = 1:3$,

$\dfrac{\text{area of } \triangle ABD}{\text{area of } \triangle SDE} = \dfrac{1}{3}$ and

$\dfrac{\text{area of } \triangle BDC}{\text{area of } \triangle EDC} = \dfrac{1}{3}$, because of the same

height.
Let the areas be k, $3k$, g, and $3g$ as shown

above. Therefore, $\dfrac{\triangle ABC}{\triangle ADC} = \dfrac{4(k+g)}{3(k+g)} = \dfrac{4}{3}$.

---

15. 2 or 10

$f\left(\dfrac{k}{2}\right) = 4 + \left(\dfrac{k}{2}\right)^2 = 4 + \dfrac{k^2}{4} = 3k - 1$,

$16 + k^2 = 12k - 4$

$k^2 - 12k + 20 = (k-10)(k-2) = 0$. Therefore,

$x = 10$ or 2.

---

16. 6

From prime factorization, $n = 2^5 \times 3^2 \times k$.

If $k = 2 \times 3$, then

$n = 2^5 \cdot 3^2 \cdot (2 \cdot 3) = 2^6 \cdot 3^3 = (2^2 \cdot 3)^3 = 12^3$

---

17. $\dfrac{3}{8}$

$\ell = \dfrac{1}{8}(2\ell + 2w) = \dfrac{\ell + w}{4}$, $4\ell = \ell + w$, $3\ell = w$

$\dfrac{w}{2\ell + 2w} = \dfrac{3\ell}{2\ell + 6\ell} = \dfrac{3\ell}{8\ell} = \dfrac{3}{8}$

Or, you can use a number. Suppose $P = 8$,
then $\ell = 1$.

Therefore, $w = \dfrac{P}{2} - \ell = 4 - 1 = 3$.

The answer $= \dfrac{3}{8}$.

Or, use convenient numbers.

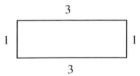

---

18. 375

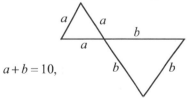

Thousands digit place have 3 possible
numbers and all the other places have 5
possible numbers each. Therefore,
$3 \times 5 \times 5 \times 5 = 375$ integers have odd integers as
digits.

---

1. (A)
The two triangles are both equilateral triangles.

$a + b = 10$,

Perimeter $= 3a + 3b = 3(a+b) = 30$

---

2. (D)
Proportion. $\dfrac{k}{1.2} = \dfrac{p}{x}$, $x = \dfrac{1.2p}{k}$

---

3. (D)
For $a > b$, $a(a-b) = 0$, the solution is $a = 0$
From the solution $a = 0$, b must be negative.
Therefore, $b < 0$ also must be true.

---

4. (E)
Substitute the numbers. (D), $|10-16| = 6 \le 4$
not true.

---

5. (D)

Geometric sequence $a_1 = 3$, $r = 3$

$a_{300} = 3 \times 3^{200-1} = 3 \times 3^{199} = 3^{200}$

---

6. (C)

These two equations are identical. Their coefficients must be equal.

$k = 5$, and $k + m = 12$. Therefore, $m = 7$.

---

7. (B)

$h = \dfrac{S}{4\pi r^2}$, $h' = \dfrac{2S}{4\pi(2r)^2} = \dfrac{2S}{4(4\pi r^2)} = \dfrac{1}{2}h$

Or, you can use numbers. When $S = 1$ and $r = 1$, $h = \dfrac{1}{4\pi}$. When $S = 2$ and $r = 2$,

$h' = \dfrac{2}{16\pi} = \dfrac{1}{8\pi}$

$\dfrac{h'}{h} = \dfrac{\frac{1}{8\pi}}{\frac{1}{4\pi}} = \dfrac{1}{2}$

---

8. (B)

$AD = 4$, $BD = 4\sqrt{3}$, and $DE = 2\sqrt{3}$

$CD = 4$, then $CE = 2$. Therefore, the area of

$\triangle CDE = \dfrac{CE \cdot DE}{2} = \dfrac{2\sqrt{3} \cdot 2}{2} = 2\sqrt{3}$

---

9. (C)

Let k = number of Black Marbles and $3k$ = Number of Red Marbles.

$\dfrac{\text{number of black marbles}}{\text{total number of marbles}} = \dfrac{k}{k+3k} = \dfrac{1}{4}$

The number of red marbles is a multiple of 3 because k is an integer.

---

10. (D)

Circumference is $2\pi r$. $2000 \times 2\pi r = 5000$

Therefore, $r = \dfrac{5000}{4000\pi} = \dfrac{5}{4\pi}$

---

11. (B)

Axis of symmetry $= \dfrac{-b}{2a} > 0$ and $a < 0$,

because the graph concaves down.
That results $b > 0$.

For $y = ax + b$, slope a is negative and y-intercept b is positive. Graph (B) is correct.

---

12. (D)

The ratio of the lengths of three circles = $a : b : c$

The ratio of the areas = $a^2 : b^2 : c^2 = a^2k : b^2k : c^2k$

The area of shaded

region $= \left(b^2 - a^2\right)k = 20$, then $k = \dfrac{20}{b^2 - a^2}$.

Therefore, the area of the largest one =

$c^2 k = \dfrac{20c^2}{b^2 - a^2}$.

---

13. (A)

$\boxed{1B}\ \boxed{2B}\ \boxed{3B}\ \boxed{4B}\ \boxed{5G}\ \boxed{6G}\ \boxed{7G}$

All possible groups are as follows.
$(1B, 2B)(1B, 3B)(1B, 4B)(1B, 5G)$
$(1B, 6G)(1B, 7G)$

$(2B, 3B)(2B, 4B)(2B, 5G)(2B, 6G)(2B, 7G)$
....
$(6G, 7G)$

Therefore, total number of all possible combination is
6+5+4+3+2+1= 21.

Desirable outcomes are as follows.
$(1B, 2B)(1B, 3B)(1B, 4B)$
$(2B, 3B)(2B, 4B)$
$(3B, 4B)$
3+2+1= 6
and
$(5G, 6G)(5G, 7G)$
$(6G, 7G)$
2+1= 3

Therefore, total number of all desirable outcomes is 6+3 = 9.

The probability is $\dfrac{9}{21} = \dfrac{3}{7}$.

Or

All possible combinations $= {_7}C_2 = \dfrac{7 \times 6}{2!} = 21$

The number of selection of two boys

$= {_4}C_2 = \dfrac{4 \times 3}{2!} = 6$

The number of selection of two girls

$= {_3}C_2 = \dfrac{3 \times 2}{2!} = 3$

Therefore,

$P(\text{two boy or two girls}) = \dfrac{3+6}{21} = \dfrac{3}{7}$

Or,

$$P = \dfrac{{_4}C_2 + {_3}C_2}{{_6}C_2} = \dfrac{6+3}{21} = \dfrac{9}{21} = \dfrac{3}{7}.$$

---

14. (C)

Sum of six numbers $= 6 \times 20 = 120$. The smallest is 15, and the five numbers will be 16, 17, 18, 19, 20. The sum of these five numbers is 90. The greatest one is $120 - 90 = 30$.

---

15. (E)

In the graph $g(k) = 2$.

When $k = -2, 0, 2,$ and 4,

$g(k) = 2$

---

16. (D)

$\pi r^2 = 16$

The radius of the circle $= \dfrac{4}{\sqrt{\pi}}$ and $d = \dfrac{8}{\sqrt{\pi}}$.

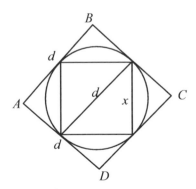

The area of the square $= \dfrac{1}{2}$ of square

$$ABCD = \dfrac{d \times d}{2} = \dfrac{d^2}{2} = \dfrac{\left(\dfrac{8}{\sqrt{\pi}} \right)^2}{2} = \dfrac{32}{\pi}$$

Or, if x is the length of a side of the square,

$x = \dfrac{d}{\sqrt{2}} = \dfrac{8}{\sqrt{2\pi}}$. Therefore, the area of the

square $= \dfrac{8}{\sqrt{2\pi}} \times \dfrac{8}{\sqrt{2\pi}} = \dfrac{64}{2\pi} = \dfrac{32}{\pi}$

$\boxed{END}$

NO MATERIAL ON THIS PAGE

Dr. John Chung's SAT Math

TEST
18

ANSWER SHEET   TEST #:

SECTION 3

1 ⒶⒷⒸⒹⒺ	11 ⒶⒷⒸⒹⒺ	21 ⒶⒷⒸⒹⒺ	31 ⒶⒷⒸⒹⒺ
2 ⒶⒷⒸⒹⒺ	12 ⒶⒷⒸⒹⒺ	22 ⒶⒷⒸⒹⒺ	32 ⒶⒷⒸⒹⒺ
3 ⒶⒷⒸⒹⒺ	13 ⒶⒷⒸⒹⒺ	23 ⒶⒷⒸⒹⒺ	33 ⒶⒷⒸⒹⒺ
4 ⒶⒷⒸⒹⒺ	14 ⒶⒷⒸⒹⒺ	24 ⒶⒷⒸⒹⒺ	34 ⒶⒷⒸⒹⒺ
5 ⒶⒷⒸⒹⒺ	15 ⒶⒷⒸⒹⒺ	25 ⒶⒷⒸⒹⒺ	35 ⒶⒷⒸⒹⒺ
6 ⒶⒷⒸⒹⒺ	16 ⒶⒷⒸⒹⒺ	26 ⒶⒷⒸⒹⒺ	36 ⒶⒷⒸⒹⒺ
7 ⒶⒷⒸⒹⒺ	17 ⒶⒷⒸⒹⒺ	27 ⒶⒷⒸⒹⒺ	37 ⒶⒷⒸⒹⒺ
8 ⒶⒷⒸⒹⒺ	18 ⒶⒷⒸⒹⒺ	28 ⒶⒷⒸⒹⒺ	38 ⒶⒷⒸⒹⒺ
9 ⒶⒷⒸⒹⒺ	19 ⒶⒷⒸⒹⒺ	29 ⒶⒷⒸⒹⒺ	39 ⒶⒷⒸⒹⒺ
10 ⒶⒷⒸⒹⒺ	20 ⒶⒷⒸⒹⒺ	30 ⒶⒷⒸⒹⒺ	40 ⒶⒷⒸⒹⒺ

SECTION 5

1 ⒶⒷⒸⒹⒺ	11 ⒶⒷⒸⒹⒺ	21 ⒶⒷⒸⒹⒺ	31 ⒶⒷⒸⒹⒺ
2 ⒶⒷⒸⒹⒺ	12 ⒶⒷⒸⒹⒺ	22 ⒶⒷⒸⒹⒺ	32 ⒶⒷⒸⒹⒺ
3 ⒶⒷⒸⒹⒺ	13 ⒶⒷⒸⒹⒺ	23 ⒶⒷⒸⒹⒺ	33 ⒶⒷⒸⒹⒺ
4 ⒶⒷⒸⒹⒺ	14 ⒶⒷⒸⒹⒺ	24 ⒶⒷⒸⒹⒺ	34 ⒶⒷⒸⒹⒺ
5 ⒶⒷⒸⒹⒺ	15 ⒶⒷⒸⒹⒺ	25 ⒶⒷⒸⒹⒺ	35 ⒶⒷⒸⒹⒺ
6 ⒶⒷⒸⒹⒺ	16 ⒶⒷⒸⒹⒺ	26 ⒶⒷⒸⒹⒺ	36 ⒶⒷⒸⒹⒺ
7 ⒶⒷⒸⒹⒺ	17 ⒶⒷⒸⒹⒺ	27 ⒶⒷⒸⒹⒺ	37 ⒶⒷⒸⒹⒺ
8 ⒶⒷⒸⒹⒺ	18 ⒶⒷⒸⒹⒺ	28 ⒶⒷⒸⒹⒺ	38 ⒶⒷⒸⒹⒺ
9 ⒶⒷⒸⒹⒺ	19 ⒶⒷⒸⒹⒺ	29 ⒶⒷⒸⒹⒺ	39 ⒶⒷⒸⒹⒺ
10 ⒶⒷⒸⒹⒺ	20 ⒶⒷⒸⒹⒺ	30 ⒶⒷⒸⒹⒺ	40 ⒶⒷⒸⒹⒺ

Grid-in answer boxes numbered 9, 10, 11, 12, 13, 14, 15, 16, 17, 18.

SECTION	1	Ⓐ Ⓑ Ⓒ Ⓓ Ⓔ	11	Ⓐ Ⓑ Ⓒ Ⓓ Ⓔ	21	Ⓐ Ⓑ Ⓒ Ⓓ Ⓔ	31	Ⓐ Ⓑ Ⓒ Ⓓ Ⓔ		
	2	Ⓐ Ⓑ Ⓒ Ⓓ Ⓔ	12	Ⓐ Ⓑ Ⓒ Ⓓ Ⓔ	22	Ⓐ Ⓑ Ⓒ Ⓓ Ⓔ	32	Ⓐ Ⓑ Ⓒ Ⓓ Ⓔ		
	3	Ⓐ Ⓑ Ⓒ Ⓓ Ⓔ	13	Ⓐ Ⓑ Ⓒ Ⓓ Ⓔ	23	Ⓐ Ⓑ Ⓒ Ⓓ Ⓔ	33	Ⓐ Ⓑ Ⓒ Ⓓ Ⓔ		
	4	Ⓐ Ⓑ Ⓒ Ⓓ Ⓔ	14	Ⓐ Ⓑ Ⓒ Ⓓ Ⓔ	24	Ⓐ Ⓑ Ⓒ Ⓓ Ⓔ	34	Ⓐ Ⓑ Ⓒ Ⓓ Ⓔ		
7	5	Ⓐ Ⓑ Ⓒ Ⓓ Ⓔ	15	Ⓐ Ⓑ Ⓒ Ⓓ Ⓔ	25	Ⓐ Ⓑ Ⓒ Ⓓ Ⓔ	35	Ⓐ Ⓑ Ⓒ Ⓓ Ⓔ		
	6	Ⓐ Ⓑ Ⓒ Ⓓ Ⓔ	16	Ⓐ Ⓑ Ⓒ Ⓓ Ⓔ	26	Ⓐ Ⓑ Ⓒ Ⓓ Ⓔ	36	Ⓐ Ⓑ Ⓒ Ⓓ Ⓔ		
	7	Ⓐ Ⓑ Ⓒ Ⓓ Ⓔ	17	Ⓐ Ⓑ Ⓒ Ⓓ Ⓔ	27	Ⓐ Ⓑ Ⓒ Ⓓ Ⓔ	37	Ⓐ Ⓑ Ⓒ Ⓓ Ⓔ		
	8	Ⓐ Ⓑ Ⓒ Ⓓ Ⓔ	18	Ⓐ Ⓑ Ⓒ Ⓓ Ⓔ	28	Ⓐ Ⓑ Ⓒ Ⓓ Ⓔ	38	Ⓐ Ⓑ Ⓒ Ⓓ Ⓔ		
	9	Ⓐ Ⓑ Ⓒ Ⓓ Ⓔ	19	Ⓐ Ⓑ Ⓒ Ⓓ Ⓔ	29	Ⓐ Ⓑ Ⓒ Ⓓ Ⓔ	39	Ⓐ Ⓑ Ⓒ Ⓓ Ⓔ		
	10	Ⓐ Ⓑ Ⓒ Ⓓ Ⓔ	20	Ⓐ Ⓑ Ⓒ Ⓓ Ⓔ	30	Ⓐ Ⓑ Ⓒ Ⓓ Ⓔ	40	Ⓐ Ⓑ Ⓒ Ⓓ Ⓔ		

Math Scoring Worksheet

A. Section 3

_____   _____
numer of correct   *number of incorrect*

+   +

B. Section 5 (1-8)

_____   _____
numer of correct   *number of incorrect*

+

C. Section 5 (9-18)

_____
numer of correct

+   +

D. Section 7

_____   _____
numer of correct   *number of incorrect*

=   =

E. Total Unrounded Raw Score

_____ − _____ ÷4 = _____
numer of correct   *number of incorrect*

F. Total Rounded Raw Score

_____ (See table)

Math Score Range = [_____ — _____]

Math Conversion Table

Raw Score	Scaled Score	Raw Score	Scaled Score
54	800	23	490-550
53	780-800	22	480-540
52	760-800	21	470-530
51	740-800	20	460-520
50	720-780	19	450-510
49	700-760	18	450-510
48	690-750	17	440-500
47	680-740	16	430-490
46	670-730	15	420-480
45	660-720	14	420-480
44	650-710	13	410-470
43	650-710	12	400-460
42	640-700	11	390-450
41	630-690	10	380-440
40	620-680	9	390-430
39	610-670	8	380-420
38	610-670	7	370-410
37	600-660	6	360-400
36	590-650	5	340-380
35	580-640	4	320-370
34	570-630	3	310-360
33	560-620	2	300-350
32	560-620	1	270-320
31	550-610	0	240-300
30	540-600	-1	200-290
29	530-590	-2	200-270
28	530-590	-3	200-260
27	520-580	-4	200-240
26	510-570	-5	200-220
25	500-560	-6 and below	200
24	500-560		

SECTION 3
Time- 25 minutes
20 Questions

Turn to Section 3 (Page 1) of your answer sheet to answer the questions in this section.

Directions: For this section, solve each problem and decide which is the best of the choices given. Fill in the corresponding circle on the answer sheet. You may use any available space for scratchwork.

Notes

1. The use of a calculator is permitted.
2. All numbers used are real numbers.
3. Figures that accompany problems in this test are intended to provide information useful in solving the problems. They are drawn as accurately as possible EXCEPT when it is stated in a specific problem that the figure is not drawn to scale. All figure lie in a plane unless other indicated.
4. Unless otherwise specified, the domain of any function f is assumed to be set of all real numbers x for which $f(x)$ is a real number.

Reference Informatiom

$A = \pi r^2$
$C = 2\pi r$
$A = \ell w$
$A = \frac{1}{2}bh$
$V = \ell w h$
$V = \pi r^2 h$
$c^2 = a^2 + b^2$
Special Right Triangles

The numbers of degrees of arc in a circle is $360°$.

The sum of the measures in degrees of the angles is $180°$.

---

1. If $x = 5 + y$ and $x = 5 - y$, what is the value of y?

(A) 0
(B) $\frac{1}{2}$
(C) 1
(D) $\frac{3}{2}$
(E) 5

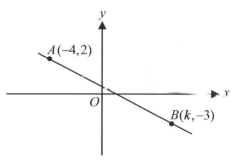

Note: Figure not drawn to scale.

2. In the *xy*-plane above, the distance between points A and B is 13. What is the value of k?

(A) 3
(B) 5
(C) 8
(D) 9
(E) 10

GO ON TO THE NEXT PAGE

3. On a number line, p represents a number within 8 units of 12. Which of the following describes the relationship?

(A) $|p-8| < 12$

(B) $|p-8| > 12$

(C) $|p-12| > 8$

(D) $|p+12| < 8$

(E) $|p-12| < 8$

4. If $x + 2y$ is 5 more than $y + 2x$, then $x - y =$

(A) -5

(B) $-\dfrac{5}{2}$

(C) $\dfrac{5}{2}$

(D) 5

(E) 7

5. If x, y, and z are positive and $x^4 y^3 z^3 > x^5 y^3 z^2$. Which of the following must be true?

    I. $x > y$

    II. $y > z$

    III. $z > x$

(A) I only

(B) II only

(C) III only

(D) I and III only

(E) II and III only

6. If $a + 2b = 20$ and $a \geq 6$, which of the following must be true?

(A) $b > 0$

(B) $b < 0$

(C) $b = 5$

(D) $b \leq 7$

(E) $b \geq 7$

GO ON TO THE NEXT PAGE

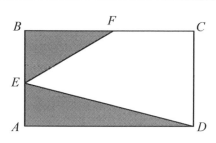

Note: Figure not drawn to scale.

7. In the rectangle $ABCD$ above, points E and F are the midpoints of $\overline{AB}$ and $\overline{BC}$ respectively. If the sum of the areas of the shaded regions is 1, what is the area of the unshaded region?

(A) 2

(B) $\dfrac{5}{3}$

(C) $\dfrac{8}{3}$

(D) $\dfrac{12}{5}$

(E) 3

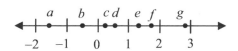

8. On the number line above, a, b, c, d, e, f, and g are coordinates of the indicated points. Which of the following is closest value to $|b - df|$?

(A) c
(B) d
(C) e
(D) f
(E) g

9. If $x^2 = 9$, what is the value of $\dfrac{4}{x+1} - \dfrac{4}{x-1}$?

(A) 3
(B) 2
(C) 1
(D) -1
(E) -2

$$2x + y = 6$$
$$mx + y = k$$

10. The graphs of the two functions above, in the xy-plane, cannot intersect at any point. Which of the following must be true?

(A) $m = 2$ and $k = 6$
(B) $m = 4$ and $k = 12$
(C) $m = 2$ and $k \neq 6$
(D) $m = 4$ and $k > 0$
(E) $m = 2$ and $k < 0$

GO ON TO THE NEXT PAGE

x	y
a	b
5	10
$a+b$	24

11. In the table above, y is directly proportional to x. Which of the following is the value of a?

(A) 3
(B) 4
(C) 5
(D) 10
(E) 12

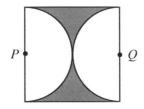

12. In the square above, two semicircles have centers at P and Q. If the area of each semicircle is 2π, what is the area of the shaded region?

(A) $4-2\pi$

(B) $8-4\pi$

(C) $16-4\pi$

(D) $\dfrac{\pi}{2}$

(E) $\dfrac{\pi}{3}$

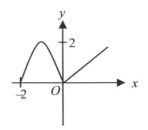

13. The graph of $y = f(x)$ is shown above. Which of the following could be the graph of

$$y = \frac{1}{2}f(x-1)?$$

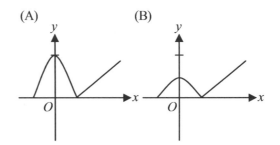

(A)   (B)

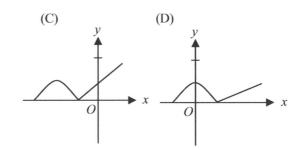

(C)   (D)

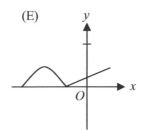

(E)

GO ON TO THE NEXT PAGE

14. In the figure above, the outer circle has diameter d . If the area of the shaded ring is equal to the area of the inner, unshaded circle, what is the radius of the inner circle, in terms of d ?

(A) $\dfrac{d}{2\sqrt{2}}$

(B) $\dfrac{d}{2}$

(C) $\dfrac{d}{2\sqrt{3}}$

(D) $\dfrac{2d}{5}$

(E) $\dfrac{2d}{3\sqrt{2}}$

15. If 6 is subtracted from the square of a positive number, the result is 24 greater than the number. What is the value of the number?

(A) 3
(B) 6
(C) 12
(D) 18
(E) 25

16. The number 0.1234 is between $\dfrac{n}{1000}$ and $\dfrac{n+1}{1000}$ for some positive integer n. Which of the following is the value of n ?

(A) 120
(B) 121
(C) 123
(D) 124
(E) 130

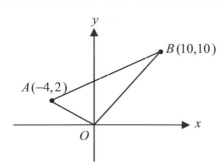

Note: Figure not drawn to scale.

17. In the figure above, what is the area of $\triangle OAB$?

(A) 30
(B) 38
(C) 40
(D) 48
(E) 70

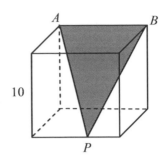

18. In the cube above, the length of an edge is 10. What is the area of $\triangle ABP$?

(A) 50

(B) $50\sqrt{2}$

(C) $50\sqrt{3}$

(D) $\dfrac{100}{\sqrt{3}}$

(E) It cannot be determined from the information given.

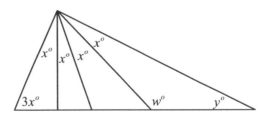

Note: Figure not drawn to scale.

19. In the figure above, $w = 120$. What is the value of y ?

(A) 10
(B) 12
(C) 20
(D) 30
(E) 40

20. Let the function $g(x)$ be defined by $g(x) = 2f(x) + k$, where $f(x)$ is a linear function. If $g(2) = 10$ and $g(5) = 18$, what is the slope of the function f ?

(A) $\dfrac{2}{3}$

(B) $\dfrac{4}{3}$

(C) 2

(D) 3

(E) $\dfrac{9}{2}$

STOP

If you finish before time is called, you may check your work on this section only.
Do not turn to any other section in the test.

SECTION 5
Time- 25 minutes
18 Questions

Turn to Section 5 (Page 1) of your answer sheet to answer the questions in this section.

Directions: For this section, solve each problem and decide which is the best of the choices given. Fill in the corresponding circle on the answer sheet. You may use any available space for scratchwork.

Reference Informatiom

$A = \pi r^2$
$C = 2\pi r$ $A = \ell w$ $A = \dfrac{1}{2}bh$ $V = \ell wh$ $V = \pi r^2 h$ $c^2 = a^2 + b^2$ Special Right Triangles

The numbers of degrees of arc in a circle is $360°$.

The sum of the measures in degrees of the angles is $180°$.

---

1. If x is 3 more than y, and y is 5 less than z, then $x + y + z =$

(A) $3z - 7$
(B) $3z - 5$
(C) $10 - 3z$
(D) $12 - 3z$
(E) $5z$

2. If $\dfrac{x}{y} = 2$, what is the value of $5 \cdot \dfrac{x^2}{y} \cdot \dfrac{3}{x}$?

(A) 15
(B) 20
(C) 24
(D) 28
(E) 30

GO ON TO THE NEXT PAGE ⟩

Dr. John Chung's SAT Math Test 18

3. If $8 \times a^{2x} = a^{2x+3}$, what is the value of a ?

(A) 1
(B) 2
(C) 3
(D) 4
(E) 5

5. What is the sum of 15 consecutive integers if the middle integer is 60?

(A) 45
(B) 90
(C) 450
(D) 500
(E) 900

$$4, \ 11, \ 18, \ 25, \ \ldots$$

4. The first four terms in a sequence are shown above. Each term after the first is found by adding 7 to the preceding term. Which term in the sequence is equal to $4 + 40 \times 7$?

(A) The 18th
(B) The 36th
(C) The 39th
(D) The 40th
(E) The 41st

6. If $x^2 - y^2 = 30$ and $x + y > 10$, where x and y are positive numbers, then which of the following must be true?

(A) $x = y$
(B) $x < y$
(C) $x - y < 30$
(D) $x - y > 3$
(E) $x - y < 3$

GO ON TO THE NEXT PAGE

7. The sum of two consecutive even integers is t. In term of t, what is the value of the greater of the two numbers?

(A) $t+2$

(B) $t+1$

(C) $\dfrac{t}{2}$

(D) $\dfrac{t+1}{2}$

(E) $\dfrac{t+2}{2}$

8. The various averages (arithmetic mean) of two of the three numbers a, b, and c are calculated, and are arranged as follows.

$P =$ The average of a and b

$Q =$ The average of b and c

$R =$ The average of c and a

If $P > Q > R$, then which of the following is true?

(A) $a = b = c$
(B) $a > b > c$
(C) $b > c > a$
(D) $c > b > a$
(E) $b > a > c$

GO ON TO THE NEXT PAGE

5 ☐☐☐ 5 ☐☐☐ 5 ☐☐ 5

Directions: For Students-Produced Response questions 9-18, use the grid at the bottom of the answer sheet page on which you have answered questions 1-8.

Each of the remaining 10 questions requires you to solve the problem and enter your answer by making the circles in the special grid, as shown in the examples below. You may use any available space for scratchwork.

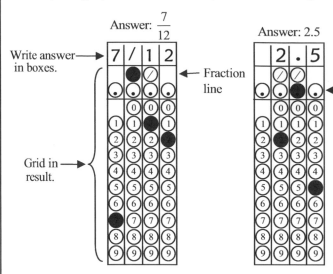

Answer: $\frac{7}{12}$

Write answer in boxes.

← Fraction line

Grid in result.

Answer: 2.5

← Decimal point

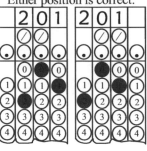

Answer: 201
Either position is correct.

Note: You may start your answers in any column, space permitting. Columns not needed should be left blank.

- Mark no more than one circle in any column.

- Because the answer sheet will be machine-scored, **you will receive credit only if the circles are filled in correctly.**

- Although not required, it is suggested that you write your answer in the boxes at the top of the columns to help you fill in the circles accurately.

- Some problems may have more than one correct answer. In such cases, grid only one answer.

- No question has a negative answer.

- **Mixed numbers** such as $3\frac{1}{2}$ must be gridded as 3.5 or 7/2. (If ☐ 3 1 / 2 ☐ is gridded, it will be interpreted as $\frac{31}{2}$, not $3\frac{1}{2}$.)

- **Decimal Answers:** If you obtain a decimal answer with more digits than the grid can accommodate, it may be either rounded or truncated, but it must fill the entire grid. For example, if you obtain an answer such as 0.6666…, you should record your result as .666 or .667. **A less accurate value such as .66 or .67 will be scored as incorrect.**

Acceptable ways to grid $\frac{2}{3}$ are:

9. The total cost to send a package with World Express is $2.00 for the first 10 ounces and $0.30 for each additional ounce. What is the total cost, in dollars, to send a package that weighs 50 ounces? (Disregard the $ sign when gridding your answer.)

10. The value of the expression $ax + 20 = 50$ when $x = 3$. What is the value of $2ax + 40$ when $x = 6$?

GO ON TO THE NEXT PAGE ⇒

11. The length of the longest side of a triangle is 10, and the remaining two sides have integer lengths. If the lengths of all three sides are different integers, what is the smallest value of the perimeter of the triangle?

13. Let $x \triangle y$ be defined by $x \triangle y = x + 2y$ for all numbers of x and y. If $(2a) \triangle b = 6$ and $a \triangle (2b) = 15$, what is the value of b?

COLOR PREFERENCES

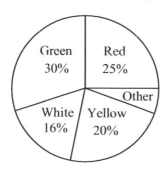

12. In the results of a survey shown above, 225 people preferred the colors yellow or red. How many people were surveyed?

14. In a biology class, microscopes, thermometers, and beakers will be assigned to the students. Each microscope is assigned to every 3 students, each thermometer is assigned to every 5 students, and each beaker is assigned to every 10 students. If the sum of the numbers of microscopes, thermometers, and beakers is 57, how many students are there in the biology class?

GO ON TO THE NEXT PAGE

15. How many integers between 10 and 100 have exactly one digit equal to 9?

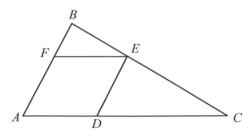

Note: Figure not drawn to scale.

17. In the figure above, $\overline{FE} \parallel \overline{AD}$, $\overline{AF} \parallel \overline{DE}$, and $\dfrac{BE}{EC} = \dfrac{1}{2}$. The area of $\square AFED$ is what fraction of the area of $\triangle ABC$?

x, 8 , $x+3$, , $2x-3$, x^2-36

16. Five different positive integers above are listed in order from least to greatest. What is the value of x?

18. How many integers between 100 and 300 are a multiple of 3 or 4?

STOP

If you finish before time is called, you may check your work on this section only.
Do not turn to any other section in the test.

SECTION 7
Time- 20 minutes
16 Questions

Turn to Section 7 (Page 2) of your answer sheet to answer the questions in this section.

Directions: For this section, solve each problem and decide which is the best of the choices given. Fill in the corresponding circle on the answer sheet. You may use any available space for scratchwork.

1. If $\dfrac{x}{5} = \dfrac{5}{x}$, what are all possible values of x?

   (A) 1

   (B) $\sqrt{5}$ and $-\sqrt{5}$

   (C) 5 and -5

   (D) $\dfrac{1}{5}$ and $-\dfrac{1}{5}$

   (E) 5 and 10

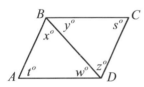

Note: Figure not drawn to scale.

2. If $AB = BC = CD = DA$, in the figure above, which of the following CANNOT be concluded?

   (A) $x = y$
   (B) $x = w$
   (C) $y = w$
   (D) $t = s$
   (E) $t = x$

GO ON TO THE NEXT PAGE

3. If 30 percent of m is 50, what is 3 percent of m?

(A) 5
(B) 10
(C) 15
(D) 20
(E) 25

4. If $p = 1 + \dfrac{1}{3} + \dfrac{1}{9} + \dfrac{1}{27} + \dfrac{1}{81} + \dfrac{1}{243}$ and $q = 1 + \dfrac{1}{3}p$, then what is the value of $q - p$?

(A) $\dfrac{1}{729}$

(B) $\dfrac{1}{243}$

(C) $\dfrac{1}{81}$

(D) $\dfrac{1}{27}$

(E) $\dfrac{1}{3}$

5. If $a - b$ is any negative number, which of the following must be positive?

(A) a
(B) b
(C) $a + b$
(D) $b - a$
(E) $a + b + 2$

MATH TEST RESULTS

Test Score	Frequency
100	3
95	7
90	10
80	5
70	17
65	3

6. The table above shows the results for 45 students on a math test. If the teacher allows students to retake the test, what is the least number of students that would have to retake the test in order to POSSIBLY establish a new and unique class median?

(A) 1
(B) 2
(C) 3
(D) 4
(E) It cannot be determined from the information given.

GO ON TO THE NEXT PAGE ⟩

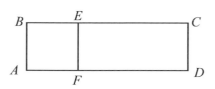

Note: Figure not drawn to scale.

7. In the figure above, the area of square $ABEF$ is $\frac{1}{3}$ of the area of rectangle $ECDF$. If the area of rectangle $ABCD$ is 256, what is the perimeter of rectangle $ECDF$?

    (A) 32
    (B) 64
    (C) 72
    (D) 80
    (E) 84

8. If $\frac{a}{b}$ is an even number, where a and b are integers, which of the following must also be an even number?

    (A) $\frac{b}{a}$

    (B) $a+b$

    (C) $a-b$

    (D) a

    (E) b

9. There are n students in a chemistry class, and only s of them are seniors. If j juniors are added to the class, what percent of the students in the class will not be seniors?

    (A) $\frac{s}{n}\times100\%$

    (B) $\frac{n-s}{n}\times100\%$

    (C) $\frac{n}{n+s}\times100\%$

    (D) $\frac{n-s}{n+j}\times100\%$

    (E) $\frac{n-s+j}{n+j}\times100\%$

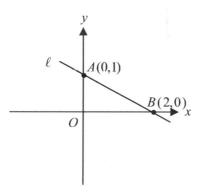

Note: Figure not drawn to scale.

10. In the xy-plane above, line m (not shown) is perpendicular to line ℓ and intersects between points A and B on line ℓ. Which of the following could be the equation of line m?

    (A) $y=x+2$
    (B) $y=-x-1$
    (C) $y=2x+2$
    (D) $y=2x-3$
    (E) $y=2x-5$

GO ON TO THE NEXT PAGE

11. If $x = \dfrac{w^2}{y}$ and $y = \dfrac{1}{w^2}$, what is the value of

$\dfrac{x}{y}$ in terms of w?

(A) w^{-4}
(B) w^{-2}
(C) w^2
(D) w^4
(E) w^6

12. If $x^2 + y^2 = (x-y)^2$, which of the following statements must also be true?

    I. $x = 0$
    II. $y = 0$
    III. $xy = 0$

(A) I only
(B) II only
(C) III only
(D) I and II only
(E) I, II, and III

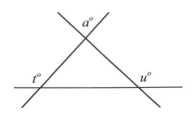

Note: Figure not drawn to scale.

13. In the figure above, what is the value of $t + u$ in terms of a?

(A) $90 + 2a$
(B) $360 - a$
(C) $180 - a$
(D) $180 + a$
(E) $180 + 2a$

14. If a number is chosen at random from the set $\{-10, -8, -7, -2, 0, 3, 5, 7, ,8, 10\}$, what is the probability that it is a member of the solution set of $|x - 2| > 6$?

(A) $\dfrac{2}{5}$

(B) $\dfrac{3}{5}$

(C) $\dfrac{1}{2}$

(D) $\dfrac{2}{3}$

(E) $\dfrac{3}{4}$

GO ON TO THE NEXT PAGE

7 ⬜ 7 ⬜ 7 ⬜ 7

$-5, -4, -3, -2, -1, 0, 1, 2, 3, 4, 5$

15. How many different sums can be obtained by adding any two different integers chosen from the integers above?

(A) 16
(B) 17
(C) 18
(D) 19
(E) 20

$4 \le x \le 10$
$2 \le y \le 5$

16. Let $A = xy$ and $B = \dfrac{x}{y}$ from the values of x and y above. What is the smallest value of $A - B$?

(A) 3
(B) 6
(C) 10
(D) 12
(E) 15

STOP

If you finish before time is called, you may check your work on this section only.
Do not turn to any other section in the test.

2Dr. John Chung's SAT Math Test 18

- 571 -

TEST 18   ANSWER KEY

#	SECTION 3	SECTION 5	SECTION 7
1	A	A	C
2	C	E	E
3	E	B	A
4	A	E	A
5	C	E	D
6	D	E	C
7	B	E	B
8	C	E	D
9	D	14	E
10	C	160	D
11	B	21	E
12	C	500	C
13	D	4	D
14	A	90	A
15	B	17	D
16	C	7	B
17	A	$\frac{4}{9}$	
18	B	99	
19	E		
20	B		

TEST 18   SECTION 3

1. (A)

$5+y = 5-y$, $\quad 2y = 0$. Therefore, $y = 0$

2. (C)

$(k-{}^-4)^2 + (-3-2)^2 = 13^2$,

$(k+4)^2 = 169-25 = 144$, then $k+4 = 12$

Therefore, $k = 8$

Or, you can solve it geometrically.

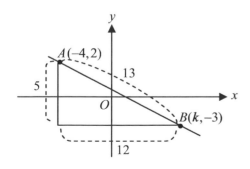

$k-(-4) = 12$, $\quad k = 8$

3. (E)

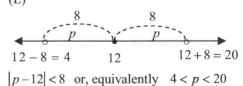

$12-8 = 4 \qquad 12 \qquad 12+8 = 20$

$|p-12| < 8$ or, equivalently $\quad 4 < p < 20$

4. (A)

$x+2y = y+2x+5$, then $x-y = -5$.

5. (C)

$\dfrac{x^4 y^3 z^3}{x^4 y^3 z^2} > \dfrac{x^5 y^3 z^2}{x^4 y^3 z^2}$ is $z > x$

6. (D)

$a = 20-2b \ge 6$, that is $-2b \ge -14$.

Therefore, $b \le 7$.

7. (B)

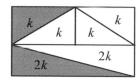

In the figure, the area of the shaded regions

$2k + k = 3k = 1$. $k = \dfrac{1}{3}$. Therefore, the area

of the unshaded regions is $5k = 5 \times \dfrac{1}{3} = \dfrac{5}{3}$.

---

8. (C)

$b \simeq -0.5$, $d \simeq 0.5$, and $f \simeq 1.8$. Therefore,

$|b - df| = |-0.5 - (0.5 \times 1.8)| = 1.4$.

---

9. (D)

$\dfrac{4}{x+1} - \dfrac{4}{x-1} = \dfrac{4(x-1) - 4(x+1)}{(x+1)(x-1)}$

$= \dfrac{-8}{x^2 - 1} = \dfrac{-8}{9 - 1} = -1$

---

10. (C)

The graphs of two lines must be parallel with different y-intercepts.

$m = 2$ but $k \neq 6$

---

11. (B)

$\dfrac{b}{a} = \dfrac{10}{5} = \dfrac{24}{a+b} = 2$, $b - 2a$ and $a + b = 12$.

$a + 2a = 3a = 12$. Therefore, $a = 4$

---

12. (C)

$\pi r^2 = 4\pi$, $r = 2$. The length of a side of the square is 4. The area of the square is 16. Therefore, the shaded region is

$16 - 2(2\pi) = 16 - 4\pi$.

---

13. (D)

$\dfrac{1}{2} f(x - 1)$: Translate to the right by 1 and shrink in half.

---

14. (A)

If the radius of the smaller circle is r, the area is πr^2. The area of the larger circle is

$\pi \left(\dfrac{d}{2} \right)^2 = \dfrac{\pi d^2}{4}$.

The shaded area = the area of the smaller one.

$\dfrac{\pi d^2}{4} - \pi r^2 = \pi r^2$. $\dfrac{\pi d^2}{4} = 2\pi r^2$.

Therefore, $2r^2 = \dfrac{d^2}{4}$,

that is

$r = \dfrac{d}{\sqrt{8}} = \dfrac{d}{2\sqrt{2}}$

---

15. (B)

Let the positive number be n.

$n^2 - 6 = n + 24$.

$n^2 - n - 30 = 0$. $(n + 5)(n - 6) = 0$.

The value of n is -5 and 6. n is positive.

$n = 6$.

---

16. (C)

$0.1234 = \dfrac{123.4}{1000}$. $\dfrac{123}{1000} < \dfrac{123.4}{1000} < \dfrac{124}{1000}$.

Therefore, $n = 123$

---

17. (A)

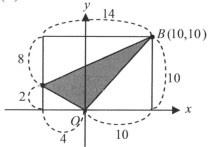

Take off the areas of the triangles from the area of the square.

The area of

$\triangle ABC = (10 \times 14) - (50 + 4 + 56) = 30$

---

18. (B)

$AB = 10$, and the height for $\overline{AB}$ is

$10\sqrt{2}$. Therefore, the area of

$\triangle ABP = \dfrac{1}{2}(10 \times 10\sqrt{2}) = 50\sqrt{2}$.

---

19. (E)

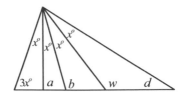

Using the exterior angle theorem,
$a = 3x + x = 4x$, $b = a + x = 5x$,
$w = b + x = 6x$. Or $3x + 3x = 120$.
$6x = 120$. Therefore, $x = 20$. Then $d = 40$.

20. (B)
$g(2) = 2f(2) + k = 10$ and
$g(5) = 2f(5) + k = 18$.

f is a linear, the slope $= \dfrac{f(5) - f(2)}{5 - 2}$.

$\quad 2f(5) + k = 18$
$-\quad 2f(2) + k = 10 \qquad f(5) - f(2) = 4$
$\overline{\quad 2f(5) - 2f(2) = 8}$

Therefore, $\dfrac{f(5) - f(2)}{5 - 2} = \dfrac{4}{3}$

TEST 18   SECTION 5

1. (A)
$x = y + 3$, $y = z - 5$, then
$x + y + z = (y + 3) + (z - 5) + z$
$\qquad\qquad = (z - 5 + 3) + (z - 5) + z$
$\qquad\qquad = 3z - 7$

2. (E)
$5 \cdot \dfrac{x^2}{y} \cdot \dfrac{3}{x} = 5 \cdot \dfrac{x}{y} \cdot 3 = 5 \cdot 2 \cdot 3 = 30$

3. (B)
$8a^{2x} = a^{2x+3} = a^{2x} \cdot a^3$, then $a^3 = 8 = 2^3$,
therefore, $a = 2$.

4. (E)
For the arithmetic sequence,
$a_n = a_1 + (n-1)d$. Then $a_n = 4 + (41-1) \times 7$.
Therefore, $n = 41$.

Or, find the pattern.
$a_1 = 4$, $a_2 = 4 + (1) \times 7$, $a_3 = 4 + (2) \times 7$,
Therefore, $4 + (40) \times 7$ is the 41st term.

5. (E)
median = average = 60. Sum = $60 \times 15 = 900$

6. (E)
$x^2 - y^2 = (x+y)(x-y) = 30$, $x + y > 10$.
Therefore, $x - y < 3$.

7. (E)
Let n = even integer. $n + n + 2 = t$. $n = \dfrac{t-2}{2}$.

Therefore, $n + 2 = \dfrac{t-2}{2} + 2 = \dfrac{t+2}{2}$

8. (E)
$\underset{(1)}{\dfrac{a+b}{2}} > \underset{(2)}{\dfrac{b+c}{2}} > \dfrac{c+a}{2}$

From (1), $a > c$. From (2), $b > a$.
Therefore, $b > a > c$

9. 14
Up to 10 ounces, it costs $2.00. For 40
additional ounces, it cost $40 \times 0.3 = \$12$.
Therefore, total cost is $14.

10. 160
When $x + 3$, $a(3) + 20 = 50$, $a = 10$. When
$x = 6$, $2ax + 40 = 2(10)(6) + 40 = 160$.

11. 21
Triangular inequality, $a \sim b < 10 < a + b$.
For the least value of the perimeter,
$a + b$ must be 11. Because a and b are
integral values, one of
$\{2,9\}$ $\{3,8\}$ $\{4,7\}$ $\{5,6\}$.

12. 500
If n is the number of people, $0.45n = 225$.
Therefore, $n = 500$

13. 4
$(2a) \triangle b = 2a + 2b = 6$ --[1],
$a \triangle (2b) = a + 4b = 15$ --[2]

$[1]+(-2)\times[2]$,

$2a+2b=6$

$\dfrac{-2a-8b=-30}{-6b=-24}$.

Therefore, $b=4$

---

14. 90

Let n be number of students.

$\dfrac{n}{3}+\dfrac{n}{5}+\dfrac{n}{10}=57$.

Therefore, $\dfrac{19n}{30}=57$. $n=90$

---

15. 17

19 , 29, ... 89: 8 numbers

$90,91,92,...98$: 9 numbers. Therefore,

$8+9=17$

---

16. 7

The values of x could be 6 or 7.

When $x=6$: 6, 8, 9, 9, 0

When $x=7$: 7, 8, 10, 11, 13

Therefore, $x=7$

---

17. $\dfrac{4}{9}$

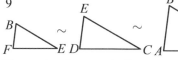

Three triangles are similar.

Ratio of the lengths $=1:2:3$

Ratio of the areas$=1:4:9=$ $k:4k:9k$

$\square AFED=9k-(k+4k)=4k$. Therefore,

$\dfrac{\square AFED}{\triangle ABC}=\dfrac{4}{9}$

---

18. 99

Let $n(k)=$ the number of multiple of k.

$n(3$ or $4)=n(3)+n(4)-n(12)$

$\left\lfloor\dfrac{299}{k}\right\rfloor-\left\lfloor\dfrac{100}{k}\right\rfloor=$ number of multiple of k.

Thus,

$\left\lfloor\dfrac{299}{3}\right\rfloor-\left\lfloor\dfrac{100}{3}\right\rfloor=99-33=66$

$\left\lfloor\dfrac{299}{4}\right\rfloor-\left\lfloor\dfrac{100}{4}\right\rfloor=74-25=49$

$\left\lfloor\dfrac{299}{12}\right\rfloor-\left\lfloor\dfrac{100}{12}\right\rfloor=24-8=16$.

Therefore, $n(3$ or $4)=66+49-16=99$

TEST 18   SECTION 7

1. (C)

Cross multiplication: $x^2=25$, $x=\pm5$

---

2. (E)

In the rhombus, diagonal bisects the angles.

$t\ne x$.

---

3. (A)

$0.3m=50$, $m=\dfrac{500}{3}$.

$0.03m=0.03\left(\dfrac{500}{3}\right)=5$

---

4. (A)

$q=1+\dfrac{1}{3}\left(1+\dfrac{1}{3}+\dfrac{1}{9}+.......+\dfrac{1}{243}\right)$

$=1+\dfrac{1}{3}+\dfrac{1}{9}+...+\dfrac{1}{729}$

Therefore, $q-p=\dfrac{1}{729}$

---

5. (D)

If $a-b<0$, then $b-a>0$

---

6. (C)

$70, 70, 80, 80, (80), 80, 80, 90, 90$

Median

If three students take higher than 80,or lower than 80, then there will be new median .

---

7. (B)

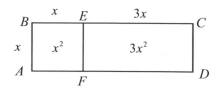

In the figure above, $4x^2=256$. $x^2=64$,

$x=8$.

The perimeter of $ECDF = 8x = 8(8) = 64$

---

8. (D)

$\dfrac{a}{b} = $ even. $a = b \times$ even $=$ even

---

9. (E)

The number of not seniors $= n - s$. After j juniors are added, the number is $n - s + j$. Also the entire number of students is $n + j$.

Therefore, the fraction will be $\dfrac{n - s + j}{n + j}$.

---

10. (D)

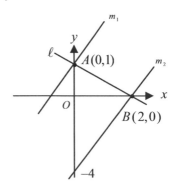

The slope of line ℓ is $-\dfrac{1}{2}$ and the slope of line m will be 2. The equation of line m is $y = 2x + b$. Substitute $(2,0)$ into the equation. $0 = 2(2) + b$, the y-intercept is -4. Therefore, the equations of line m must have slope 2 and y-intercept between 1 and -4.

---

11. (E)

$x = \dfrac{w^2}{y} = \dfrac{w^2}{1/w^2} = w^4$ and $y = \dfrac{1}{w^2}$. Therefore,

$\dfrac{x}{y} = \dfrac{w^4}{1/w^2} = w^6$

---

12. (C)

$x^2 + y^2 = (x - y)^2$, $x^2 + y^2 = x^2 - 2xy + y^2$.
Therefore, $xy = 0$

Choice I and II are the answer for "**could be true?**"

---

13. (D)

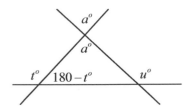

Exterior angle $u = a + 180 - t$. Therefore, $t + u = 180 + a$.

---

14. (A)

The solution set of $|x - 2| > 6$ are as follows.
$x - 2 > 6$ or $x - 2 < -6$. That is
$x > 8$ or $x < -4$. Therefore, four numbers of $-10, -8, -7, 10$ are the solutions. Probability $P = \dfrac{4}{10} = \dfrac{2}{5}$.

---

15. (D)

The sums of two different integers are as follows.
$-9, -8, -7, -6, \ldots \ldots 0$
$-7, -6, -5, -4, \ldots \ldots 0, 1$
$-5, -4, -3, -2, \ldots \ldots 0, 1, 2$
$\ldots \ldots \ldots \ldots \ldots \ldots \ldots, 3$
$\ldots \ldots \ldots \ldots \ldots \ldots \ldots, 4$
$\ldots \ldots \ldots \ldots \ldots \ldots \ldots, 5$
$\ldots \ldots \ldots \ldots \ldots \ldots \ldots, 6$
$\ldots \ldots \ldots \ldots \ldots \ldots \ldots, 7$
$\ldots \ldots \ldots \ldots \ldots \ldots \ldots, 8$
$\ldots \ldots \ldots \ldots \ldots \ldots \ldots, 9$
Therefore, the number of the different sums is 19.

---

16. (B)

When $x = 4$ and $y = 2$, $A = 4 \times 2 = 8$, and $B = \dfrac{4}{2} = 2$. Therefore $8 - 2 = 6$.

Or, $A - B = xy - \dfrac{x}{y} = \dfrac{x^2 y - x}{y}$

$= \dfrac{x(y^2 - 1)}{y}$. To have smallest, $x = 4$ and $y = 2$. Therefore, $\dfrac{x(y^2 - 1)}{y} = \dfrac{4(2^2 - 1)}{2} = 6$.

$\boxed{END}$

Dr. John Chung's SAT Math

TEST
19

ANSWER SHEET   TEST #:

SECTION 3

1 ⒶⒷⒸⒹⒺ	11 ⒶⒷⒸⒹⒺ	21 ⒶⒷⒸⒹⒺ	31 ⒶⒷⒸⒹⒺ
2 ⒶⒷⒸⒹⒺ	12 ⒶⒷⒸⒹⒺ	22 ⒶⒷⒸⒹⒺ	32 ⒶⒷⒸⒹⒺ
3 ⒶⒷⒸⒹⒺ	13 ⒶⒷⒸⒹⒺ	23 ⒶⒷⒸⒹⒺ	33 ⒶⒷⒸⒹⒺ
4 ⒶⒷⒸⒹⒺ	14 ⒶⒷⒸⒹⒺ	24 ⒶⒷⒸⒹⒺ	34 ⒶⒷⒸⒹⒺ
5 ⒶⒷⒸⒹⒺ	15 ⒶⒷⒸⒹⒺ	25 ⒶⒷⒸⒹⒺ	35 ⒶⒷⒸⒹⒺ
6 ⒶⒷⒸⒹⒺ	16 ⒶⒷⒸⒹⒺ	26 ⒶⒷⒸⒹⒺ	36 ⒶⒷⒸⒹⒺ
7 ⒶⒷⒸⒹⒺ	17 ⒶⒷⒸⒹⒺ	27 ⒶⒷⒸⒹⒺ	37 ⒶⒷⒸⒹⒺ
8 ⒶⒷⒸⒹⒺ	18 ⒶⒷⒸⒹⒺ	28 ⒶⒷⒸⒹⒺ	38 ⒶⒷⒸⒹⒺ
9 ⒶⒷⒸⒹⒺ	19 ⒶⒷⒸⒹⒺ	29 ⒶⒷⒸⒹⒺ	39 ⒶⒷⒸⒹⒺ
10 ⒶⒷⒸⒹⒺ	20 ⒶⒷⒸⒹⒺ	30 ⒶⒷⒸⒹⒺ	40 ⒶⒷⒸⒹⒺ

SECTION 5

1 ⒶⒷⒸⒹⒺ	11 ⒶⒷⒸⒹⒺ	21 ⒶⒷⒸⒹⒺ	31 ⒶⒷⒸⒹⒺ
2 ⒶⒷⒸⒹⒺ	12 ⒶⒷⒸⒹⒺ	22 ⒶⒷⒸⒹⒺ	32 ⒶⒷⒸⒹⒺ
3 ⒶⒷⒸⒹⒺ	13 ⒶⒷⒸⒹⒺ	23 ⒶⒷⒸⒹⒺ	33 ⒶⒷⒸⒹⒺ
4 ⒶⒷⒸⒹⒺ	14 ⒶⒷⒸⒹⒺ	24 ⒶⒷⒸⒹⒺ	34 ⒶⒷⒸⒹⒺ
5 ⒶⒷⒸⒹⒺ	15 ⒶⒷⒸⒹⒺ	25 ⒶⒷⒸⒹⒺ	35 ⒶⒷⒸⒹⒺ
6 ⒶⒷⒸⒹⒺ	16 ⒶⒷⒸⒹⒺ	26 ⒶⒷⒸⒹⒺ	36 ⒶⒷⒸⒹⒺ
7 ⒶⒷⒸⒹⒺ	17 ⒶⒷⒸⒹⒺ	27 ⒶⒷⒸⒹⒺ	37 ⒶⒷⒸⒹⒺ
8 ⒶⒷⒸⒹⒺ	18 ⒶⒷⒸⒹⒺ	28 ⒶⒷⒸⒹⒺ	38 ⒶⒷⒸⒹⒺ
9 ⒶⒷⒸⒹⒺ	19 ⒶⒷⒸⒹⒺ	29 ⒶⒷⒸⒹⒺ	39 ⒶⒷⒸⒹⒺ
10 ⒶⒷⒸⒹⒺ	20 ⒶⒷⒸⒹⒺ	30 ⒶⒷⒸⒹⒺ	40 ⒶⒷⒸⒹⒺ

Student-produced response grids numbered 9, 10, 11, 12, 13, 14, 15, 16, 17, 18.

Math Scoring Worksheet

A. Section 3

      _____   _____
      numer of correct   *number of incorrect*

      +   +

B. Section 5 (1-8)

      _____   _____
      numer of correct   *number of incorrect*

      +

C. Section 5 (9-18)

      _____
      numer of correct

      +   +

D. Section 7

      _____   _____
      numer of correct   *number of incorrect*

      =   =

E. Total Unrounded Raw Score

      _____ $-$ _____ $\div 4$ $=$ _____
      numer of correct   *number of incorrect*

F. Total Rounded Raw Score   _____ (See table)

Math Score Range = | _____ |

Math Conversion Table

Raw Score	Scaled Score	Raw Score	Scaled Score
54	800	23	490-550
53	780-800	22	480-540
52	760-800	21	470-530
51	740-800	20	460-520
50	720-780	19	450-510
49	700-760	18	450-510
48	690-750	17	440-500
47	680-740	16	430-490
46	670-730	15	420-480
45	660-720	14	420-480
44	650-710	13	410-470
43	650-710	12	400-460
42	640-700	11	390-450
41	630-690	10	380-440
40	620-680	9	390-430
39	610-670	8	380-420
38	610-670	7	370-410
37	600-660	6	360-400
36	590-650	5	340-380
35	580-640	4	320-370
34	570-630	3	310-360
33	560-620	2	300-350
32	560-620	1	270-320
31	550-610	0	240-300
30	540-600	-1	200-290
29	530-590	-2	200-270
28	530-590	-3	200-260
27	520-580	-4	200-240
26	510-570	-5	200-220
25	500-560	-6 and below	200
24	500-560		

SECTION 3
Time- 25 minutes
20 Questions

Turn to Section 3 (Page 1) of your answer sheet to answer the questions in this section.

Directions: For this section, solve each problem and decide which is the best of the choices given. Fill in the corresponding circle on the answer sheet. You may use any available space for scratchwork.

Notes

1. The use of a calculator is permitted.
2. All numbers used are real numbers.
3. Figures that accompany problems in this test are intended to provide information useful in solving the problems. They are drawn as accurately as possible EXCEPT when it is stated in a specific problem that the figure is not drawn to scale. All figure lie in a plane unless other indicated.
4. Unless otherwise specified, the domain of any function f is assumed to be set of all real numbers x for which $f(x)$ is a real number.

Reference Informatiom

$A = \pi r^2$
$C = 2\pi r$ $\quad A = \ell w \quad A = \dfrac{1}{2}bh \quad V = \ell wh \quad V = \pi r^2 h \quad c^2 = a^2 + b^2 \quad$ Special Right Triangles

The numbers of degrees of arc in a circle is $360°$.

The sum of the measures in degrees of the angles is $180°$.

---

1. If k is an integer and $|1 - k| < 1$, which of the following could be the value of k?

(A) -2
(B) -1
(C) 0
(D) 1
(E) 2

2. In the cube above, its exterior faces were painted blue. If the cube is cut into 27 smaller cubes, how many of the smaller cubes have unpainted faces?

(A) 1
(B) 3
(C) 6
(D) 8
(E) 9

GO ON TO THE NEXT PAGE

3 ☐ ☐ 3 ☐ ☐ 3 ☐ ☐ 3

3. Jane started her postage-stamp collection with 30 stamps, and then she increased her collection by adding 5 stamps per week. Which of the following represents the total number of stamps in her collection at the end of w weeks?

(A) $30 + w$
(B) $5(1 + 5w)$
(C) $5(5 + w)$
(D) $5(6 + w)$
(E) $6(5 + w)$

4. If $\frac{a}{b} = 3$, $\frac{b}{c} = \frac{1}{2}$, and $\frac{c}{d} = \frac{2}{3}$, what is the value of $\frac{2a}{d}$?

(A) 1

(B) $\frac{3}{2}$

(C) 2

(C) $\frac{5}{2}$

(D) 3

5. If seven different lines lie in a plane, what is the greatest possible number of intersections?

(A) 18
(B) 20
(C) 21
(D) 22
(E) 30

6. If $m^a \cdot m^b = m^5$ and $\left(m^{2a}\right)^b = m^{12}$, which of the following could be a value of a?

(A) 3
(B) 4
(C) 5
(D) 6
(E) 8

GO ON TO THE NEXT PAGE

7. At a grocery store, an apple and a pear cost $0.75 , a pear and a banana cost $0.55 , and an apple and a banana cost $0.50 . What is the cost of an apple?

(A) $0.15
(B) $0.30
(C) $0.35
(D) $0.40
(E) $0.45

8. If $p > 0$, then 5 percent of 7 percent of $4p$ equals what percent of p ?

(A) 14%
(B) 7%
(C) 1.4%
(D) 0.14%
(E) 0.07%

9. Ashley is twice as old as Carlos and Bobby is 5 years younger than Ashley. If the sum of the ages of the three children is s, how old is Carlos, in terms of s ?

(A) $\dfrac{s}{5}$

(B) $\dfrac{s+1}{5}$

(C) $\dfrac{s}{5}+1$

(D) $\dfrac{2s}{5}$

(E) $\dfrac{2s}{5}+1$

10. A list consists of 20 positive numbers. A new list of 20 numbers is formed by adding 5 to each of the original numbers. If m is the average (arithmetic mean) of the original list, and k is the average of the new list, which of the following must be true?

(A) $k = m$
(B) $k = 2m$
(C) $k = m + 5$
(D) $k = m^2$
(E) $k = m + 100$

GO ON TO THE NEXT PAGE

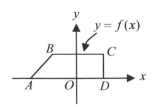

11. The figure above shows the graph of $y = f(x)$.

If function g is defined by $g(x) = \dfrac{1}{2} f(x)$ and
the area of the enclosed region $ABCD$ of
$f(x)$ and the x-axis is 20, what is the area of
the enclosed region of $g(x)$ and the x-axis?

(A) 5
(B) 8
(C) 10
(D) 12
(E) It cannot be determined from the
     information given.

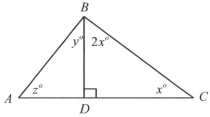

Note: Figure not drawn to scale.

12. In $\triangle ABC$ above, $y = z$ and $BC = 8$. What is
the area of $\triangle ABC$?

(A) 12
(B) 18
(C) $8 + 8\sqrt{3}$
(D) $16\sqrt{3}$
(E) $12 + 12\sqrt{2}$

13. How many integers greater than 199, when
doubled, will yield a 3-digit integer?

(A) 200
(B) 300
(C) 400
(D) 1000
(E) More than 1000

14. If f is a linear function for which
$f(10) - f(5) = 10,$ what is the value of
$f(20) - f(15)$?

(A) 10
(B) 12
(C) 15
(D) 20
(E) 30

GO ON TO THE NEXT PAGE

$a,\ ar,\ ar^2,\ ar^3,...$

15. In the sequence above, the first term is a and each term after the first is obtained by multiplying the preceding term by r. If the third and fifth terms in the sequence are 2 and 8, what is the value of the 7th term?

(A) 16
(B) 24
(C) 32
(D) 48
(E) 96

16. Let $\Phi(x)$ be defined as the sum of the digits of the positive integer x. For example, $\Phi(7) = 7$ and $\Phi(123) = 6$. For how many two-digit values of x does $\Phi(\Phi(x)) = 2$?

(A) 8
(B) 9
(C) 10
(D) 12
(E) 13

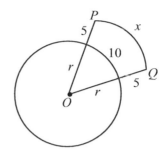

Note: Figure not drawn to scale.

17. In the circle above, the length of an arc is 10, and there are two radii shown extended by 5 units outside the circle. If the length of arc PQ is x, what must x equal?

(A) 12

(B) $10 + r$

(C) $\dfrac{10(r+5)}{r}$

(D) $\dfrac{10(r+10)}{r}$

(E) $\dfrac{15(r+10)}{r+5}$

GO ON TO THE NEXT PAGE

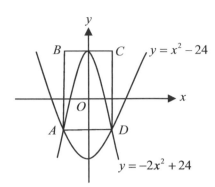

Note: Figure not drawn to scale.

18. The figure above shows the graphs of $y = x^2 - 24$ and $y = -2x^2 + 24$, and points A and D of the intersection of the graphs. What is the area of rectangle $ABCD$?

(A) 84
(B) 128
(C) 168
(D) 192
(E) 256

19. If $(a+b)(a^2 - b^2) = 0$, which of the following must be true?

    I. $a = b$
    II. $a = -b$
    III. $a^2 = b^2$

(A) I only
(B) II only
(C) III only
(D) I and III only
(E) I, II, and III

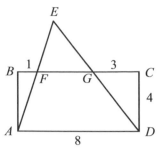

Note: Figure not drawn to scale.

20. In the figure above, $ABCD$ is a rectangle, $AD = 8$, and $CD = 4$. If points F and G are on $\overline{BC}$ so that $BF = 1$ and $GC = 3$, and line AF and DG intersect at point E, what is the area of $\triangle AED$?

(A) 28
(B) 30
(C) 32
(D) 36
(E) 48

STOP

If you finish before time is called, you may check your work on this section only
Do not turn to any other section in the test.

SECTION 5
Time- 25 minutes
18 Questions

Turn to Section 5 (Page 1) of your answer sheet to answer the questions in this section.

Directions: For this section, solve each problem and decide which is the best of the choices given. Fill in the corresponding circle on the answer sheet. You may use any available space for scratchwork.

Notes

1. The use of a calculator is permitted.
2. All numbers used are real numbers.
3. Figures that accompany problems in this test are intended to provide information useful in solving the problems. They are drawn as accurately as possible EXCEPT when it is stated in a specific problem that the figure is not drawn to scale. All figure lie in a plane unless other indicated.
4. Unless otherwise specified, the domain of any function f is assumed to be set of all real numbers x for which $f(x)$ is a real number.

Reference Information

$A = \pi r^2$
$C = 2\pi r$ $A = \ell w$ $A = \dfrac{1}{2}bh$ $V = \ell wh$ $V = \pi r^2 h$ $c^2 = a^2 + b^2$ Special Right Triangles

The numbers of degrees of arc in a circle is 360°.

The sum of the measures in degrees of the angles is 180°.

---

1. If $|x| < 5$ and $|x| > 2$, how many integers satisfy both of the two inequalities?

(A) 1
(B) 2
(C) 3
(D) 4
(E) More than 4

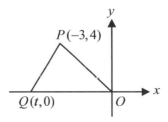

Note: Figure not drawn to scale.

2. In the figure above, if $OP = OQ$, what is the value of t?

(A) −5
(B) −4
(C) −3
(D) 4
(E) 5

GO ON TO THE NEXT PAGE

3. If $\dfrac{1}{k} > \dfrac{1}{2}$, then which of the following must be true?

(A) $k > 2$
(B) $k \le 2$
(C) $1 < k < 2$
(D) $0 < k < 2$
(E) $0 \le k \le 2$

4. If the average (arithmetic mean) of six consecutive odd integers is m, which of the following is the expression for the smallest odd integer?

(A) $6m - 6$
(B) $2m - 4$
(C) $m - 6$
(D) $m - 5$
(E) $m - 4$

5. Let the function f be defined by
$f(x) = ax + b$, where a and b are constants.
If $ab < 0$, which of the following graphs could be the graph of f?

(A)

(B)

(C)

(D)

(E)

GO ON TO THE NEXT PAGE

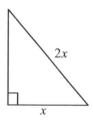

Note: Figure not drawn to scale.

6. In the figure above, the perimeter of the triangle is $9 + 3\sqrt{3}$. What is the value of x?

(A) 3
(B) 6
(C) $3\sqrt{3}$
(D) $3\sqrt{2}$
(E) $3 + \sqrt{3}$

7. If two coins are selected from a collection consisting of 2 pennies, 2 nickels, 2 dimes, and 2 quarters, how many different sums are possible?

(A) 10
(B) 16
(C) 24
(D) 36
(E) 72

NUMBER OF MEMBERS OF A FAMILY
AT WORLD COMPUTER COMPANY

Number of Members of a Family	Number of Employees
7	5
6	15
5	35
4	55
3	30
Fewer than 3	60

8. The table above shows the number of members of a family for 200 employees at World Computer Company this year. Which of the following can be determined from the information in the table?

   I. The average (arithmetic mean) number of members of a family of all employees
   II. The median number of members of a family of all employees
   III. The mode of the number of members of a family of all employees

(A) I only
(B) II only
(C) III only
(D) II and III only
(E) I, II, and III

GO ON TO THE NEXT PAGE

Directions: For Students-Produced Response questions 9-18, use the grid at the bottom of the answer sheet page on which you have answered questions 1-8.

Each of the remaining 10 questions requires you to solve the problem and enter your answer by making the circles in the special grid, as shown in the examples below. You may use any available space for scratchwork.

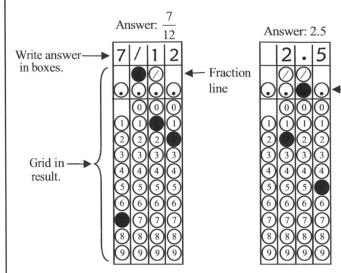

Answer: $\frac{7}{12}$

Write answer in boxes.

← Fraction line

Grid in result.

Answer: 2.5

← Decimal point

Answer: 201

Either position is correct.

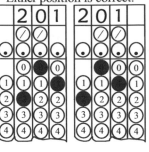

Note: You may start your answers in any column, space permitting. Columns not needed should be left blank.

- Mark no more than one circle in any column.

- Because the answer sheet will be machine-scored, **you will receive credit only if the circles are filled in correctly.**

- Although not required, it is suggested that you write your answer in the boxes at the top of the columns to help you fill in the circles accurately.

- Some problems may have more than one correct answer. In such cases, grid only one answer.

- No question has a negative answer.

- **Mixed numbers** such as $3\frac{1}{2}$ must be gridded as 3.5 or 7/2. (If ☐3☐1☐/☐2☐ is gridded, it will be interpreted as $\frac{31}{2}$, not $3\frac{1}{2}$.)

- **Decimal Answers:** If you obtain a decimal answer with more digits than the grid can accommodate, it may be either rounded or truncated, but it must fill the entire grid. For example, if you obtain an answer such as 0.6666..., you should record your result as .666 or .667. **A less accurate value such as .66 or .67 will be scored as incorrect.**

Acceptable ways to grid $\frac{2}{3}$ are:

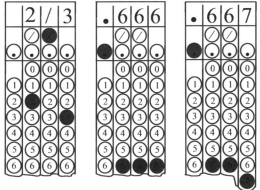

9. In the figure above, how many different routes are possible from A to B and back to A, but not taking the same route?

10. If $(x-5)y = x^2 - 10x + 25$, what is the value of y, when $x = 10$?

GO ON TO THE NEXT PAGE

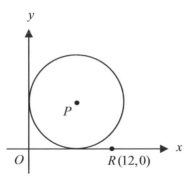

Note: Figure not drawn to scale.

11. In the figure above, a circle with a radius of 8 is tangent to the *x*- and *y*-axis. If line *k* (not shown) contains points *P* and *R* , at what point does *k* intersect the *y*-axis?

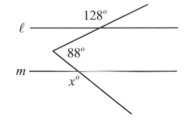

12. In the figure above, if $\ell \parallel m$, what is the value of *x* ?

13. A jar contains 10 red, 10 blue, 10 green, and 10 yellow marbles. If marbles are drawn out and arranged on the table, how many marbles must be drawn out to ensure that there are five marbles of the same color on the table?

14. A work crew of four people requires two hours and thirty minutes to do $\frac{1}{2}$ of a certain job. How long, in minutes, will it take a crew of ten people to do $\frac{1}{3}$ of the same job if every one works at the same rate?

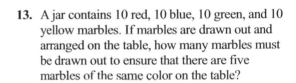

GO ON TO THE NEXT PAGE

15. If $a^x = b$, $b^y = c$, and $c^z = 125$, what is the value of xyz, when $a = 5$?

16. If $\dfrac{16}{k} + \dfrac{8}{k} = P$, where P is an integer, how many positive integer values of k are possible?

17. In $\triangle ABC$, $\angle A > 90^o$, the length of $\overline{AC}$ is 6, and the length of $\overline{AB}$ is 8. What is one possible length of $\overline{BC}$?

18. Let the function g be defined by $g(x) = 2f(x) - 3$, where $f(x)$ is a linear function. If $g(3) = 9$ and $g(0) = 3$, what is the slope of the graph of the linear function?

STOP

If you finish before time is called, you may check your work on this section only.
Do not turn to any other section in the test.

SECTION 7
Time- 20 minutes
16 Questions

Turn to Section 7 (Page 2) of your answer sheet to answer the questions in this section.

Directions: For this section, solve each problem and decide which is the best of the choices given. Fill in the corresponding circle on the answer sheet. You may use any available space for scratchwork.

Notes

1. The use of a calculator is permitted.
2. All numbers used are real numbers.
3. Figures that accompany problems in this test are intended to provide information useful in solving the problems. They are drawn as accurately as possible EXCEPT when it is stated in a specific problem that the figure is not drawn to scale. All figure lie in a plane unless other indicated.
4. Unless otherwise specified, the domain of any function f is assumed to be set of all real numbers x for which $f(x)$ is a real number.

Reference Informatiom

       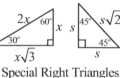

$A = \pi r^2$
$C = 2\pi r$
$A = \ell w$
$A = \dfrac{1}{2}bh$
$V = \ell wh$
$V = \pi r^2 h$
$c^2 = a^2 + b^2$
Special Right Triangles

The numbers of degrees of arc in a circle is $360°$.

The sum of the measures in degrees of the angles is $180°$.

---

1. If $81,000 = 1,000(9k + 9)$, then k?

(A) 4
(B) 6
(C) 8
(D) 10
(E) 100

2. If 2 less than m is negative and if 2 more than m is positive, which of the following could be the value of m?

(A) -3
(B) -2
(C) 0
(D) 2
(E) 3

GO ON TO THE NEXT PAGE

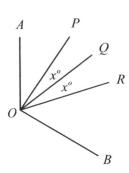

Note: Figure not drawn to scale.

3. In the figure above, $\angle AOB = 125^o$, $\angle AOR = 70^o$, and $\angle BOP = 85^o$. If $\angle POQ = \angle QOR = x^o$, what is the value of x?

(A) 10
(B) 15
(C) 18
(D) 20
(E) 25

4. A train takes one hour longer to travel d miles at its normal speed than it would if it were traveling 10 miles faster. Which of the following equations could represent the equation to find the train's normal speed s, in miles per hour?

(A) $\dfrac{d}{s+10} = s+1$

(B) $\dfrac{d}{s} = \dfrac{d}{s+1}$

(C) $s(s+10) = d$

(D) $\dfrac{d}{s+10} - \dfrac{d}{s} = 1$

(E) $\dfrac{d}{s} - \dfrac{d}{s+10} = 1$

5. If $k(a^2 - b^2) = 0$, which of the following could be true?

 I. $k = 0$
 II. $a = 5$ and $b = 5$
 III. $a = -b$

(A) I only
(B) II only
(C) III only
(D) I and III only
(E) I, II, and III

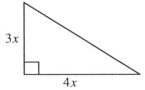

Note: Figures not drawn to scale.

6. In the figures above, the length of a side of the square is $2x + 4$, and the lengths of two sides of the triangle are $3x$ and $4x$. If the perimeters of these two figures are equal, what is the length of a side of the square?

(A) 4
(B) 8
(C) 12
(D) 16
(E) 24

GO ON TO THE NEXT PAGE

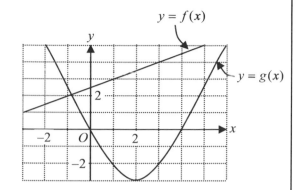

$y = f(x)$

$y = g(x)$

7. The figure above shows the graphs of the functions f and g. If $g(k) = 0$, what is one possible value of $f(k)$?

(A) -2

(B) $2\frac{1}{2}$

(C) 3

(D) $3\frac{1}{2}$

(E) 4

8. If $|2 - x| < x$, which of the following is a possible value of x?

(A) -2
(B) -1
(C) 0
(D) 1
(E) 2

9. In a group of men and women, 20 are right-handed, 14 are men, 10 are right-handed men, and 2 are left-handed women. How many women are in the group?

(A) 26
(B) 20
(C) 14
(D) 12
(E) 6

10. Which of the following CANNOT be written as the sum of three consecutive integers?

(A) 6
(B) 15
(C) 21
(D) 39
(E) 56

GO ON TO THE NEXT PAGE

11. If $3^{x^2+y^2} = \dfrac{1}{9^{xy}}$, what is the value of $x+y$?

(A) −2
(B) −1
(C) 0
(D) 1
(E) 2

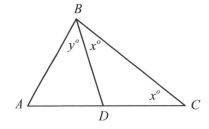

Note: Figure not drawn to scale.

12. In $\triangle ABC$ above, $AD = CD$. What is the value of $x+y$?

(A) 90
(B) 85
(C) 80
(D) 75
(E) 60

13. Which of the following is equivalent to $|k| < 10$?

  I. $k^2 < 10$
  II. $-10 < k < 10$
  III. $(k+10)(k-10) < 0$

(A) I only
(B) II only
(C) III only
(D) I and II only
(E) II and III only

14. If the average (arithmetic mean) of 9 different positive integers is 15 and the median is 20, what is a possible value of the greatest of the 9 integers?

(A) 99
(B) 75
(C) 39
(D) 30
(E) 28

GO ON TO THE NEXT PAGE

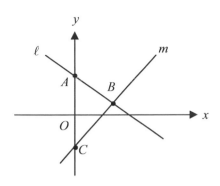

Note: Figure not drawn to scale.

15. In the figure above, line ℓ is given by the equation $y = -x + 8$, and line m is given by $y = 2x - 7$. If line ℓ intersects line m at point B, what is the area of $\triangle ABC$?

(A) 16

(B) 24

(C) 32

(D) $37\dfrac{1}{2}$

(E) 50

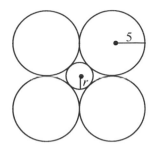

Note: Figure not drawn to scale.

16. In the figure above, a circle of radius r is surrounded by four circles of radius 5. What is the value of r?

(A) $\sqrt{2}$

(B) $10\sqrt{2} + 1$

(C) $5(\sqrt{2} - 1)$

(D) $5(\sqrt{2} - 5)$

(E) $10(\sqrt{2} - 1)$

STOP

If you finish before time is called, you may check your work on this section only.
Do not turn to any other section in the test.

TEST 19　　ANSWER KEY

#	SECTION 3	SECTION 5	SECTION 7
1	D	D	C
2	A	A	C
3	D	D	B
4	C	D	E
5	C	E	E
6	A	A	C
7	C	A	B
8	C	B	E
9	C	12	D
10	C	5	E
11	C	24	C
12	C	144	A
13	B	17	E
14	A	40	C
15	C	3	D
16	C	8	C
17	C	11, 12, or 13	
18	E	1	
19	C		
20	C		

TEST 19　　SECTION 3

1. (D)

$|1-k| = |k-1| < 1$.

Algebraically, $-1 < k-1 < 1$,

$0 < k < 2$. Therefore, $k = 1$

Or, test the choices.

2. (A)

Except the outer cubes, only one inside is unpainted.

3. (D)

The number of stamps $= 30 + 5 \times w = 5(6+w)$

4. (C)

$\dfrac{a}{b} \times \dfrac{b}{c} \times \dfrac{c}{d} = \dfrac{3}{1} \times \dfrac{1}{2} \times \dfrac{2}{3} = 1$. Therefore,

$2\dfrac{a}{d} = 2 \cdot 1 = 2$

Or, use convenient numbers. If $c = 2$, $d = 3$,

then $b = 1$ and $a = 3$. Therefore

$\dfrac{2a}{d} = \dfrac{2(3)}{3} = 2$.

---

5. (C)

The first two lines have one intersection. The third line will have two intersections and the fourth line will have three intersections, so on…

Therefore, the total intersections will be

$1+2+3+4+5+6 = 21$

---

6. (A)

$m^a \cdot m^b = m^{a+b} = m^5$.　$a+b = 5$ --(1)

$\left(m^{2a}\right)^b = m^{2ab} = m^{12}$.　$2ab = 12$, $ab = 6$ --(2)

From (1) and (2)　$a = 2$ or 3.

---

7. (C)

Let $a, b,$ and p are the number of apples, bananas, and pears respectively.

$a + p = \$0.75$, $b + p = \$0.55$

and $a + b = \$0.5$. Add both sides.

$2(a + b + p) = \$1.80$. That is

$a + b + p = \$0.90$.

The price of an apple is, $\$0.90 - \$0.55 = \$0.35$

---

8. (E)

$0.05 \times 0.07 \times 4p = 0.014p$

$0.014p = 1.4\%$ of p

---

9. (C)

$A = 2C$, $B = A - 5 = 2C - 5$.

Then $A + B + C =$

$2C + 2C - 5 + C = 5C - 5 = s$.

Therefore, $C = \dfrac{s+5}{5} = \dfrac{s}{5} + 1$.

---

10. (C)

$m = \dfrac{S}{20}$ and $k = \dfrac{S + 20 \times 5}{20}$. Because

$S = 20m$, then $k = \dfrac{20m + 100}{20} = m + 5$.

---

11. (C)

The area will be half of the original region.

$\dfrac{1}{2} \times 20 = 10$.

---

12. (C)

The area of $\triangle ABC = \dfrac{\left(4 + 4\sqrt{3}\right) \cdot 4}{2} = 8 + 8\sqrt{3}$.

---

13. (B)

$200, 201, 202, 203, \ldots\ldots\ldots 499$. When doubled,

$400, 402, 404, 406, \ldots\ldots\ldots 998$. They are all 3 digit-numbers. Therefore, $499 - 199 = 300$.

---

14. (A)

Because f is a linear, the slope between two points arc constant.

$\dfrac{f(10) - f(5)}{10 - 5} = \dfrac{f(20) - f(15)}{20 - 15}$.

Therefore, $\dfrac{10}{5} = \dfrac{f(20) - f(15)}{5}$.

Then $f(20) - f(15) = 10$.

---

15. (C)

$ar^2 = 2$ and $ar^4 = 8$.

Then $\dfrac{ar^4}{ar^2} = r^2 = 4$, $a = \dfrac{1}{2}$

The 7th term

$a_7 = ar^6 = a\left(r^2\right)^3 = \dfrac{1}{2}(4)^3 = 32$

---

16. (C)

If $\Phi(\Phi(x)) = 2$, the possible values of $\Phi(x)$ are only 2 and 11, less than or equal to 18. The maximum of $\Phi(99)$ is 18. Therefore, the possible two-digit values of x are 11, 20, 29, 38, 47, 56, 65, 74, 83, 92. (10 numbers)

---

17. (C)

The sectors are similar. The ratio of the lengths are constant. $\dfrac{10}{x} = \dfrac{r}{r+5}$.

Therefore, $x = \dfrac{10(r+5)}{r}$.

---

18. (E)

The coordinates of point D can be obtained from the two graphs. $x^2 - 24 = -2x^2 + 24$, $3x^2 = 48$. Then $x = 4$ and $y = -8$. The length of AD is 8 and the length of CD is $24 - (-8) = 32$. Therefore, the area of $ABCD$ is $8 \times 32 = 256$.

19. (C)

$(a-b)(a^2-b^2)=(a-b)(a-b)(a+b)=0$.

The solutions are $a=b$ or $a=-b$. Therefore, $a^2=b^2$ is the answer for " must be true?"

---

20. (C)

$\triangle EFG$ is similar to $\triangle AED$.

 ~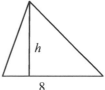

$\dfrac{h-4}{4}=\dfrac{h}{8}$, $h=8$. The area of

$\triangle AED = \dfrac{8\times 8}{2} = 32$.

Or,

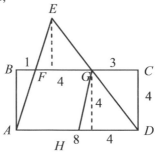

$\triangle EFG \cong \triangle GDH$, therefore the height of $\triangle EFG$ is 4. The height of $\triangle AED$ is 8.

TEST 19   SECTION 5

1. (D)

$|x|<5$ is equivalent to $-5<x<5$.

$|x|>2$ is equivalent to $x>2$ or $x<-2$.

Therefore, $-4,-3,3,4$ satisfy both of the inequalities.

---

2. (A)

$OP=\sqrt{3^2+4^2}=5=OQ$. Therefore, $t=-5$.

---

3. (D)

From $\dfrac{1}{k}>\dfrac{1}{2}$, k must be positive. Multiply k to both sides. $1>\dfrac{1}{2}k$, that is $k<2$.

Therefore, $0<k<2$ is the answer.

---

4. (D)

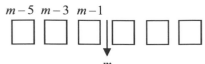

m is the median and also the average. In the figure above, the smallest number is $m-5$.

---

5. (E)

$ab<0$ has two possible cases, (1) $a>0$ and $b<0$, or (2) $a<0$ and $b>0$. Only graph (E) has a positive slope and negative y-intercept.

---

6. (A)

For the special triangle, $3x+x\sqrt{3}=9+3\sqrt{3}$. Therefore, $x=3$.

---

7. (A)

10 possible combinations, (1,1) (5,5) (10,10) (25,25) (1,5) (1,10) (1,25) (5,10) (5,25) (10,25)

---

8. (B)

The average cannot be determined, since the number of families with one member or two members cannot be determined. The median can be obtained from the middle number, which is 4. The mode cannot be determined from the information given, because all 60 families have one or two members. Therefore, the mode can be 1, or 2, or 3.

---

9. 12

The product of choices is $(3\times 2)\times(1\times 2)=12$

---

10. 5

$(x-5)y = (x-5)^2$. Therefore,
$y = x-5 = 10-5 = 5$.
Or, substitute 10 into x.
$(10-5)y = 100-100+25 = 25$,
$5y = 25$, $y = 5$.

---

11. 24
Algebraically, the coordinates of point P is $(8, 8)$.

The slope of line $k = \dfrac{8-0}{8-12} = -2$. The equation of line k is $y = -2x + b$. At the point $P(8,8)$, $8 = -2(8) + b$. Therefore, $b = 24$.

Geometrically, the two triangles are similar.

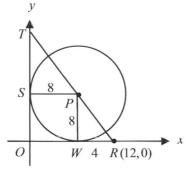

$\triangle TOR \sim \triangle PWR$. Therefore, $\dfrac{TO}{12} = \dfrac{8}{4}$, $TO = 24$.

---

12. 144

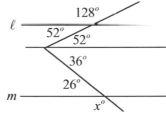

In the figure above, $x = 180 - 36 = 144$.

---

13. 17
The worst case is as follows.
    RRRR   BBBB   GGGG   YYYY
The next marble will be the one that can ensure the same colors for five.

---

14. 40
Four people need 5 hrs to do the entire job. That is an inverse proportion. $4 \times 5 = 10 \times x$, $x = 2$. Now 10 people need 2 hrs to do the

---

same job. But to do $\dfrac{1}{3}$ of the job, they need
$\dfrac{1}{3}(2) = \dfrac{2}{3}$ hrs = 40minutes.

---

15. 3
$\left(\left(a^x\right)^y\right)^z = a^{xyz} = 5^3$. If $a = 5$, $xyz = 3$.

---

16. 8
$\dfrac{16}{k} + \dfrac{8}{k} = \dfrac{24}{k}$ = integer. k is factors of 24.
Therefore, $k = 1, 2, 3, 4, 6, 8, 12, 24$.
Eight numbers.

---

17. 11, 12, or 13

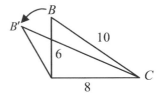

$B'C$ of the obtuse triangle is greater than 10, but less than 14. $10 < B'C < 6+8$.
Therefore, 11, 12, or 13 is the answer.

---

18. 1
$g(3) = 2f(3) - 3 = 9$, that is $f(3) = 6$
$g(0) = 2f(0) - 3 = 3$, that is $f(0) = 3$
The slope between the two points, $(3,6)$ and $(0,3)$ is
$m = \dfrac{6-3}{3-0} = 1$.

TEST 19   SECTION 7

1. (C)
$81,000 = 1,000(9k+9)$, $9k+9 = 81$.
Therefore, $k = 8$.

---

2. (C)
$m-2 < 0$ and $m+2 > 0$ is $-2 < m < 2$.
Therefore, (C) is the answer.

---

3. (B)

$125 = 70 + 85 - 2x$, $2x = 30$. Therefore, $x = 15$.

4. (E)

The traveling time at normal speed s is $\dfrac{d}{s}$.

The traveling time at speed of $(s+10)$ is

$$\dfrac{d}{(s+10)}$$

Therefore, $\dfrac{d}{s} - \dfrac{d}{s+10} = 1$ hour

5. (E)

The solutions are $k = 0$, $a = b$, or $a = -b$.
Therefore, I ,II, and III is the answer.

6. (C)

The length of the hypotenuse of the triangle
$= 5x$.
$4(2x+4) = 3x + 4x + 5x$, $x = 4$.
Therefore, the length of a side of the square
is $2(4) + 4 = 12$.

7. (B)

At $k = 0$ and 4, $g(k) = 0$. Therefore,
$f(0) = 2.5$ and $f(4) = 4.5$

8. (E)

Substitute the choices into $|2 - x| < x$

(A) $|2 - (-2)| = 4 < -2$,False.

(B) $|2 - (-1)| = 3 < -1$, False

(C) $|2 - 0| = 2 < -1$, False

(D) $|2 - 1| = 1 < -1$,False

(E) $|2 - 2| = 0 < 2$,True

9. (D)

	Men	Women	
RH	10		20
LH		2	
	14		

Fill out the table to get the number of people
in the group. 12 is the answer.

10. (E)

$6 = 1 + 2 + 3$, $15 = 4 + 5 + 6$, ... (E) cannot
be written as the sum of three consecutive
integers.
Or, the middle number is the median and also

the average. $\dfrac{6}{2} = 3$, 3 is the median.

Therefore, the integers are 2,3,4 . But, when
56 is divided by 3, it is not integer.

11. (C)

$3^{x^2+y^2} = 9^{-xy} = \left(3^2\right)^{-xy} = 3^{-2xy}$. That is

$x^2 + y^2 = -2xy$

Therefore, $(x+y)^2 = 0$, $x + y = 0$

12. (A)

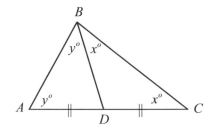

$AD = CD = BD$. $\triangle ABD$ and $\triangle CBD$ are
isosceles triangles. Therefore, $2(x + y) = 180$,
$x + y = 90$

13. (E)

$|k| < 10$ is equivalent to $k^2 < 100$,

$k^2 - 100 < 0$

$(k - 10)(k + 10) < 0$, and $-10 < k < 10$

14. (C)

☐ ☐ ☐ ☐ 20 ☐ ☐ ☐ ☐
1 2 3 4   21 22 23 ?

When assigned possibly the minimum
numbers as shown above, the last one will be
the greatest one.
The sum of the nine numbers is
$9 \times 15 = 135$. Therefore,
$135 - (1 + 2 + 3 + 4 + 20 + 21 + 22 + 23) = 39$

15. (D)

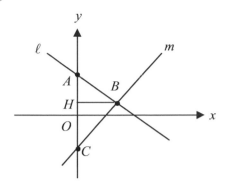

The y-intercept of line m, $y = -7$ and the y-intercept of line ℓ, $y = 8$, $x = 8$

The intersection of the two lines can be obtained from the two equations.

$2x - 7 = -x + 8$, $x = 5$. Then

$AC = 8 - (-7) = 15$. $BH = 5$. Therefore, the area of $\triangle ABC$ is $\dfrac{1}{2}(15)(5) = 37\dfrac{1}{2}$.

---

16. (C)

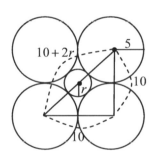

The triangle in the figure is an isosceles right triangle.

$10 + 2r = 10\sqrt{2}$.

Therefore, $r = \dfrac{10\sqrt{2} - 10}{2} = 5(\sqrt{2} - 1)$

$\boxed{END}$

NO MATERIAL ON THIS PAGE

Dr. John Chung's SAT Math

TEST
20

ANSWER SHEET   TEST #:

SECTION 3

1 ⒶⒷ©ⒹⒺ	11 ⒶⒷ©ⒹⒺ	21 ⒶⒷ©ⒹⒺ	31 ⒶⒷ©ⒹⒺ
2 ⒶⒷ©ⒹⒺ	12 ⒶⒷ©ⒹⒺ	22 ⒶⒷ©ⒹⒺ	32 ⒶⒷ©ⒹⒺ
3 ⒶⒷ©ⒹⒺ	13 ⒶⒷ©ⒹⒺ	23 ⒶⒷ©ⒹⒺ	33 ⒶⒷ©ⒹⒺ
4 ⒶⒷ©ⒹⒺ	14 ⒶⒷ©ⒹⒺ	24 ⒶⒷ©ⒹⒺ	34 ⒶⒷ©ⒹⒺ
5 ⒶⒷ©ⒹⒺ	15 ⒶⒷ©ⒹⒺ	25 ⒶⒷ©ⒹⒺ	35 ⒶⒷ©ⒹⒺ
6 ⒶⒷ©ⒹⒺ	16 ⒶⒷ©ⒹⒺ	26 ⒶⒷ©ⒹⒺ	36 ⒶⒷ©ⒹⒺ
7 ⒶⒷ©ⒹⒺ	17 ⒶⒷ©ⒹⒺ	27 ⒶⒷ©ⒹⒺ	37 ⒶⒷ©ⒹⒺ
8 ⒶⒷ©ⒹⒺ	18 ⒶⒷ©ⒹⒺ	28 ⒶⒷ©ⒹⒺ	38 ⒶⒷ©ⒹⒺ
9 ⒶⒷ©ⒹⒺ	19 ⒶⒷ©ⒹⒺ	29 ⒶⒷ©ⒹⒺ	39 ⒶⒷ©ⒹⒺ
10 ⒶⒷ©ⒹⒺ	20 ⒶⒷ©ⒹⒺ	30 ⒶⒷ©ⒹⒺ	40 ⒶⒷ©ⒹⒺ

SECTION 5

1 ⒶⒷ©ⒹⒺ	11 ⒶⒷ©ⒹⒺ	21 ⒶⒷ©ⒹⒺ	31 ⒶⒷ©ⒹⒺ
2 ⒶⒷ©ⒹⒺ	12 ⒶⒷ©ⒹⒺ	22 ⒶⒷ©ⒹⒺ	32 ⒶⒷ©ⒹⒺ
3 ⒶⒷ©ⒹⒺ	13 ⒶⒷ©ⒹⒺ	23 ⒶⒷ©ⒹⒺ	33 ⒶⒷ©ⒹⒺ
4 ⒶⒷ©ⒹⒺ	14 ⒶⒷ©ⒹⒺ	24 ⒶⒷ©ⒹⒺ	34 ⒶⒷ©ⒹⒺ
5 ⒶⒷ©ⒹⒺ	15 ⒶⒷ©ⒹⒺ	25 ⒶⒷ©ⒹⒺ	35 ⒶⒷ©ⒹⒺ
6 ⒶⒷ©ⒹⒺ	16 ⒶⒷ©ⒹⒺ	26 ⒶⒷ©ⒹⒺ	36 ⒶⒷ©ⒹⒺ
7 ⒶⒷ©ⒹⒺ	17 ⒶⒷ©ⒹⒺ	27 ⒶⒷ©ⒹⒺ	37 ⒶⒷ©ⒹⒺ
8 ⒶⒷ©ⒹⒺ	18 ⒶⒷ©ⒹⒺ	28 ⒶⒷ©ⒹⒺ	38 ⒶⒷ©ⒹⒺ
9 ⒶⒷ©ⒹⒺ	19 ⒶⒷ©ⒹⒺ	29 ⒶⒷ©ⒹⒺ	39 ⒶⒷ©ⒹⒺ
10 ⒶⒷ©ⒹⒺ	20 ⒶⒷ©ⒹⒺ	30 ⒶⒷ©ⒹⒺ	40 ⒶⒷ©ⒹⒺ

9   10   11   12   13

14   15   16   17   18

1	Ⓐ Ⓑ Ⓒ Ⓓ Ⓔ	11	Ⓐ Ⓑ Ⓒ Ⓓ Ⓔ	21	Ⓐ Ⓑ Ⓒ Ⓓ Ⓔ	31	Ⓐ Ⓑ Ⓒ Ⓓ Ⓔ
2	Ⓐ Ⓑ Ⓒ Ⓓ Ⓔ	12	Ⓐ Ⓑ Ⓒ Ⓓ Ⓔ	22	Ⓐ Ⓑ Ⓒ Ⓓ Ⓔ	32	Ⓐ Ⓑ Ⓒ Ⓓ Ⓔ
3	Ⓐ Ⓑ Ⓒ Ⓓ Ⓔ	13	Ⓐ Ⓑ Ⓒ Ⓓ Ⓔ	23	Ⓐ Ⓑ Ⓒ Ⓓ Ⓔ	33	Ⓐ Ⓑ Ⓒ Ⓓ Ⓔ
4	Ⓐ Ⓑ Ⓒ Ⓓ Ⓔ	14	Ⓐ Ⓑ Ⓒ Ⓓ Ⓔ	24	Ⓐ Ⓑ Ⓒ Ⓓ Ⓔ	34	Ⓐ Ⓑ Ⓒ Ⓓ Ⓔ
5	Ⓐ Ⓑ Ⓒ Ⓓ Ⓔ	15	Ⓐ Ⓑ Ⓒ Ⓓ Ⓔ	25	Ⓐ Ⓑ Ⓒ Ⓓ Ⓔ	35	Ⓐ Ⓑ Ⓒ Ⓓ Ⓔ
6	Ⓐ Ⓑ Ⓒ Ⓓ Ⓔ	16	Ⓐ Ⓑ Ⓒ Ⓓ Ⓔ	26	Ⓐ Ⓑ Ⓒ Ⓓ Ⓔ	36	Ⓐ Ⓑ Ⓒ Ⓓ Ⓔ
7	Ⓐ Ⓑ Ⓒ Ⓓ Ⓔ	17	Ⓐ Ⓑ Ⓒ Ⓓ Ⓔ	27	Ⓐ Ⓑ Ⓒ Ⓓ Ⓔ	37	Ⓐ Ⓑ Ⓒ Ⓓ Ⓔ
8	Ⓐ Ⓑ Ⓒ Ⓓ Ⓔ	18	Ⓐ Ⓑ Ⓒ Ⓓ Ⓔ	28	Ⓐ Ⓑ Ⓒ Ⓓ Ⓔ	38	Ⓐ Ⓑ Ⓒ Ⓓ Ⓔ
9	Ⓐ Ⓑ Ⓒ Ⓓ Ⓔ	19	Ⓐ Ⓑ Ⓒ Ⓓ Ⓔ	29	Ⓐ Ⓑ Ⓒ Ⓓ Ⓔ	39	Ⓐ Ⓑ Ⓒ Ⓓ Ⓔ
10	Ⓐ Ⓑ Ⓒ Ⓓ Ⓔ	20	Ⓐ Ⓑ Ⓒ Ⓓ Ⓔ	30	Ⓐ Ⓑ Ⓒ Ⓓ Ⓔ	40	Ⓐ Ⓑ Ⓒ Ⓓ Ⓔ

Math Scoring Worksheet

A. Section 3

_____   _____
numer of correct   number of incorrect

+   +

B. Section 5 (1-8)

_____   _____
numer of correct   number of incorrect

+

C. Section 5 (9-18)

_____
numer of correct

+   +

D. Section 7

_____   _____
numer of correct   number of incorrect

=   =

E. Total Unrounded Raw Score

_____ − _____ ÷4 = _____
numer of correct   number of incorrect

F. Total Rounded Raw Score   _____ (See table)

Math Score Range = | ——— |

Math Conversion Table

Raw Score	Scaled Score	Raw Score	Scaled Score
54	800	23	490-550
53	780-800	22	480-540
52	760-800	21	470-530
51	740-800	20	460-520
50	720-780	19	450-510
49	700-760	18	450-510
48	690-750	17	440-500
47	680-740	16	430-490
46	670-730	15	420-480
45	660-720	14	420-480
44	650-710	13	410-470
43	650-710	12	400-460
42	640-700	11	390-450
41	630-690	10	380-440
40	620-680	9	390-430
39	610-670	8	380-420
38	610-670	7	370-410
37	600-660	6	360-400
36	590-650	5	340-380
35	580-640	4	320-370
34	570-630	3	310-360
33	560-620	2	300-350
32	560-620	1	270-320
31	550-610	0	240-300
30	540-600	-1	200-290
29	530-590	-2	200-270
28	530-590	-3	200-260
27	520-580	-4	200-240
26	510-570	-5	200-220
25	500-560	-6 and below	200
24	500-560		

SECTION 3
Time- 25 minutes
20 Questions

Turn to Section 3 (Page 1) of your answer sheet to answer the questions in this section.

Directions: For this section, solve each problem and decide which is the best of the choices given. Fill in the corresponding circle on the answer sheet. You may use any available space for scratchwork.

Notes

1. The use of a calculator is permitted.
2. All numbers used are real numbers.
3. Figures that accompany problems in this test are intended to provide information useful in solving the problems. They are drawn as accurately as possible EXCEPT when it is stated in a specific problem that the figure is not drawn to scale. All figure lie in a plane unless other indicated.
4. Unless otherwise specified, the domain of any function f is assumed to be set of all real numbers x for which $f(x)$ is a real number.

Reference Informatiom

$A = \pi r^2$
$C = 2\pi r$
$A = \ell w$
$A = \dfrac{1}{2}bh$
$V = \ell wh$
$V = \pi r^2 h$
$c^2 = a^2 + b^2$
Special Right Triangles

The numbers of degrees of arc in a circle is $360°$.

The sum of the measures in degrees of the angles is $180°$.

---

1. If $|t| = 2$, what is the value of $(t+3)(t-3)$?

    (A) −5
    (B) 0
    (C) 5
    (D) 7
    (E) 9

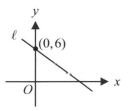

Note: Figure not drawn to scale.

2. In the figure above, the slope of line ℓ is $-\dfrac{2}{3}$.

    What is the x-intercept of the line?

    (A) 4
    (B) 6
    (C) 9
    (D) 10
    (E) 12

GO ON TO THE NEXT PAGE

3. In a certain chess club of n members, each of the members played a game with each of the other members. If there were 36 games, how many members are in the chess club?

(A) 12
(B) 9
(C) 8
(D) 7
(E) 6

4. The sum of the first 25 positive odd integers is subtracted from the sum of the first 25 positive even integers. What is the value of the result?

(A) 0
(B) 25
(C) 50
(D) 75
(E) 100

5. If the graph of the function g in the xy-plane contains the points $(0,4)$, $(-2,0)$, and $(2,0)$, which of the following could be true?

I. The graph of g is a parabola.
II. The graph of g has a maximum value.
III. The graph of g is a line

(A) I only
(B) II only
(C) III only
(D) I and II only
(E) I, II, and III

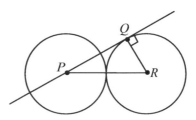

6. In the figure above, two identical circles have centers at points P and R with radius of 5, and are tangent to each other. If a line is tangent to the circle, what is the area of $\triangle PQR$?

(A) 50

(B) $25\sqrt{3}$

(C) 25

(D) $\dfrac{25\sqrt{3}}{2}$

(E) $\dfrac{25}{2}$

GO ON TO THE NEXT PAGE ⇨

7. A right circular cylinder has a volume of 150π and a height of 6. What is the circumference of the base?

   (A) 25π
   (B) 15π
   (C) 10π
   (D) 5π
   (E) 3π

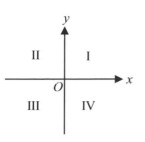

9. In the figure above, which quadrants contain points (x, y) that satisfy the condition $\dfrac{x}{y} < 0$?

   (A) I only
   (B) I and II only
   (C) I and III only
   (D) II and IV only
   (E) III and IV only

8. The angles of a quadrilateral $ABCD$ satisfy $\angle A = 2\angle B = 4\angle C = 8\angle D$. How many degrees are in $\angle D$?

   (A) 192°
   (B) 96°
   (C) 84°
   (D) 48°
   (E) 24°

10. If x, y, and z are all integers where $-5 < x < y < z < 4$, which of the following is the greatest possible value of xz?

    (A) 2
    (B) 3
    (C) 4
    (D) 5
    (E) 8

GO ON TO THE NEXT PAGE

t	3	b	9
$f(t)$	b	12	k

11. The table above gives the values of $f(t)$ for the selected values of $t.$ If $f(t)$ varies directly proportional to $t,$ which of the following could be the value of k?

(A) 12
(B) 18
(C) 36
(D) 54
(E) 108

12. If $rst < 0$ and $r^3s^2t > 0$, which of the following must be true?

(A) $r > 0$
(B) $r < 0$
(C) $t > 0$
(D) $s < 0$
(E) $s > 0$

GO ON TO THE NEXT PAGE

(D)

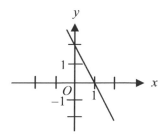

13. The figure above shows the graph of
$y = mx + b$, where m and b are constants.
Which of the following represents the graph of
$y = -2mx - 2b$?

(E)

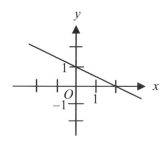

(A)

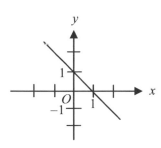

(B)

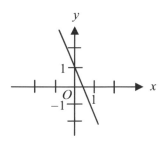

(C)

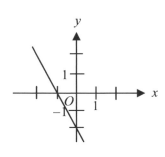

GO ON TO THE NEXT PAGE

14. For the function f, $f(x)$ is inversely proportional to x. If $f(k) = a$, what is the value of $f\left(\dfrac{k}{3}\right)$?

(A) $\dfrac{a}{3}$

(B) $\dfrac{2a}{3}$

(C) $3a$

(D) $6a$

(E) $9a$

16. If $\sqrt{x+p} = \sqrt{x-q}$, which of the following could be true?

    I. $p = 0$ and $q = 0$
    II. $pq = 0$
    III. $p + q = 0$

(A) I only
(B) II only
(C) III only
(D) I and III only
(E) II and III only

15. If it takes 12 people 8 hours to do a certain job, how many hours would it take 3 people, working at the same rate, to do $\dfrac{1}{4}$ of the same job?

(A) 4
(B) 8
(C) 12
(D) 18
(E) 36

17. Let the function f and g be defined by $f(x) = x + 3$, and $g(2x) = 2f(x)$. If $g(2m) = g(m) + 10$, what is the value of m?

(A) 4
(B) 6
(C) 8
(D) 10
(E) 20

GO ON TO THE NEXT PAGE

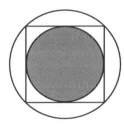

18. In the figure above, a square is inscribed in a circle, and a small circle is inscribed in the square. If a point is chosen from the figure, what is the probability that the point will be in the shaded region?

(A) $\dfrac{1}{4}$

(B) $\dfrac{1}{3}$

(C) $\dfrac{1}{2}$

(D) $\dfrac{2}{3}$

(E) $\dfrac{3}{4}$

19. If $0 < a < b < c < 6$, where a, b, and c are integers, and b is the average (arithmetic mean) of a and c, how many values of $\{a, b, c\}$ are possible?

(A) 2
(B) 3
(C) 4
(D) 5
(E) 6

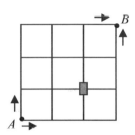

20. In the figure above, a path from point A to point B is determined by moving upward or to the right along the paths. How many different paths are possible from A to B that do not pass through the point ▧?

(A) 12
(B) 11
(C) 10
(D) 9
(E) 8

STOP

**If you finish before time is called, you may check your work on this section only.
Do not turn to any other section in the test.**

SECTION 5
Time- 25 minutes
18 Questions

Turn to Section 5 (Page 1) of your answer sheet to answer the questions in this section.

Directions: For this section, solve each problem and decide which is the best of the choices given. Fill in the corresponding circle on the answer sheet. You may use any available space for scratchwork.

Notes

1. The use of a calculator is permitted.
2. All numbers used are real numbers.
3. Figures that accompany problems in this test are intended to provide information useful in solving the problems. They are drawn as accurately as possible EXCEPT when it is stated in a specific problem that the figure is not drawn to scale. All figure lie in a plane unless other indicated.
4. Unless otherwise specified, the domain of any function f is assumed to be set of all real numbers x for which $f(x)$ is a real number.

Reference Informatiom

   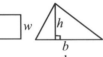 

$A = \pi r^2$
$C = 2\pi r$
$A = \ell w$
$A = \dfrac{1}{2}bh$
$V = \ell wh$
$V = \pi r^2 h$
$c^2 = a^2 + b^2$
Special Right Triangles

The numbers of degrees of arc in a circle is $360°$.

The sum of the measures in degrees of the angles is $180°$.

---

1. If $a + b < 0$ and $c < 0$, which of the following must be true?

(A) $a + b < c$

(B) $\dfrac{a}{c} + \dfrac{b}{c} < 0$

(C) $ac + bc > 0$

(D) $a + b + c > 0$

(D) $a + b - c > 0$

2. Which of the following could be the value of n^3, where n is an integer?

(A) 0.8×10^{14}
(B) 0.8×10^{15}
(C) 0.8×10^{16}
(D) 0.8×10^{17}
(E) 0.8×10^{18}

GO ON TO THE NEXT PAGE

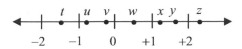

3. On the number line above, which of the following could be the point of $\left|\dfrac{u}{w}\right|$?

(A) t
(B) u
(C) x
(D) y
(E) z

4. If the product of two positive integers is greater than 20, and the sum of the two integers is 10, which of the following cannot be one of the integers?

(A) 3
(B) 5
(C) 6
(D) 7
(E) 8

5. If k is a negative number, which of the following must be a positive number?

(A) $k - 5$
(B) $k + 5$
(C) $k^2 - 5$
(D) $(k + 5)(k - 5)$
(E) $k(k - 5)$

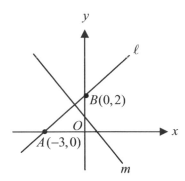

Note: Figure not drawn to scale.

6. In the figure above, line m is perpendicular to line ℓ. Which of the following equations could describe line m?

(A) $y = -\dfrac{2}{3}x - 1$

(B) $y = \dfrac{2}{3}x + 1$

(C) $y = -\dfrac{3}{2}x + 1$

(D) $y = \dfrac{3}{2}x + 1$

(E) $y = -\dfrac{3}{2}x - 1$

GO ON TO THE NEXT PAGE

7. If $x > 0$ and $\dfrac{x^2}{x^{-2m}} = \dfrac{1}{x^2}$, what is the value of m?

(A) -2

(B) $-\dfrac{3}{2}$

(C) $-\dfrac{1}{2}$

(D) $\dfrac{3}{2}$

(E) 3

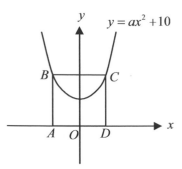

Note: Figure not drawn to scale.

8. The graph of the function $y = ax^2 + 10$ is shown above in the xy-plane. If the area of square $ABCD$ is 400, what is the value of a?

(A) 100

(B) 10

(C) 1

(D) $\dfrac{1}{10}$

(E) $\dfrac{1}{100}$

GO ON TO THE NEXT PAGE

Directions: For Students-Produced Response questions 9-18, use the grid at the bottom of the answer sheet page on which you have answered questions 1-8.

Each of the remaining 10 questions requires you to solve the problem and enter your answer by making the circles in the special grid, as shown in the examples below. You may use any available space for scratchwork.

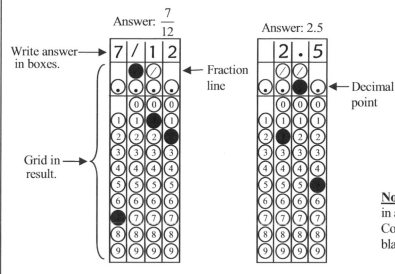

Answer: $\frac{7}{12}$

Write answer in boxes. ← Fraction line

Grid in result.

Answer: 2.5 ← Decimal point

Answer: 201
Either position is correct.

Note: You may start your answers in any column, space permitting. Columns not needed should be left blank.

- Mark no more than one circle in any column.

- Because the answer sheet will be machine-scored, **you will receive credit only if the circles are filled in correctly.**

- Although not required, it is suggested that you write your answer in the boxes at the top of the columns to help you fill in the circles accurately.

- Some problems may have more than one correct answer. In such cases, grid only one answer.

- No question has a negative answer.

- **Mixed numbers** such as $3\frac{1}{2}$ must be gridded as 3.5 or 7/2. (If [3 1 / 2] is gridded, it will be interpreted as $\frac{31}{2}$, not $3\frac{1}{2}$.)

- **Decimal Answers:** If you obtain a decimal answer with more digits than the grid can accommodate, it may be either rounded or truncated, but it must fill the entire grid. For example, if you obtain an answer such as 0.6666..., you should record your result as .666 or .667. **A less accurate value such as .66 or .67 will be scored as incorrect.**

Acceptable ways to grid $\frac{2}{3}$ are:

9. If p is inversely proportional to q^2 and if $p = 10$ when $q = 4$, what is the value of p when $q = 10$?

10. How many different pairs of positive even integers have a product of 120?

GO ON TO THE NEXT PAGE

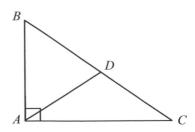

Note: Figure not drawn to scale.

11. In the figure above, $\triangle ABD$ is an equilateral triangle. If the area of $\triangle ABD$ is 8, what is the area of $\triangle ABC$?

12. For all x, the function f is defined by $f(x) = x^2 - 5$, and the function g is defined by $g(x) = f(x) - 13$. If $g(k) = 7$, what is the positive value of k?

13. If a ball is thrown straight up at an initial speed of 40 feet per second, its height h, in feet, after t seconds is given by the formula $h = 40t - 10t^2$. What is the maximum height, in feet, that the ball can reach?

14. The nth term of a sequence is given by $a_n = a + nk$, where a and k are constants. If the 5th term is 20 and the 10th term is 30 in the sequence, what is the value of a_1?

GO ON TO THE NEXT PAGE

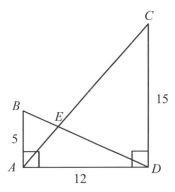

Note: Figure not drawn to scale.

15. In the figure above, $AB = 5$, $AD = 12$, and $CD = 15$. What is the area of $\triangle CED$?

16. When a positive integer m is divided by 7, the remainder is 4. What is the remainder when $20m$ is divided by 35?

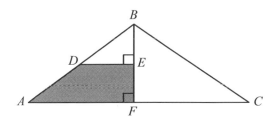

Note: Figure not drawn to scale.

17. In $\triangle ABC$ above, $AB = BC$ and D is the midpoint of $\overline{AB}$. If the area of the shaded region is 1, what is the area of $\triangle ABC$?

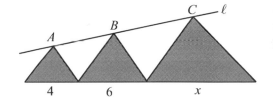

Note: Figure not drawn to scale.

18. The figure above shows three equilateral triangles with sides of length 4, 6, and x, respectively. If $A, B,$ and C lie on line ℓ, what is the value of x?

STOP

If you finish before time is called, you may check your work on this section only.
Do not turn to any other section in the test.

SECTION 7
Time- 20 minutes
16 Questions

Turn to Section 7 (Page 2) of your answer sheet to answer the questions in this section.

Directions: For this section, solve each problem and decide which is the best of the choices given. Fill in the corresponding circle on the answer sheet. You may use any available space for scratchwork.

Notes

1. The use of a calculator is permitted.
2. All numbers used are real numbers.
3. Figures that accompany problems in this test are intended to provide information useful in solving the problems. They are drawn as accurately as possible EXCEPT when it is stated in a specific problem that the figure is not drawn to scale. All figure lie in a plane unless other indicated.
4. Unless otherwise specified, the domain of any function f is assumed to be set of all real numbers x for which $f(x)$ is a real number.

Reference Informatiom

  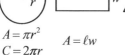

$A = \pi r^2$
$C = 2\pi r$
$A = \ell w$
$A = \frac{1}{2}bh$
$V = \ell w h$
$V = \pi r^2 h$
$c^2 = a^2 + b^2$
Special Right Triangles

The numbers of degrees of arc in a circle is $360°$.

The sum of the measures in degrees of the angles is $180°$.

---

1. If a six-pack of soda costs k cents, how many individual cans can be purchased for d dollars?

(A) $\dfrac{d}{k}$

(B) $\dfrac{6d}{k}$

(C) $\dfrac{100d}{k}$

(D) $\dfrac{600d}{k}$

(E) $\dfrac{kd}{600}$

2. If the average (arithmetic mean) of 10, a, and b is 20, what is the average of a and b?

(A) 25
(B) 28
(C) 30
(D) 40
(E) 50

GO ON TO THE NEXT PAGE

3. Abby, Bernard, and Cabin made a total of 95 candy bars. Bernard made 5 times as many as Cabin, and Cabin made three times as many as Abby. How many candy bars did Abby make?

(A) 5
(B) 10
(C) 15
(D) 45
(E) 75

4. If $0.1k$ percent of n is 5, what is $3k$ percent of n?

(A) 15
(B) 50
(C) 75
(D) 120
(E) 150

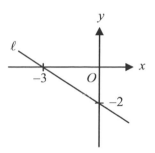

Note: Figure not drawn to scale.

5. In the figure above, what is the equation of line ℓ?

(A) $y = -\dfrac{2}{3}x - 2$

(B) $y = -\dfrac{3}{2}x - 2$

(C) $y = -\dfrac{3}{2}x - 3$

(D) $y = \dfrac{2}{3}x - 2$

(E) $y = \dfrac{3}{2}x - 3$

6. If $x < 0$, which of the following could have solution?

I. $|x| = -x$
II. $|x| = x$
III. $|x| = \sqrt{x^2}$

(A) I only
(B) II only
(C) III only
(D) I and III only
(E) I, II, and III

GO ON TO THE NEXT PAGE

7. For positive integers x and y, $x + y = 20$. Which of the following cannot be a possible value of $x - y$?

(A) −8
(B) −2
(C) 3
(D) 4
(E) 8

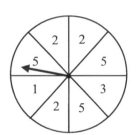

8. In the figure above, the spinner is spun 24 times. How many times would it be expected to land on 5?

(A) 3 times
(B) 9 times
(C) 10 times
(D) 12 times
(E) 14 times

Questions 9-10 refer to the following figures and information.

Movies	Number of people
X	25
Y	30
Neither	5

Figure 1

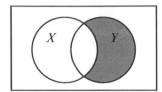

Figure 2

Note: Figure 2 not drawn to scale.

In figure 1 above, 50 people were asked if they had seen two particular movies X and Y. Figure 2 above is another way to which movies they have seen.

9. How many people have seen both two movies?

(A) 5
(B) 10
(C) 15
(D) 20
(E) 25

10. What is the total number of people represented by the shaded region in figure 2?

(A) 5
(B) 10
(C) 15
(D) 20
(E) 30

GO ON TO THE NEXT PAGE

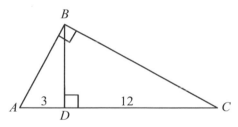

Note: Figure not drawn to scale.

11. In $\triangle ABC$, $AD = 3$ and $DC = 12$. What is the length of $\overline{AB}$?

(A) $\sqrt{40}$ (approximately 6.32)
(B) $\sqrt{43}$ (approximately 6.56)
(C) $\sqrt{45}$ (approximately 6.71)
(D) $\sqrt{48}$ (approximately 6.93)
(E) $\sqrt{50}$ (approximately 7.07)

12. Twice the sum of two consecutive integers is k less than five times the smaller integer, where k is a positive integer. What is the larger integer, in terms of k?

(A) $k + 2$
(B) $k + 3$
(C) $k + 5$
(D) $2k + 2$
(E) $2k + 6$

PROFIT OF SMITH COMPANY
(in thousand-dollars)

Year	1	2	3	4	5	6	7	8
Profit	10	12	16	24	28	30	31	31.2

13. Which of the following graphs best represents the information in the table above?

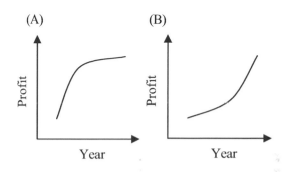

(A)   (B)

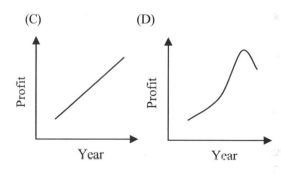

(C)   (D)

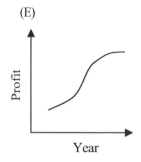

(E)

GO ON TO THE NEXT PAGE

14. If $\dfrac{a-2b}{b-2a} = \dfrac{3}{2}$, what is the value of $\dfrac{2a}{b}$?

(A) $\dfrac{2}{7}$

(B) $\dfrac{3}{8}$

(C) $\dfrac{7}{8}$

(D) $\dfrac{7}{4}$

(E) $\dfrac{8}{3}$

15. If $3^k + 3^k + 3^k = (27)^5$, what is the value of k?

(A) 15
(B) 14
(C) 9
(D) 6
(E) 3

16. In the xy-coordinate plane, the graph of $y = -x^2 + 12$ intersects line ℓ at $(p, 3)$ and $(t, -4)$. What is the greatest possible value of the slope of ℓ?

(A) 9
(B) 7
(C) 1
(D) −1
(E) −9

STOP

If you finish before time is called, you may check your work on this section only.
Do not turn to any other section in the test.

NO MATERIAL ON THIS PAGE

TEST 20   SECTION 3

1. (A)

$|t| = 2$, $t^2 = 4$. Therefore,

$(t+3)(t-3) = t^2 - 9 = 4 - 9 = -5$.

Or, substitute $t = 2$ or -2 into the expression.

2. (C)

The equation of line is $y = -\dfrac{2}{3}x + 6$.

Therefore, $0 = -\dfrac{2}{3}x + 6$, $x = 9$.

Or, the answer can be graphed as follows.

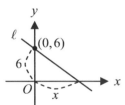

Slope $= -\dfrac{6}{x} = -\dfrac{2}{3}$, that is $x = 9$

3. (B)

2 members have one game. 3 members have 3 games.

#of members	#of games
2	1
3	1+2
4	1+2+3
9	1+2+3+4+5+6+7+8=36

In the table above, there are 9 members in the club. Or, $_nC_2 = \dfrac{n(n+1)}{2} = 36$, then $n = 9$.

---

4. (B)

$S_{even} = 2+4+6+......+48+50$

$\dfrac{S_{odd} = 1+3+5+.......+47+49}{S = 1+1+1+1+...\ +1+\ 1}$

Therefore, $S = 1 \times 25 = 25$

Or, $S_{even} = \dfrac{(2+50) \times 25}{2} = 650$

$S_{odd} = \dfrac{(1+49) \times 25}{2} = 625$

Therefore, $650 - 625 = 25$.

---

5. (D)

The possible graph is as follows.

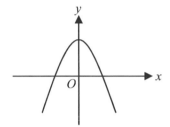

---

6. (D)

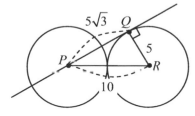

$\triangle PQR$ is a special right triangle.

Therefore, the area of $\triangle PQR$

$= \dfrac{5\sqrt{3} \times 5}{2} = \dfrac{25\sqrt{3}}{2}$.

---

7. (C)

The volume of a cylinder $V = \pi r^2 h = 150\pi$.

For $h = 6$, $\pi r^2 (6) = 150\pi$, that is $r = 5$.

Therefore, $C = 2\pi r = 10\pi$.

---

8. (E)

If $\angle D = x$, then $\angle C = 2x$, $\angle B = 4x$, and $\angle A = 8x$.

Therefore, $x + 2x + 4x + 8x = 15x = 360$.

$x = 24$

---

9. (D)

$\dfrac{x}{y} < 0$ implies $x > 0$ and $y < 0$ (**IV**) or

$x < 0$ and $y > 0$ (**II**)

---

10. (E)

x and z must be both positive or negative.
There are two possible outcomes as follows.
(1) $-5, < 1 < 2 < 3 < 4$, $xz = 3$
(2) $-5 < -4 < -3 < -2 < 4$, $xz = 8$
Therefore, the greatest one is 8.

---

11. (B)

$\dfrac{b}{3} = \dfrac{12}{b}$, $b^2 = 36$, $b = 6$ or -6.

Therefore, $\dfrac{12}{\pm 6} = \dfrac{k}{9}$, $k = 18$ or -18.

---

12. (D)

$r^3 s^2 t > 0$ is equivalent to $rt > 0$. (Divided by positive $r^2 s^2$). From $rt > 0$ and $rst < 0$, s must be negative.

---

13. (D)

In the graph, $m = 1$ and $b = -1$. Therefore, the graph of $y = -2(1)x - 2(-1) = -2x + 2$.
(D) is the answer.

---

14. (C)

Inverse proportion is $xy = K$ (constant).

$f(k) = a$ implies (k, a) and

$f\left(\dfrac{k}{3}\right) = p$ implies $\left(\dfrac{k}{3}, p\right)$. By definition,

$k \times a = \dfrac{k}{3} \times p$. Therefore, $p = 3a$.

---

15. (B)

By inverse proportion, $12 \times 8 = 3 \times h$,

$h = 32$ hrs.

Therefore, $\dfrac{1}{4}(32) = 8$hrs.

---

16. (D)

$\sqrt{x+p} = \sqrt{x-q}$, $x+p = x-q$, that is
$p+q = 0$.

$p = q = 0$ is also true.

Or, for the question " could be true", you can check I, II, and III

I. $\sqrt{x} = \sqrt{x}$ II. If $p = 0$, $\sqrt{x} = \sqrt{x-q}$

III. $p = -q$, $\sqrt{x-q} = \sqrt{x-q}$

But for the question " must be true", you have to solve it algebraically.

---

17. (D)

$g(2x) = 2f(x) = 2x+6$. $g(2m) = 2m+6$

and $g(m) = m+6$. Because,

$g(2m) = g(m)+10$

$2m+6 = m+6+10$, that is , $m = 10$.

---

18. (C)

Use a convenient number. If the radius of the smaller circle is 1, the radius of the larger circle is $\sqrt{2}$.

The ratio of the areas of smaller circle to the larger circle $= \dfrac{\pi(1)^2}{\pi(\sqrt{2})^2} = \dfrac{1}{2}$. Or, if the ratio of

the lengths is $\dfrac{1}{\sqrt{2}}$, the ratio of the areas is

$\dfrac{1^2}{\left(\sqrt{2}\right)^2} = \dfrac{1}{2}$.

---

19. (C)

$0 < a < b < c < 6$, $b = \dfrac{a+c}{2}$, therefore

$2b = a+c$.

If $b = 2$, $a+c = 4$. $(a,b,c) \rightarrow (1,2,3)$

If $b = 3$, $a+c = 6$.

   $(a,b,c) \rightarrow (1,3,5),(2,3,4)$

If $b = 4$, $a+c = 8$. $(a,b,c) \rightarrow (3,4,5)$

If $b = 5$, $a+c = 10$. Not working.

---

20. (B)

Add numbers at intersections.

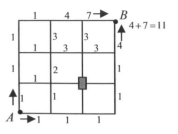

---

1. (C)

Negative number $\times$ Negative number $=$ Positive.

$(a+b)c > 0$, that is, $ab+ac > 0$

---

2. (C)

$0.8 \times 10^{16} = 2^3 \times 10^{15} = (2 \cdot 10^5)^3$

---

3. (D)

$u \simeq -0.8$ and $w \simeq 0.5$. $\left|\dfrac{-0.8}{0.5}\right| = 1.6$.

Point y is the number.

---

4. (E)

The possible outcomes which satisfy $a+b = 10$ is as follows. $(1,9)$ $(2,8)$ $(3,7)$ $(4,6)$ $(5,5)$. But $(3,7)$ $(4,6)$ $(5,5)$ have the product greater than 20. Therefore, 8 cannot be one of the pairs.

---

5. (E)

$k(k-5) = negative \times negative = positive$. Or, you can use -1 and -10 to check the choices.

---

6. (C)

The slope of line ℓ is $\dfrac{2}{3}$ and y-intercept is 2.

The equation of line m is, $y = -\dfrac{3}{2}x+b$,

where, $0 < b < 2$. Therefore, (C) is the correct.

---

7. (A)

$\dfrac{x^2}{x^{-2m}} = x^{2-(-2m)} = x^{2+2m}$ and $\dfrac{1}{x^2} = x^{-2}$.

Therefore, $2+2m = -2$, $m = -2$.

8. (D)

The length of a side of the square is 20. The coordinates of point C is $(10, 20)$. Therefore,

$$20 = a(10)^2 + 10 = 100a + 10, \quad a = \frac{1}{10}.$$

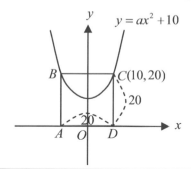

9. 1.6 or $\dfrac{8}{5}$

$p \times q^2 = K$ (constant).

$(10)(4^2) = p(10^2) = K$

Therefore, $p = \dfrac{160}{100} = 1.6$ or $\dfrac{8}{5}$.

10. 4

$(2, 60)$ $(4, 30)$ $(6, 20)$ $(10, 12)$ have the product of 120. $(8, 15)$ is not even pair.

11. 16

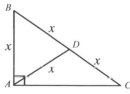

$\triangle ABD$ is an equilateral triangle and $\triangle ADC$ is an isosceles triangle. The areas of the two triangles are equal, because they have the same lengths of bases and the same height. Therefore, the area of $\triangle ABC$ is 16.

12. 5

$g(k) = f(k) - 13 = (k^2 - 5) - 13 = k^2 - 18$.

Therefore, $k^2 - 18 = 7$, that is, $k = \pm 5$, the positive one is 5.

13. 40

The axis of symmetry is, $t = \dfrac{-40}{-20} = 2$,

therefore, the maximum height is

$h(2) = 40(2) - 10(2^2) = 40$

Or, use graph as follows.

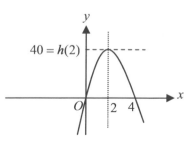

The x-intercepts are 0 and 4. when $x = 2$, the graph has a maximum height of 40.

14. 12

$a_{10} = a + 10k = 30$ ---(1) and

$a_5 = a + 5k = 20$ ---(2)

From (1) − (2), $k = 2$ and $a = 10$. Therefore,

$a_1 = a + k = 10 + 2 = 12$

15. 67.5

$\triangle ABE$ and $\triangle DEC$ are similar. The ratio of lengths is $5 : 15$ or $1 : 3$. In the figure below,

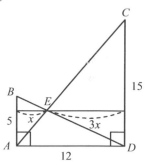

$x + 3x = 12$, $x = 3$, $3x = 9$. Therefore, the

area of $\triangle CED$ is, $\dfrac{1}{2}(15 \times 9) = 67.5$

16. 10

Choose $m = 11$, 18,…. If $m = 11$, then

$20m = 220$.

When 220 is divided by 35, the remainder is 10.

Or, algebraically, let $m = 7q + 4$, where q is a quotient. Then
$20m = 20(7q+4) = 140q + 80$, When divided by 35, $20m = 35(4q+2)+10$.
Therefore, the remainder is 10.

---

17. $\dfrac{8}{3}$

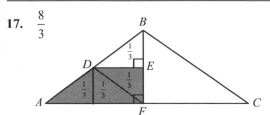

In the figure, each triangle has an area of $\dfrac{1}{3}$.

Therefore, the area of $\triangle ABC$ is $\dfrac{8}{3}$.

---

18. 9

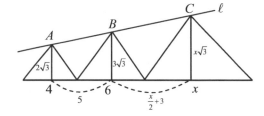

The heights of the three equilateral triangles are $2\sqrt{3}$, $3\sqrt{3}$, and $\dfrac{x\sqrt{3}}{2}$, respectively. The slope of line ℓ between any two points are equal. That is,

$$\frac{3\sqrt{3}-2\sqrt{3}}{5} = \frac{x\sqrt{3}-3\sqrt{3}}{\dfrac{x}{2}+3}$$

$$\frac{\sqrt{3}}{5} = \frac{\sqrt{3}\left(\dfrac{x}{2}-3\right)}{\dfrac{x}{2}+3} \quad \text{or} \quad \frac{1}{5} = \frac{\dfrac{x}{2}-3}{\dfrac{x}{2}+3}$$

Cross multiplication gives
$\dfrac{5x}{2} - 15 = \dfrac{x}{2} + 3$ or $2x = 18$. Therefore, $x = 9$.

If you use $x = 2k$, $\dfrac{\sqrt{3}}{5} = \dfrac{k\sqrt{3}-3\sqrt{3}}{k+3}$, you can get the answer in easier way. $k = 4.5$, then $x = 9$.

Or

The ratio of lengths of the triangles= $4 : 6 : x$.
The ratio of areas is $4^2 : 6^2 : x^2 = 16 : 36 : x^2$.
The areas increase proportionally. Therefore,
$\dfrac{16}{36} = \dfrac{36}{x^2}$. $16x^2 = 36 \times 36 \Rightarrow x^2 = 81$.

$x = \sqrt{81} = 9$.

TEST 20	SECTION 7

1. (D)

Proportion. $\dfrac{k}{6} = \dfrac{100d}{x}$ or $x = \dfrac{600d}{k}$

---

2. (A)

$\dfrac{10+a+b}{3} = 20$, $a+b = 50$. Therefore, the average is, $\dfrac{50}{2} = 25$.

---

3. (A)

If the number of candy bars for Abby is x. Cabin's is $3x$ and Bernard's is $15x$. Therefore, $x + 3x + 15x = 19x = 95$, or $x = 5$.

---

4. (E)

$\dfrac{0.1k}{100} \times n = 5$, $0.1kn = 500$, $kn = 5000$.

$\dfrac{3k}{100} \times n = \dfrac{3kn}{100} = \dfrac{3(5000)}{100} = 150$

---

5. (A)

Slope is $-\dfrac{2}{3}$ and y-intercept is -2. Therefore, the equation is, $y = -\dfrac{2}{3}x - 2$

---

6. (D)

I. $|x| = -x$, any negative value f x will be the solution.

II. $|x| = x$, for $x < 0$, no solution.

III. $|x| = \sqrt{x^2}$, for any $x < 0$ will be the solution.

---

7. (C)

The pairs of (x, y) are as follows.

$(1,19)\ (2,18)\ (3,17)....(10,10)$. The difference between two numbers must be even integers. Or, Let $x - y = k$ ---- (1) , $x + y = 20$ --- (2)

From (1)+(2) $2x = k + 20$, that is,

$x = \dfrac{k + 20}{2}$.

Because x is an integer, k should be even integers.

---

8. (B)

The probability that the arrow lands on 5 is $\dfrac{3}{8}$.

The expected times $= 24 \times \dfrac{3}{8} = 9$ times.

---

9. (B)

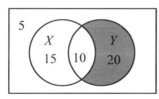

In the figure above,
$n(X \cup Y) = n(X) + n(Y) - n(X \cap Y)$
$45 = 25 + 30 - k$, $k = 10$

---

10. (D)

Only number of movie $Y = 30 - 10 = 20$

---

11. (C)

$AB^2 = AD \times AC$, $AB^2 = 3 \times 15 = 45$.
$AB = \sqrt{45}$

---

12. (B)

$2(n + (n+1)) = 5n - k$, $n = k + 2$. Therefore,
The larger one is, $n + 1 = k + 3$

---

13. (E)

Year	1	2	3	4	5	6	7	8
Profit	10	12	16	24	28	30	31	31.2
Increased by		2	4	8	4	2	1	0.2

In the table, increase faster, but later on slow. (E) is the best representing the table.

---

14. (D)

From cross-multiplication,
$2(a - 2b) = 3(b - 2a)$

$8a = 7b$, $a = \dfrac{7b}{4}$. Therefore,

$\dfrac{2a}{b} = \dfrac{2\left(\dfrac{7b}{8}\right)}{b} = \dfrac{7}{4}$.

---

15. (B)

$3^k + 3^k + 3^k = 3 \cdot 3^k = 3^{1+k}$, and
$(27)^5 = (3^3)^5 = 3^{15}$

Therefore, $1 + k = 15$, that is, $k = 14$.

---

16. (B)

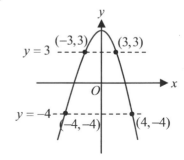

In the graph, the value of p and t can be obtained from the equation. $3 = -p^2 + 12$.

$p^2 = 9$, then $p = \pm 3$. you can find the value of t. The greatest slope will be between $(-3, 3)$ and $(-4, -4)$.

Therefore, the greatest slope $= \dfrac{3 - (-4)}{-3 - (-4)} = 7$.

$\boxed{END}$

NO MATERIAL ON THIS PAGE

NO MATERIAL ON THIS PAGE

NO MATERIAL ON THIS PAGE

NO MATERIAL ON THIS PAGE

NO MATERIAL ON THIS PAGE

17549655R00342